Handbook of
Ecological Indicators for Assessment of Ecosystem Health

Handbook of
Ecological Indicators for Assessment of Ecosystem Health

Edited by

Sven E. Jørgensen
Robert Costanza
Fu-Liu Xu

Taylor & Francis
Taylor & Francis Group

Boca Raton   London   New York   Singapore

A CRC title, part of the Taylor & Francis imprint, a member of the
Taylor & Francis Group, the academic division of T&F Informa plc.

Library of Congress Cataloging-in-Publication Data

Handbook of ecological indicators for assessment of ecosystem health / edited by Sven E. Jørgensen, Robert Costanza, Fu-Liu Xu.
    p. cm.
    Includes bibliographical references and index.
    ISBN 1-56670-665-3
    1. Ecosystem health. 2. Environmental indicators.   I. Jørgensen, Sven Erik, 1934. II. Costanza, Robert. III. Xu, Fu-Liu. IV. Title.

QH541.15.E265H36 2005
577.27--dc22   2004015982

Visit the CRC Press Web site at www.crcpress.com

© 2005 by CRC Press

No claim to original U.S. Government works
International Standard Book Number 1-56670-665-3
Library of Congress Card Number 2004015982
Printed in the United States of America 1 2 3 4 5 6 7 8 9 0
Printed on acid-free paper

The Editors

Sven Erik Jørgensen is professor of environmental chemistry at the Danish University of Pharmaceutical Sciences. He has doctorates in engineering from Karlsruhe University and sciences from Copenhagen University. He has been editor in chief of *Ecological Modelling* since the journal started in 1975. He is chairman of the International Lake Environment Committee. He has edited or authored 58 books in Danish and English and written 300 papers of which two-thirds have been published in peer-reviewed international journals. He was the first person to receive the Prigogine Award in 2004 for his outstanding work in the use thus far of equilibrium thermodynamics on ecosystems. He has also received the prestigious Stockholm Water Prize for his outstanding contribution to a global dissemination of ecological modeling and ecological management of aquatic ecosystems, mainly lakes and wetlands.

Robert Costanza is Gordon Gund professor of ecological economics and director of the Gund Institute for Ecological Economics in the Rubenstein School of Environment and Natural Resources at the University of Vermont. His research interests include: landscape-level integrated spatial simulation modeling; analysis of energy and material flows through economic and ecological systems; valuation of ecosystem services, biodiversity, and natural capital; and analysis of dysfunctional incentive systems and ways to correct them. He is the author or co-author of over 350 scientific papers and 18 books. His work has been cited in more than 2000 scientific articles since 1987 and more than 100 interviews and reports on his work have appeared in various popular media.

Fu-Liu Xu is an associate professor at the College of Environmental Sciences, Peking University, China. He was a guest professor at the Research Center for Environmental Quality Control (RCRQC), Kyoto University, from August 2003 to January 2004; and at the Research Center for Environmental Sciences, Chinese University of Hong Kong (CUHK), from August to October 2001. He is a member of the editorial boards for two international journals. He received his Ph.D. from Royal Danish University of Pharmacy in 1998. His research fields include system ecology and ecological modeling, ecosystem health and ecological indicators, ecosystem planning and management.

Contributors

M. Austoni
University of Parma
Parma, Italy

S. Bargigli
University of Siena
Siena, Italy

Simone Bastianoni
University of Siena
Siena, Italy

Paul Bertram
U.S. Environmental Protection
    Agency
Chicago, Illinois

Mark T. Brown
University of Florida
Gainesville, Florida

Villy Christiansen
University of British Columbia
Vancouver, Canada

Philippe Cury
Centre de Recherche Halieutique
    Méditerranéenne et Tropicale
Sète, France

Guilio A. De Leo
University of Parma
Parma, Italy

Robert Deal
Shawnee State University
Portsmouth, Ohio

Christina Forst
Oak Ridge Institute for Science
    and Education, on appointment
    to U.S. Environmental
    Protection Agency
Oak Ridge, Tennessee

G. Giordani
University of Parma
Parma, Italy

Paul Horvatin
U.S. Environmental Protection
    Agency
Chicago, Illinois

Sven E. Jørgensen
Danish University of
    Pharmaceutical Sciences
Copenhagen, Denmark

Nadia Marchettini
University of Siena
Siena, Italy

Joao C. Marques
University of Coimbra
Coimbra, Portugal

William J. Mitsch
Ohio State University
Columbus, Ohio

Felix Müller
University of Kiel
Kiel, Germany

Miguel A. Pardal
University of Coimbra
Coimbra, Portugal

Jaciro M. Patrício
University of Coimbra
Coimbra, Portugal

Charles Perrings
University of York
York, United Kingdom

I. Petrosillo
University of Lecce
Lecce, Italy

Martin Plus
Ifremer-Station d'Arcachon
    Département Environnement
    Littoral
Quai du Cdt Silhouette
Arcachon, France

Federico Maria Pulselli
University of Siena
Siena, Italy

Dave Raffaelli
University of York
York, United Kingdom

M. Raugei
University of Siena
Siena, Italy

Anna Renwick
University of York
York, United Kingdom

Marco Rosini
University of Siena
Siena, Italy

F. Salas
University of Coimbra
Coimbra, Portugal

Harvey Shear
Environment Canada
Downsview, Ontario, Canada

Jim Smart
University of York
York, United Kingdom

Yuri M. Svirezhev
Potsdam Institute for Climate
    Impact Research
Potsdam, Germany

Sergio Ulgiati
University of Siena
Siena, Italy

P. Viaroli
University of Parma
Parma, Italy

Naiming Wang
Ohio State University
Columbus, Ohio

P.G. Wells
Environment Canada
Dartmouth, Nova Scotia, Canada

Piran White
University of York
York, United Kingdom

Xixyuan Wu
Texas A&M University
College Station, Texas

Fu-Liu Xu
Peking University
Beijing, China

N. Zaccarelli
University of Lecce
Lecce, Italy

Jose Manuel Zaldívar-Comenges
European Commission, Joint
 Research Center
Institute for Environment and
 Sustainability
Inland and Marine Water Unit
Ispra, Italy

Li Zhang
Ohio State University
Columbus, Ohio

Giovanni Zurlini
University of Lecce
Lecce, Italy

Andy Zuwerink
Ohio State University
Columbus, Ohio

Contents

Chapter 9
Using Ecological Indicators in a Whole-Ecosystem Wetland Experiment   213
W.J. Mitsch, N. Wang, L. Zhang, R. Deal, X. Wu, and A. Zuwerink

Chapter 10
The Joint Use of Exergy and Emergy as Indicators of Ecosystems
Performances 239
S. Bastianoni, N. Marchettini, F.M. Pulselli, and M. Rosini

CHAPTER 1

Introduction

S.E. Jørgensen

1.1 THE ROLE OF ECOSYSTEM HEALTH ASSESSMENT IN ENVIRONMENTAL MANAGEMENT

The idea to apply an assessment of ecosystem health to environmental management emerged in the late 1980s. The parallels with the assessment of human health are very obvious. We go to the doctor to get a diagnosis (to determine what is wrong) and hopefully initiate a cure to bring us back to normal. The doctor will take various measurements and make examinations (pulse, blood pressure, sugar in the urine etc.) before making a diagnosis and suggesting a cure.

The idea behind the assessment of ecosystem health is similar (see Figure 1.1). If we observe that an ecosystem is not healthy, we want a diagnosis. What is wrong? What caused this unhealthy condition? What can we do to bring the ecosystem back to normal? To answer these questions, and also to come up with a cure, ecological indicators are applied.

Since ecosystem health assessment (EHA) emerged in the late 1980s, numerous attempts have been made to use the idea in practice, and again and again environmental managers and ecologists have asked the question: Which ecological indicators should we apply? It is clear today that it is not possible to find one indicator or even a few indicators that can be used generally, as some naively thought when EHA was introduced. Of course there are general

1-56670-665-3/05/$0.00 + $1.50

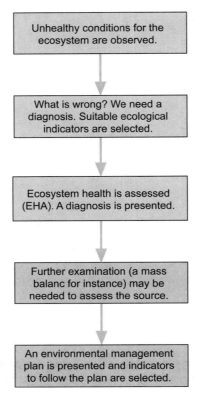

Figure 1.1   How ecological indicators are used for EHA and how to follow the effect of the environmental management plan.

ecological indicators that are used almost every time we have to assess ecosystem health; but they are never sufficient to present a complete diagnosis — the general indicators always have to be supplemented by other indicators. Our doctor has also general indicators. He will always take the patient's pulse, temperature, and blood pressure — very good general indicators — but he also has also always to supplement these general indicators with others that he selects according to the description of the problem as given by the patient. The same is true for the ecological doctor. If he observes dead fish but clear water, he will suspect the presence of a toxic substance in the ecosystem, while he will associate dead fish and very muddy water with oxygen depletion. In these two cases he will use two different sets of indicators, although some general indicators may be used in both cases.

The first international conference on the application of ecological indicators for the assessment of ecosystem health was held in Fort Lauderdale, Florida, in October 1900. Since then there have been several national and international conferences on ecological indicators and on EHA. In 1992 a book entitled *Ecosystem Health* was published by Island Press. Blackwell published a book with the same title in 1998 and also launched a journal entitled *Ecosystem*

Health in the mid-1990s with Rapport as the editor-in-chief. Elsevier launched a journal with the title *Ecological Indicators* in 2000 with Eric Hyatt and Felix Mueller as editors-in-chief. It can therefore be seen from this short overview of the development of the use of EHA and ecological indicators to perform the EHA that there has been significant interest in EHA and ecological indicators.

Some may have expected that EHA would replace ecological modeling to a certain extent, as it was a new method to quantify the disease of an ecosystem. It is also possible (as will be discussed in the next chapter) to assess ecosystem health based solely upon observations. On the other hand, EHA cannot be used to make prognoses and does not give the overview of the ecological components and their interactions like a model does. EHA and ecological modeling are two rather different but complementary tools that together give a better image of the environmental management possibilities than if either were used independently. Today, models are used increasingly (as will also demonstrated in this volume) as a tool to perform an EHA. The models are, furthermore, used to give prognoses of the development of the EHA-applied ecological indicators when a well-defined environmental management plan is followed.

A number of ecological indicators have been applied during the last 15 years or so to assess ecosystem health. As already stressed, general ecological indicators do not exist (or at least have not been discovered yet). A review of the literature published over the last 15 years regarding EHA and a selection of ecological indicators will reveal that it is also not possible to generate a set of indicators that can be used for specific problems or specific ecosystems. There are general indicators and there are problem- and ecosystem-specific indicators, which will be used again and again for the same problems or the same type of ecosystems; but because all ecosystems are different, even ecosystems of the same type are very different, and there are always some case-specific indicators that are selected on the basis of sound theoretical considerations. We can therefore not simply give, let us say, 300 lists of ecological indicators, with each list valid for a specific problem in a specific ecosystem (we presume for instance 20 different problems and 15 different type of ecosystems, totaling 300 combinations). Our knowledge about human health is much more developed than our knowledge about ecosystem health, and there is still no general procedure on how to assess a diagnosis for each of the several hundred possible cases a doctor will meet in his practice. We will, however, attempt to give an overview of the most applied ecological indicators for different ecosystems in the next chapter. It is possible to give such an overview, but not to give a general applicable procedure with a general valid list of indicators. This does of course not mean that we have nothing to learn from case studies. Because the selection of indicators is difficult and varies from case to case, it is of course possible to expand one's experience by learning about as many case studies as possible. This is the general idea behind this volume. By presenting a number of different case studies representing different ecosystems and different problems, an overview of the applicable indicators should be obtained.

1.2 THE CONCEPTUAL FLOW IN THIS VOLUME

Chapters 3 to 15 present different case studies focusing on different ecosystems and different problems. Chapter 2 has tried to give an overview of the other chapters (chapters 3 to 15) by presenting:

1. A discussion of the selection of ecological indicators for assessment of ecosystem health.
2. A classification of indicators.
3. The definition of some important holistic indicators.
4. An overview of all the applied ecological indicators with indication of where they have been applied and where they could be applied.
5. Three different procedures which can be applied for EHA.
6. A short presentation of a recently developed consistent ecosystem theories that can explain the close relationship between E.P. Odum's attributes (1969 and 1971) and the presented holistic indicators rooted in thermodynamics. The presented ecosystem theory is based on integration of several different approaches, that are consistent to a high extent (Jørgensen, 2002).

REFERENCES

1. Jørgensen, S.E. *Integration of Ecosystem Theories: A Pattern*, 3rd edition. Kluwer Scientific Publ. Company, Dordrecht, The Netherlands, 2002, 428 p.
2. Odum, E.P. The strategy of ecosystem development. *Science*, 164, 262–270, 1969.
3. Odum, E.P. *Fundamentals of Ecology*. W.B. Saunders Co., Philadelphia, 1974, 354 p.

CHAPTER 2

Application of Indicators for the Assessment of Ecosystem Health

S.E. Jørgensen, F.-L. Xu, F. Salas, and J.C. Marques

This chapter provides a comprehensive overview of the wide spectrum of indicators applicable for the assessment of ecosystem health. The applied indicators are classified in seven levels: (1) application of specific species; (2) ratio between classes of organisms; (3) specific chemical compounds; (4) trophic levels; (5) rates; (6) composite indicators included E.P. Odum's attributes and various indices; (7) holistic indicators as, for instance, biodiversity and resistance; (8) thermodynamic indicator. The chapter shows by several examples (based on case studies) that the application of the seven levels are consistent, at least to a certain extent, i.e., that indicators in level 1 and 2, for instance, would give the same indication as indicators from for instance level 6 and 7. The chapter presents furthermore an ecosystem theory that is shown to be applicable as fundamental for the ecological indicators, particularly the indicators from level 6 and 7.

1-56670-665-3/05/$0.00 + $1.50

2.1 CRITERIA FOR THE SELECTION OF ECOLOGICAL INDICATORS FOR EHA

Von Bertalanffy characterized the evolution of complex systems in terms of four major attributes:[1]

1. Progressive integration (which entails the development of integrative linkages between different species of biota and between biota, habitat, and climate).
2. Progressive differentiation (progressive specialization as systems evolve biotic diversity to take advantage of abilities to partition resources more finely and so forth).
3. Progressive mechanization (covers the growing number of feedbacks and regulation mechanisms).
4. Progressive centralization (which does probably not refer to a centralization in the political meaning, as ecosystems are characterized by short and fast feedbacks and decentralized control, but to the more and more developed cooperation among the organisms (the "Gaia" effect) and the growing adaptation to all other component in the ecosystem).

Costanza summarizes the concept definition of ecosystem health as:[2]

1. Homeostasis
2. Absence of disease
3. Diversity or complexity
4. Stability or resilience
5. Vigor or scope for growth
6. Balance between system components.

He emphasizes that it is necessary to consider all or least most of the definitions simultaneously. Consequently, he proposes an overall system health index, $HI = V \times O \times R$, where V is system vigor, O is the system organization index and R is the resilience index. With this proposal, Costanza touches on probably the most crucial ecosystem properties to cover ecosystem health.

Kay uses the term "ecosystem integrity" to refer to the ability of an ecosystem to maintain its organization.[3] Measures of integrity should therefore reflect the two aspects of the organizational state of an ecosystem: function and structure. Function refers to the overall activities of the ecosystem. Structure refers to the interconnection between the components of the system. Measures of function would indicate the amount of energy being captured by the system. Measures of structure would indicate the way in which exergy is moving through the system, therefore the exergy stored in the ecosystem could be a reasonable indicator of the structure.

Kay (1991) presents the fundamental hypothesis that ecosystems will organize themselves to maximize the degradation of the available work (exergy) in incoming energy[3] and that material flows will tend to close, which is necessary to ensure a continuous supply of material for the energy degrading processes. Maximum degradation of exergy is a consequence of the development of ecosystems from the early to the mature state, but as ecosystems cannot degrade more energy than that corresponding to the incoming solar radiation, maximum degradation may not be an appropriate goal function for *mature* ecosystems. This is discussed further in section 4 of this chapter. It should, however, be underlined here that the use of satellite images to indicate where an ecosystem may be found on a scale from an early to a mature system, is a very useful method to assess ecosystem integrity. These concepts have been applied by Akbari to analyze a nonagricultural and an agricultural ecosystem.[4] He found that the latter system, representing an ecosystem at an early stage, has a higher surface-canopy air temperature (less exergy is captured) and less biomass (less stored exergy) than the nonagricultural ecosystem, which represents the more mature ecosystem.

O'Connor and Dewling proposed five criteria to define a suitable index of ecosystem degradation, which we think can still be considered up-to-date.[5] The index should be:

1. Relevant
2. Simple and easily understood by laymen
3. Scientifically justifiable
4. Quantitative
5. Acceptable in terms of costs.

On the other hand, from a more scientific point of view, we may say that the characteristics defining a good ecological indicator are:

1. Ease of handling
2. Sensibility to small variations of environmental stress
3. Independence of reference states
4. Applicability in extensive geographical areas and in the greatest possible number of communities or ecological environments
5. Possible quantification.

It is not easy to fulfill all of these five requirements. In fact, despite the panoply of bio-indicators and ecological indicators that can be found in the literature, very often they are more or less specific for a given kind or stress or applicable to a particular type of community or scale of observation, and rarely will its wider validity have actually been proved conclusively. As will be seen through this volume, the generality of the selected indicators is only limited.

2.2 CLASSIFICATION OF ECOSYSTEM HEALTH INDICATORS

The ecological indicators applied today in different contexts, for different ecosystems, and for different problems can be classified on six levels from the most reductionistic to the most holistic indicators. Ecological indicators for EHA do not include indicators of climatic conditions, which in this context are considered entirely natural conditions.

2.2.1 Level 1

Level 1 covers the presence or absence of specific species. The best-known application of this type of indicator is the saprobien system,[6] which classifies streams into four classes according to their pollution by organic matter causing oxygen depletion:

1. Oligosaprobic water (unpolluted or almost unpolluted)
2. Beta-mesosaprobic (slightly polluted)
3. Alpha-mesosaprobic (polluted)
4. Poly-saprobic (very polluted).

This classification was originally based on observations of species that were either present or absent. The species that were applied to assess the class of pollution were divided into four groups:

1. Organisms characteristic of unpolluted water
2. Species dominating in polluted water
3. Pollution indicators
4. Indifferent species.

Records of fish in European rivers have been used to find by artificial neural network (ANN) a relationship between water quality and presence (and absence) of fish species. The result of this examination has shown that present or absent of fish species can be used as strong ecological indicators for the water quality.

2.2.2   Level 2

Level 2 uses the ratio between classes of organisms. A characteristic example is Nyggard algae index.

2.2.3   Level 3

Level 3 is based on concentrations of chemical compounds. Examples are assessment of the level of eutrophication on the basis of the total phosphorus concentration (assuming that phosphorus is the limiting factor for eutrophication). When the ecosystem is unhealthy due to too high concentrations of specific toxic substances, the concentration of one or more focal toxic compounds is, of course, a very relevant indicator. Chapter 4 gives an example where the PCB contamination of the Great North American Lakes has been followed by recording the concentrations of PCB in birds and in water. It is often important to find a concentration in a medium or in organisms where the concentration can be easily determined and has a sufficiently high value that is magnitudes higher than the detection limit, in order to facilitate a clear indication.

2.2.4   Level 4

Level 4 applies concentration of entire trophic levels as indicators; for instance, the concentration of phytoplankton (as chlorophyll-a or as biomass per m^3) is used as indicator for the eutrophication of lakes. A high fish concentration has also been applied as indicator for a good water quality or birds as indicator for a healthy forest ecosystem.

2.2.5 Level 5

Level 5 uses process rates as indicators. For instance, primary production determinations are used as indicators for eutrophication, either as maximum gC/m^2 day or gC/m^3 day or gC/m^2 year or gC/m^3 year. A high annual growth of trees in a forest is used as an indicator for a healthy forest ecosystem and a high annual growth of a selected population may be used as an indicator for a healthy environment. A high mortality in a population can, on the other hand, be used as indication of an unhealthy environment. High respiration may indicate that an aquatic ecosystem has a tendency towards oxygen depletion.

2.2.6 Level 6

Level 6 covers composite indicators, for instance, those represented by many of E.P. Odum's attributes (see Table 2.1). Examples are biomass,

Table 2.1 Differences between initial stage and mature stage are indicated; a few attributes are added to those published by Odum[7,8]

Properties	Early stages	Late or mature stage
A: Energetic		
P/R	$\gg 1$ or $\ll 1$	Close to 1
P/B	High	Low
Yield	High	Low
Specific entropy	High	Low
Entropy production per unit of time	Low	High
Exergy	Low	High
Information	Low	High
B: Structure		
Total biomass	Small	Large
Inorganic nutrients	Extrabiotic	Intrabiotic
Diversity, ecological	Low	High
Diversity, biological	Low	High
Patterns	Poorly organized	Well organized
Niche specialization	Broad	Narrow
Size of organisms	Small	Large
Life cycles	Simple	Complex
Mineral cycles	Open	Closed
Nutrient exchange rate	Rapid	Slow
Life span	Short	Long
C: Selection and homeostasis		
Internal symbiosis	Undeveloped	Developed
Stability (resistance to external perturbations)	Poor	Good
Ecological buffer capacity	Low	High
Feedback control	Poor	Good
Growth form	Rapid growth	Feedback
Growth types	R strategists	K strategists

respiration/biomass, respiration/production, production/biomass, and ratio of primary producer to consumers. E.P. Odum uses these composite indicators to assess whether an ecosystem is at an early stage of development or a mature ecosystem.

2.2.7 Level 7

Level 7 encompasses holistic indicators such as resistance, resilience, buffer capacity, biodiversity, all forms of diversity, size and connectivity of the ecological network, turnover rate of carbon, nitrogen, and energy. As will be discussed in the next section, high resistance, high resilience, high buffer capacity, high diversity, a big ecological network with a medium connectivity, and normal turnover rates, are all indications of a healthy ecosystem.

2.2.8 Level 8

Level 8 indicators are thermodynamic variables, which can be called super-holistic indicators as they try to see the forest through the trees and capture the total image of the ecosystem without the inclusion of details. Such indicators are exergy, energy, exergy destruction, entropy production, power, mass, and energy system retention time. The economic indicator cost/benefit (which includes all ecological benefits, not only the economic benefits of the society) also belong to this level.

Section 2.4 gives an overview of the application of the eight levels in chapters 3 to 15.

2.3 INDICES BASED ON INDICATOR SPECIES

When talking about indicator species, it is important to distinguish between two cases: indicator species and bioaccumulative species (the latter is more appropriate in toxicological studies).

The first case refers to those species whose appearance and dominance is associated with an environmental deterioration, as being favored for such fact, or for its tolerance of that type of pollution in comparison to other less resistant species. In a sense, the possibility of assigning a certain grade of pollution to an area in terms of the present species has been pointed out by a number of researchers including Bellan[9] and Glemarec and Hily[10], mainly in organic pollution studies.

Following the same policy some authors have focused on the presence/absence of such species to formulate biological indices, as detailed below.

Indices such as the Bellan (based on polychaetes) or the Bellan–Santini (based on amphipods) attempt to characterize environmental conditions by analyzing the dominance of species, indicating some type of pollution in relation to the species considered to indicate an optimal environmental situation.[11–12] Several authors do not advise the use of these indicators because often such indicator species may occur naturally in relatively high densities. The point is that there is no reliable methodology to know at which level one of those indicator species can be well represented in a community that is not really affected by any kind of pollution, which leads to a significant exercise of subjectivity.[13] Roberts et al.[16] also proposed an index based on macrofauna species which accounts for the ratio of each species abundance in control vs. samples proceeding from stressed areas. It is, however, semiquantitative as well as specific to site and pollution type. In the same way, the benthic response index[17] is based upon the type (pollution tolerance) of species in a sample, but its applicability is complex as it is calculated using a two-step process in which ordination analysis is employed to quantify a pollution gradient within a calibration data set.

The AMBI index, for example, which accounts for the presence of species indicating a type of pollution and of species indicating a nonpolluted situation, has been considered useful in terms of the application of the European Water Framework Directive to coastal ecosystems and estuaries. In fact, although this index is very much based on the paradigm of Pearson and Rosenberg[18] which emphasizes the influence of organic matter enrichment on benthic communities, it was shown to be useful for the assessment of other anthropogenic impacts, such as physical alterations in the habitat, heavy metal inputs, etc. What is more, it has been successfully applied to Atlantic (North Sea; Bay of Biscay; and south of Spain) and Mediterranean (Spain and Greece) European coasts.[14]

Regarding submarine vegetation, there is a series of genera that universally appear when pollution situations occur. Among them, there are the green algae: *Ulva*, *Enteromorpha*, *Cladophora* and *Chaetomorpha*; and the red algae: *Gracilaria*, *Porphyra* and *Corallina*.

High structural complexity species, such as Phaeophyta (belonging to Fucus and Laminaria orders), are seen worldwide as the most sensitive to any kind of pollution, with the exception of certain species of the Fucus order that can cope with moderate pollution.[19] On the other hand, marine Spermatophytae are considered indicator species of good water quality.

In the Mediterranean Sea, for instance, the presence of Phaeophyta *Cystoseira* and *Sargassum* or meadows of *Posidonia oceanica* indicate good water quality. Monitoring population density and distribution of such species allows detecting and evaluating the impact whatever activity.[20] *Posidonia oceanica* is possibly the most commonly used indicator of water quality in the Mediterranean Sea[21,22] and the conservation index,[23] based on the named marine Spermatophyta, is used in such littoral.

The description of above-mentioned indices is given below.

2.3.1 Bellan's Pollution Index[11]

$$IP = \sum \frac{\text{Dominance of pollution indicator species}}{\text{Dominance of pollution/clear water indicators}}$$

Species considered as pollution indicators by Bellan are *Platenereis dumerilli, Theosthema oerstedi, Cirratulus cirratus* and *Dodecaria concharum*.

Species considered as clear-water indicators by Bellan are *Syllis gracillis, Typosyllis prolifera, Typosyllis* sp. and *Amphiglena mediterranea*.

Index values over 1 show that the community is pollution disturbed. As organic pollution increases, the value of the index goes higher, which is why (in theory) different pollution grades can be established, although the author does not fix them.

This index was designed in principle to be applied to rocky superficial substrates. Nevertheless, Ros et al. modified it in terms of the used indicator species in order to be applicable to soft bottoms.[24] In this case, the pollution indicator species are *Capitella capitata, Malococerus fuliginosus* and *Prionospio malmgremi*, and the clear water indicator species is *Chone duneri*.

2.3.2   Pollution Index Based on Ampiphoids[12]

This index follows the same formulation and interpretation as Bellan's, but is based on the amphipods group.

The pollution indicator species are *Caprella acutrifans* and *Podocerus variegatus*. The clear-water indicator species are *Hyale*. sp., *Elasmus pocllamunus* and *Caprella liparotensis*.

2.3.3   AMBI[14]

For the development of the AMBI, the soft bottom macrofauna is divided into five groups according to their sensitivity to an increasing stress:

    I. Species very sensitive to organic enrichment and present under unpolluted conditions.

    II. Species indifferent to enrichment, always in low densities with nonsignificant variations with time.

    III. Species tolerant to an excess of organic matter enrichment. These species may occur under normal conditions, but their populations are stimulated by organic enrichment.

    IV. Second-order opportunist species, mainly small-sized polychaetes.

    V. First-order opportunist species, essentially deposit-feeders.

The formula is as follows:

$$AMBI = \frac{(0 \times \%GI) + (1.5 \times \%GII) + (3 \times \%GIII) + (4.5 \times \%GIV) + (6 \times \%GV)}{100}$$

The index results are classified as:

- Normal: 0.0–1.2
- Slightly polluted: 1.2–3.2

- Moderately polluted: 3.2–5.0
- Highly polluted: 5.0–6.0
- Very highly polluted: 6.0–7.0.

For the application of this index, nearly 2000 taxa have been classified, which are representative of the most important soft-bottom communities present in European estuarine and coastal systems. The marine biotic index can be applied using the AMBI software[14] (freely available at < http://www. azti.es >).

2.3.4 Bentix[15]

This index is based on AMBI index but lies in the reduction of the ecological groups involved in the formulae in order to avoid errors in the grouping of the species and reduce effort in calculating the index:

$$\text{Bentix} = \frac{(6 \times \%\text{GI}) + 2(\%\text{GII} + \%\text{GIII})}{100}$$

> Group I: This group includes species sensitive to disturbance in general.
> Group II: Species tolerant to disturbance or stress whose populations may respond to enrichment or other source of pollution.
> Group III: This group includes the first order opportunistic species (pronounced unbalanced situation), pioneer, colonizers, or species tolerant to hypoxia.

A compiled list of indicator species in the Mediterranean Sea was made, each assigned a score ranging from 1–3 corresponding to each one of the three ecological groups:

- Normal: 4.5–6.0
- Slightly polluted: 3.5–4.5
- Moderately polluted: 2.5–3.5
- Highly polluted: 2.0–2.5
- Very highly polluted: 0.

2.3.5 Macrofauna Monitoring Index[16]

The authors developed an index for biological monitoring of dredge spoil disposal. Each of the 12 indicator species is assigned a score, based primarily on the ratio of its abundance in control versus impacted samples. The index value is the average score of those indicator species present in the sample.

Index values of < 2, 2–6 and > 6 are indicative of severe, patchy, and no impact, respectively.

The index is site- and impact-specific but the process of developing efficient monitoring tools from an initial impact study should be widely applicable.[16]

2.3.6 Benthic Response Index[17]

The benthic response index (BRI) is the abundance weighted average pollution tolerance of species occurring in a sample, and is similar to the weighted average approach used in gradient analysis.[25,26] The index formula is:

$$I_s = \frac{\sum_{i=1}^{n} p_i \sqrt[3]{a_{si}}}{\sum_{i=1}^{n} \sqrt[3]{a_{si}}}$$

where I_s is the index value for sample s, n is the number of species for sample s, p_i is the position for species i on the pollution gradient (pollution tolerance score), and a_{si} is the abundance of species i in sample s.

According to the authors, determining the pollutant score (p_i) for the species involves four steps:

1. Assembling a calibration infaunal data set.
2. Conducting an ordination analysis to place each sample in the calibration set on a pollution gradient.
3. Computing the average position of each species along the gradient.
4. Standardizing and scaling the position to achieve comparability across depth zones.

The average position of species $i(p_i)$ on the pollution gradient defined in the ordination is calculated as:

$$p_i = \frac{\sum_{j=1}^{t} g_j}{t}$$

where t is the number of samples to be used in the sum, with only the highest t species abundance values included in the sum. The g_j is the position on the pollution gradient in the ordination space for sample j.

This index only has been applied for assessing benthic infaunal communities on the Mayland shelf of southern California employing a 717-sample calibration data set.

2.3.7 Conservation Index[23]

$$CI = \frac{L}{L + D}$$

where L is the meadow of living *Posidonia oceanica* and D the dead meadow coverage.

Authors applied the index near chemical industrial plants. Results led them to establish four grades of *Posidonia* meadow conservation, which allow identification of increasing impact zones, as changes in the industry activity can be detected by the conservation status in a certain location.

Also, there are species classified as bioaccumulative, defined as those capable of resisting and accumulating diverse pollutant substances in their tissues, which allows their detection when they are present in the environment at such low levels (and are therefore difficult to detect using analytical techniques).[27]

The disadvantage of using accumulator indicator species in the detection of pollutants arises from the fact that a number of biotic and abiotic variables may affect the rate at which the pollutant is accumulated, and therefore both laboratory and field tests need to be undertaken so that the effects of extraneous parameters can be identified.

Molluscs, specifically the bivalve class, have been one of the most commonly used species in determining the existence and quantity of a toxic substance.

Individuals of the genres *Mytilus*,[28–37] *Cerastoderma*,[38–40] *Ostrea*[35,41] and *Donax*[42,43] are considered to be ideal for research involving the detection of the concentration of a toxic substance in the environment, due to their sessile nature, wide geographical distribution, and capability to detoxify when pollution ceases. In that sense, Goldberg et al.[29] introduced the concept of "mussel watch" when referring to the use of molluscs in the detection of polluting substances, due to their wide geographical distribution and their capability of accumulating those substances in their tissues. The National Oceanic and Atmospheric Agency (NOAA) in the U.S. has developed a "mussel watch" program since 1980 focusing on pollution control along the North American coasts. There are programs similar to the North American one in Canada,[31,44] the Mediterranean Sea,[45] the North Sea[46] and on the Australian coast.[47–49]

Likewise, certain species of the amphipods group are considered capable of accumulating toxic substances,[50,51] as well as species of the polychaetes group like *Nereis diversicolor*,[52,53] *Neanthes arenaceodentata*,[54] *Glycera alba*, *Tharix marioni*,[55] or *Nephtys hombergi*.[56]

Some fish species have also been used in various work focusing on the effects of toxic pollution of the marine environment, due to their bioaccumulative capability[57–59] and the existing relationship among pathologies suffered by any benthic fishes and the presence of polluting substances.[60–62]

Other authors such as Levine,[63] Maeda and Sakaguchi,[64] Newmann et al.,[65] and Storelli and Marcotrigiano[66] have looked into algae as indicators for the presence of heavy metals, pesticides and radionuclides. *Fucus*, *Ascophyllum* and *Enteromorpha* are the most utilized.

For reasons of comparison, the concentrations of substances in organisms must be translated into uniform and comparable units. This is done through the ecologic reference index (ERI), which represents a potential for environmental effects. This index has only been applied using blue mussels:

$$ERI = \frac{\text{Measured concentration}}{\text{BCR}}$$

Table 2.2 Upper limit of BCR for hazardous substances in blue mussel according to OSPAR/MON (1998)

Substance	Upper limit of BCR value (ng/g dry weight)
Cadmium	550
Mercury	50
Lead	959
Zinc	150,000

where BCR is the value of the background/reference concentration (see Table 2.2).

Few indices (such as the latter) based on the use of bioaccumulative species have been formulated, most of which involve the simple measurement of the effects (e.g., percentage incidence or percentage mortality) of a certain pollutant on those species, or the use of biomarkers (which can be useful to scientists evaluating the specificity of the responses to natural or anthropogenic changes). However, it is very difficult for the environmental manager to interpret increasing or decreasing changes in biomarker data.

The Working Group on Biological Effects of Contaminants (WGBEC) in 2002 recommended different techniques for biological monitoring programs (see Table 2.3).

2.4 INDICES BASED ON ECOLOGICAL STRATEGIES

Some indices try to assess environmental stress effects accounting for the ecological strategies followed by different organisms. That is the case of trophic indices such as the infaunal index proposed by Word,[67] which are based on the different feeding strategies of the organisms. Another example is the nematodes/copepods index[68] which account for the different behavior of two taxonomic groups under environmental stress situations. However, several authors have rejected them due to their dependence on parameters such as depth and sediment particle size, as well as because of their unpredictable pattern of variation depending on the type of pollution.[69,70] More recently, other proposals have appeared, such as the polychaetes/amphipods ratio index, or the index of r/K strategies, which considers all benthic taxa although the difficulty of scoring exactly each species through the biological trait analysis has been emphasized.

Feldman's R/P index, based on marine vegetation, is often used in the Mediterranean Sea. It was established as a biogeographical index and it is based on the fact that *Rodophyceae* sp. number decreases from the tropics to the poles. Its application as an indicator holds on the higher or lower sensitivity of *Phaeophyceae* and *Rhodophyceae* to disturbance.

Table 2.3 Review of different techniques for biological monitoring

Method	Organism	Issues addressed	Biological significance	Threshold value
Bulky DNA adduct formation	Fish	PAHs, other synthetic organics	Measures genotoxic effects. Sensitive indicator of past and present exposure	2 × reference site or 20% change
AChE	Fish	Organophosphates and carbonates or similar molecules	Measures exposures	−2.5 × reference site
Metallothionein induction	Bivalve molluscs Fish *Mytilus* sp.	Measures induction of metallothionein protein by certain metals	Measures exposure and disturbance of copper and zinc metabolism	2.0 × reference site
EROD or P4501A induction	Fish	Measures induction of enzymes with metabolise planar organic contaminants		2.5 × reference site
ALA-D inhibition	Fish	Lead	Index of exposure	2.0 × reference site
PAH bile metabolites	Fish	PAHs	Measures exposure to and metabolism PAHs	2.0 × reference site
Lysososmal stability	Fish	Not contaminant specific but responds to a wide variety of xenobiotics contaminants and metals	Provides a link between exposure and pathological endpoints	2.5 × reference site
Lysosomal neutral red retention	*Mytilus* sp. *Mytilus* sp.	Not contaminant specific but responds to a wide variety of xenobiotics contaminants and metals	Provides a link between exposure and pathological endpoints	2.5 × reference site

(*Continued*)

Table 2.3 Continued

Method	Organism	Issues addressed	Biological significance	Threshold value
Early toxicopathic lesions, pre-neoplastic and neoplastic liver histopathology	Fish	PAHs	Measures pathological changes associated with exposure to genotoxic and non-genotoxic carcinogens	2.0 × reference site or 20% change
Scope for growth	Bivalve molluscs	Responds to a wide variety of contaminants	Integrative response which is a sensitive and sublethal measure of energy available for growth	
Shell thickening	Crassostea gigas	Specific to organotins	Disruption to pattern of shell growth	
Vitellogenin induction	Male and juvenile fish	Oestrogenic substances	Measures feminization of male fish and reproductive impairment	
Imposex	Neogastropod molluscs	Specific to organotins	Reproductive interference	2.0 × reference site or 20% change
Intersex	Littorina littorina	Specific to reproductive effects of organotins	Reproductive interference in coastal waters	2.0 × reference site or 20% change
Reproductive success in fish	Zoarces viviparous	Not contaminant-specific, will respond to a wide of environmental contaminants	Measures reproductive output and survival of eggs and fry in relation to contaminants	

2.4.1   Nematodes/Copepods Index[68]

This index is based on the ratio between abundances of nematodes and copepods:

$$I = \frac{\text{Nematode abundance}}{\text{Copepod abundance}}$$

Values of such a ratio can increase or decrease according to levels of organic pollution. This happens by means of a different response of those groups to the input of organic matter to the system. Values of over 100 demonstrate a high organic pollution.

According to the authors, the index application should be limited to certain intertidal zones. In infralittoral areas at certain depths, despite the absence of pollution, the values obtained were very high. The explanation for this is the absence of copepods at such depths, possibly due to a change in the optimal interstitial habitat for that taxonomic group (see Reference 68).

2.4.2   Polychaetes/Amphipods Index

This index is similar to the nematodes/copepods, but is applied to the macrofauna level using the polychaetes and amphipods groups. The index was formerly designed to measure the effects of crude oil pollution:

$$I = \log_{10}\left(\frac{\text{Polychaetes abundance}}{\text{Amphipods abundance}}\right) + 1$$

The index values are classified as: $I \leq 1 = $ nonpolluted and $I > 1 = $ polluted.

2.4.3   Infaunal Index[67]

The macrozoobenthos species can be divided into:

1. Suspension feeders
2. Interface feeders
3. Surface deposit feeders
4. Subsurface deposit feeders.

Based on this division, the trophic structure of macrozoobenthos can be determined using the following equation:

$$ITI = 100 - (100/3) \times \frac{(0n_1 + 1n_2 + 2n_3 + 3n_4)}{(n_1 + n_2 + n_3 + n_4)}$$

in which n_1, n_2, n_3 and n_4 are the number of individuals sampled in each of the above mentioned groups. ITI values near 100 mean that suspension feeders are dominant and that the environment is not disturbed. Near a value of 0 subsurface, feeders are dominant, meaning that the environment is probably heavily disturbed due to human activities.

One of the disadvantages of a trophic index is the determination of the diet of the organisms, which can be developed through the study of the stomach content or in laboratory experiments. Generally, the real diet (i.e., the one studied observing the stomach content) is difficult to establish, and can vary from one population to another among the same taxonomic entity. For example, *Nereis virens* is an omnivore species along the European coast but a herbivore along the North American coast.[71]

Another aspect to be considered when determining the trophic category of many polychaetes species, is their alternative feeding behavior that can appear under certain circumstances. Buhr (1976) determined, through laboratory experiments, that the terebellid *Lanice conchylega*, considered as a detritivore, changes into a filterer when a certain concentration of phytoplankton is present in the water column Taghon et al. (1980)[72] observed that some species of the Spionidae family, usually taken for a detritivore, could change into a filterer, modifying the mandibular palps into a characteristic helicoidal shape.

On the other hand, some species of the Sabellidae and Owenidae families can change from filterers to detritivores. Some limnivores and detritivores can be considered carnivores when they consume the remains of other animals.[73]

Those facts nowadays lead to doubts about the existence of a clear separation among such diverse feeding strategies. This is why other characteristics such as the grade of individual's mobility and the morphology of the mouth apparatus intervene in the definition of the trophic category of polychaetes.[74] The different combinations of that set of characteristics are what Fauchald and Jaumars term "feeding guilds."[71]

Authors such as Maurer et al.[75] and Pires and Múniz[76] have tried the use of the classification of the different polychaetes species in feeding guilds when studying the structure of the benthic system and when identifying the different impacts, both with good results.

The main problem when using such a classification is without doubt the difficulty that carries the determination of each one of those combinations for each species. According to a study by Dauer, many families hold more than one combination depending on the type of feeding they follow, their grade of mobility, and the morphology of their mouth apparatus; being monospecific every combination (Dauer et al., 1981). This leads us to believe that such a classification very often does not make much sense from a practical point of view.

2.4.4 Feldman Index

$$I = \frac{N^\circ \text{ Species of } Rhodophyceae}{N^\circ \text{ Species of } Phaeophyceae}$$

Cormaci and Furnari detected values of over 8 in polluted areas in southern Italy,[77] when normal values in a balanced community oscillate between 2.5 and 4.5. Verlaque studied the effects of a thermal power station,[78] and also found higher values of those the index, but considers due to the presence of communities with higher optimum temperature.

However, Belsher and Bousdouresque analyzed vegetation in small harbors and found that as the Phaeophyceae increases, the index decreases.[79]

2.5 INDICES BASED ON THE DIVERSITY VALUE

Diversity is the other mostly used concept, focusing on the fact that the relationship between diversity and disturbances can be seen as a decrease in the diversity when the disturbances increase.

Magurran divides the diversity measurements into three main categories:[80]

1. Indices that measure the enrichment of the species, such as Margalef, which are, in essence, a measurement of the number of species in a defined sampling unit.

2. Models of the abundance of species, as the K-dominance curves[70] or the log-normal model,[81] which describe the distribution of their abundance, going from those that represent situations in which there is a high uniformity to those that characterize cases in which the abundance of the species is very unequal. However, the log-normal model deviation was rejected once ago by several authors due to the impossibility of finding any benthic marine sample that clearly responded to the log-normal distribution model.[70,82,83]

3. Indices based on the proportional abundance of species that pretend to solve enrichment and uniformity in a simple expression. Such indices can also be divided into those based on statistics, information theory, and dominance indices. Indices derived from the information theory, such as the Shannon–Wiener, are based on something logical: diversity or information in a natural system can be measured in a similar way as information contained in a code or message. On the other hand, dominance indices such as Simpson or Berger–Parker are referred to as measurements that mostly ponder the abundance of common species instead of the enrichment of the species.

Meanwhile, average taxonomic diversity and distinctness measures has been used in some research to evaluate biodiversity in the marine environment,[84–86] as it takes into account taxonomic, numerical, ecological, genetic, and filogenetic aspects of diversity. These measures address some of the problems identified with species richness and the other diversity indices.[85]

2.5.1 Shannon–Wiener Index[87]

This index is based on the information theory. It assumes that individuals are sampled at random, out of an "indefinitely large" community, and that all the species are represented in the sample.

The index takes the form:

$$H' = - \sum p_i \log 2p_i$$

where p_i is the proportion of individuals found in the species i. In the sample, the real value of p_i is unknown, but it is estimated through the ratio N_i/N, where N_i is the number of individuals of the species i and N is the total number of individuals.

The units for the index depend on the log used. So, for log 2, the unit is bits/individual; "natural bels" and "nat" for log e; and "decimal digits" and "decits" for log 10.

The index can take values between 0 and 5. Maximum values are rarely over 5 bits per individual. Diversity is a logarithmic measurement which makes it, to a certain extent, a sensitive index in the range of values next to the upper limit.[88]

As an ordinary basis, in the literature, low index values are considered to be indication of pollution.[89–98]

However, one of the problems arising with its use is the lack of objectivity when establishing as a precise manner from what value it should start detecting the effects of such pollution.

Molvaer et al.,[99] established the following relationship between the indices and the different ecological levels according to what is recommended by the Water Framework Directive:

- High status: >4 bits/individual
- Good status: 4–3 bits/individual
- Moderate status: 3–2 bits/individual
- Poor status: 2–1 bits/individual
- Bad status: 1–0 bits/individual.

Detractors of Shannon index base their criticisms on its lack of sensitivity when it comes to detecting the initial stages of pollution.[18, 100–101]

Gray and Mirza,[102] in a study on the effects of a cellulose paste factory waste, set out the uselessness of this index as it responses to such obvious changes that there is no need of a tool to detect them.

Ros and Cardell,[103] in their study on the effects of great industrial and human domestic pollution, consider the index as a partial approach to the knowledge of pollution effects on marine benthic communities and, without any explanation to that statement, set out a new structural index proposal, the lack of applicability of which has already been demonstrated by Salas.[104]

2.5.2 Pielou Evenness Index

$$J' = H'/H'_{max} = H'/\log S$$

where H'_{max} is the maximum possible value of Shannon diversity and S is the number of species.

The index oscillates from 0 to 1.

2.5.3　Margalef Index

The Margalef index quantifies the diversity relating specific richness to the total number of individuals:

$$D = (S - 1)/\log_2 N$$

where S is the number of species and N is the total number of individuals. The author did not establish reference values.

The main problem that arises when applying this index is the absence of a limit value, therefore it is difficult to establish reference values. Ros and Cardell[103] consider values below 4 as typical of polluted. Bellan-Santini,[12] on the contrary, established that limit when the index takes values below 2.05.

2.5.4　Berger–Parker Index

The index expresses the proportional importance of the most abundant species, and takes this shape:

$$D = n_{max}/N$$

where n_{max} is the number of individuals of the one most abundant species and N is the total number of individuals. The index oscillates from 0 to 1 and, in contrast with the other diversity indices, high values show a low diversity.

2.5.5　Simpson Index

Simpson defined their index on the probability that two individuals randomly extracted from an infinitely large community could belong to the same species:[105]

$$D = \sum p_i^2$$

where p_i is the individuals proportion of the species i. To calculate the index for a finite community use:

$$D = \sum [n_i(n_i-1)/N(N-1)]$$

where n_i is the number of individuals in the species i and N is the total number of individuals.

Like the Berger–Parker index, this one oscillates from 0 to 1, it has no dimensions and similarly, the high values imply a low diversity.

2.5.6　Deviation from the Log-Normal Distribution[102]

This method, proposed by Gray and Mirza in 1979, is based on the assumption that when a sample is taken from a community, the distribution of the individuals tends to follow a log-normal model.

The adjustment to a logarithmical normal distribution assumes that the population is ruled by a certain number of factors and it constitutes a community in a steady equilibrium; meanwhile, the deviation from such distribution implies that any perturbation is affecting it.

2.5.7 K-Dominance Curves[106]

The *K*-dominance curve is the representation of the accumulated percentage of abundance vs. the logarithm of the sequence of species ordered in a decreasing order. The slope of the straight line obtained allows the valuation of the pollution grade. The higher the slope is, the higher the diversity is too.

2.5.8 Average Taxonomic Diversity[84]

This measure, equal to taxonomic distinctness, is based on the species abundances (denoted by x_i, the number of individuals of species i in the sample) and on the taxonomic distance (ω_{ij}) through the classification tree, between every pair of individuals (the first from species and the second from species j).

It is the average taxonomic distance apart of every pair of individuals in the sample, or the expected path length between any two individuals chosen at random:

$$\Delta = \left[\sum \sum_{i<j} \omega_{ij} x_i x_j \right] / [N(N-1)/2]$$

where the double summation is over all pairs of species i and j $(i, j = 1, 2, \ldots S;$ $i < j)$, and $N = \sum_i x_i$, the total number of individuals in the sample.

2.5.9 Average Taxonomic Distinctness[84]

To remove the dominating effect of the species abundance distribution, Warwick and Clarke[84] proposed to divide the average taxonomic diversity index by the Simpson index, giving the average taxonomic distinctness index:

$$\Delta^* = \left[\sum \sum_{i<j} \omega_{ij} x_i x_j \right] / \left[\sum \sum_{i<j} x_i x_j \right]$$

when quantitative data is not available and the sample consists simply of a species list (presence/absence data) the average taxonomic distinctness takes the following form:

$$\Delta^+ = \left[\sum \sum_{i<j} \omega_{ijj} \right] / [S(S-1)/2]$$

where S, as usual, is the observed number of species in the sample and the double summation ranges over all pairs i and j of the species $(i < j)$.

Taxonomic distinctness is reduced in respect to increasing environmental stress and this response of the community lies at the base of this index concept. Nevertheless, it is most often very complicated to meet certain requirements to apply it, as having a complete list of the species present in the area under study in pristine situations. Moreover, some works, have shown that in fact taxonomic distinctness is not more sensitive than other diversity indices usually applied when detecting disturbances,[107] and consequently this measure has not been widely used on marine environment quality assessment and management studies.

2.6 INDICATORS BASED ON SPECIES BIOMASS AND ABUNDANCE

Other approaches account for the variation of organism's biomass as a measure of environmental disturbances. Along these lines, there are methods such as SAB,[18] consisting of a comparison between the curves resulting from ranking the species as a function of their representativeness in terms of both their abundance and biomass. The use of this method is not advisable because it is purely graphical, which leads to a high degree of subjectivity that impedes relating it quantitatively to the various environmental factors. The ABC method[108] also involves the comparison between the cumulative curves of species biomass and abundance, from which Warwick and Clarke[85] derived the W-statistic index.

2.6.1 ABC Method[108]

This method is based on the idea that the distribution of a number of individuals for the different species in the macrobenthos communities is different to the biomass distribution.

It is adapted from the K-dominance curve already mentioned, showing in one graphic the K-dominance and biomass curves. The graphics are made up comparing the interval of species (in the abscise axis), decreasingly arranged and in logarithmical scale, to the accumulated dominance (in the ordinate axis).

According to the range of disturbance, three different situations can be given:

1. In a system with no disturbances, a relatively low number of individuals contribute to the major part of the biomass, and at the same time, the distribution of the individuals among the different species is similar. The representations would show the biomass curve above the dominance one, indicating higher numeric diversity than biomass.
2. Under moderate disturbances, there is a decrease in the dominance as regards biomass; however, abundances increase. The graphic shows both curves intersected.
3. In the case of intense disturbances, the situation is totally the opposite, and only a few species monopolize the greater part of the individuals,

which are of a small size, which is why the biomass is low and is more equally shared. It can be seen in the representation how the curve of the number of individuals is placed above the biomass curve, indicating a higher diversity in the biomass distribution.

Some studies have tried to lead this method into a measurable index,[109–112] with the study by Clarke being the most commonly accepted one:[110]

$$W = \sum_{i=1}^{S} (B_i - A_i) / 50(S - 1)$$

where B_i is the biomass of species i, A_i the abundance of specie i, and S is the number of species.

The index can take values from $+1$, indicating a nondisturbed system (high status) to -1, which defines a polluted situation (low status). Values close to 0 indicate a moderate level of pollution (moderate status).

The method is specific of organic pollution and it has been applied, with satisfactory results, to soft-bottom tropical communities,[113,114] to experiments,[115] to fish-factoring disturbed areas,[116] and on coastal lagoons.[117,118] However, several studies obtained confusing results after applying that technique to estuarine zones,[109,119–122] induced by the appearance of dominant species in normal conditions and favored by different environmental factors.

Although it is a method designed to be applied to benthic macrofauna, Abou-Aisha et al.[123] used it to detect the impact of phosphorus waste in macroalgae, in three areas of the Red Sea. In spite of that, the problem when applying it to marine vegetation lies on the difficulty of counting the number of individuals in the vegetal species.

2.7 INDICATORS INTEGRATING ALL ENVIRONMENT INFORMATION

From a more holistic point of view, some studies proposed indices capable of at least trying to integrate the whole environmental information. A first approach for application in coastal areas was developed by Satmasjadis,[124] relating sediment particles size to benthic organisms diversity. Wollenweider et al.[125] developed a trophic index (TRIX) integrating chlorophyll-a, oxygen saturation, total nitrogen, and phosphorus to characterize the trophic state of coastal waters.

In a progressively more complex way, other indices such as the index of biotic integrity (IBI) for coastal systems,[126] the benthic index of environmental condition,[96] or the Chesapeake Bay B-BI index[127] included physicochemical factors, diversity measures, specific richness, taxonomical composition, and the trophic structure of the system.

Similarly, a set of specific indices of fish communities has been developed to measure the ecological status of estuarine areas. The estuarine biological health index (BHI) combines two separate measures (health and importance) into a single index. The estuarine fish health index (FHI) is based on both qualitative and quantitative comparisons with a reference fish community.[128] The estuarine biotic integrity index (EBI)[129] reflects the relationship between anthropogenic alterations in the ecosystem and the status of higher trophic levels, and the estuarine fish importance rating (FIR) is based on a scoring system of seven criteria that reflect the potential importance of estuaries to the associated fish species. This index is able to provide a ranking based on the importance of each estuary and helps to identify the systems with major importance for fish conservation.

Nevertheless, these indicators are rarely used in a generalized way because they have usually been developed for application in a particular system or area, which makes them dependent on seasonality and the type of habitat. On the other hand, they are difficult to apply as they need a large amount of data of different nature.

2.7.1 Trophic Index[125]

$$TRIX = \frac{k}{n} \times \sum (M_i - L_i)/(U_i - L_i)$$

in which $k = 10$ (scaling the result between 0 and 10), $n = 4$ (number of variables are integrated, M_i = measured value of variable i, U_i = upper limit of variable i, L_i = lower limit of value i.

The resulting TRIX values are dependent on the upper and the lower limit chosen and indicate how close the current state is to the natural state. However, comparing TRIX values of different areas becomes more difficult. When a wider, more general range is used for the limits, TRIX values for different areas can more easily be compared to each other.

2.7.2 Coefficient of Pollution[124]

Calculation of the index is based on several integrated equations. These equations are:

$$S' = s + t/(5 + 0.2s)$$

$$i_0 = (-0.0187s'2 + 2.63s' - 4)(2.20 - 0.0166h)$$

$$g' = I/(0.0124i + 1.63)$$

$$P = g'/[g(i/i_0)^{1/2}]$$

where P is the coefficient of pollution, S' is the sand equivalent, s is the percent sand, t is the percent silt, i_0 is the theoretical number of individuals, i is the number of individuals, h is the station depth, g' is the theoretical number of species, and g is the number of species.

2.7.3 Benthic Index of Environmental Condition[96]

Benthic index $= (2.3841 \times$ Proportion of expected diversity$) + (-1.6728 \times$ Proportion of total abundance as tubifids$) + (0.6683 \times$ Proportion of total abundance as bivalves$)$.

The expected diversity is calculated throughout Shannon–Wiener index adjusted for salinity:

$$\text{Expected diversity} = 0.75411 + (0.00078 \times \text{salinity}) + (0.00157 \times \text{salinity}^2)$$
$$+ (-0.00078 \times \text{salinity}^3)$$

This index was developed for estuarine macrobenthos in the Gulf of Mexico in order to discriminate between areas with degraded environmental conditions and areas with nondegraded or reference conditions.

The final development of the index involved calculating discriminating scores for all samples sites and normalizing calculated scores to a scale of 0 to 10, setting the break point between degraded and nondegraded reference sites at 4.1. So the index values lower than 4.1 indicate degraded conditions, higher values than 6.1 indicate nondegraded situations, and values between 6.1 and 4.1 reveal moderate disturbance.

2.7.4 B-IBI[127]

Eleven metrics are used to calculate the B-IBI[127]

1. Shannon–Wiener species diversity index
2. Total species abundance
3. Total species biomass
4. Percent abundance of pollution-indicative taxa
5. Percent abundance of pollution-sensitive taxa
6. Percent biomass of pollution-indicative taxa
7. Percent biomass of pollution-sensitive taxa
8. Percent abundance of carnivore and omnivores
9. Percent abundance of deep-deposit feeders
10. Tolerance Score
11. Tanypodinae to Chironomidae percent abundance ratio.

The scoring of metrics to calculate the B-IBI is done by comparing the value of a metric from the sample of unknown sediment quality to thresholds established from reference data distributions.

This index was developed to establish ecological status of Chesapeake Bay and it is specific to habitat type and seasonality, its use advisable only during spring.

2.7.5 Biotic Integrity (IBI) for Fishes

A fish index of biotic integrity (IBI) was developed for tidal fish communities of several small tributaries to the Chesapeake Bay.[130,131]

Nine metrics are used to calculate the index having in account species richness, trophic structure and abundance:

1. Number of species
2. Number of species comprising 90% of the catch
3. Number of species in the bottom trawl
4. Proportion of carnivores
5. Proportion of planktivores
6. Proportion of benthivores
7. Number of estuarine fish
8. Number of anadromous fish
9. Total fish with Atlantic menhaden removed.

The scoring of metrics to calculate the index is done by comparing the value of a metric from the sample of unknown water quality to thresholds established from reference data distributions.

2.7.6   Fish Health Index (FHI)[128]

This index is based on the community degradation index (CDI), which measures the degree of dissimilarity (degradation) between a potential fish assemblage and the actual measured fish assemblage.

FHI provide a measure of the similarity (health) between the potential and actual fish assemblages and is calculated using the formula:

$$\text{FHI} = 10 \, (J)[\ln (P)/ \ln (P_{\max})]$$

where J is the number of species in the system divided by the number of species in the reference community, $P =$ is the potential species richness (number of species) of each reference community, and $P_{\max}$ is the maximum potential species richness from all the reference communities. The index ranges from 0 (poor) to 10 (good).

The FHI was used to assess the state of South Africa's estuaries.[128] Although the index has proved to be a useful tool in condensing information of estuarine fish assemblages into a single numerical value, the index is only based on presence/absence data and does not take into account the relative proportions of the various species present.

2.7.7   Estuarine Ecological Index (EBI)[129]

The EBI includes the following eight metrics:

1. Total number of species
2. Dominance
3. Fish abundance
4. Number of nursery
5. Number of estuarine spawning species
6. Number of resident species
7. Proportion of benthic associated species
8. Proportion of abnormal or diseased fishes.

The usefulness of this index requires it to reflect not only the current status of fish communities but also to be applicable over a wide range of estuaries, although this is not entirely achieved.[132]

2.7.8   Estuarine Fish Importance Rating (FIR)[133]

This index is constructed from seven weighted measures of species and estuarine importance and is designed to work on a presence/absence data set where species are only considered to be present if they constituted more than 1% of any catch by number.

Measures of species importance:

- Number of exploitable species
- Number of estuarine-dependent species
- Number of endemic species.

Measures of estuarine importance:

- Type
- Size
- Condition
- Isolation.

This index is able to provide a ranking, based on the importance of each estuary and helps to identify the systems with major importance for fish conservation.

2.8   PRESENTATION AND DEFINITION OF LEVEL 7 AND 8 INDICATORS — HOLISTIC INDICATORS

An ecological network is often drawn as a conceptual diagram that is used as the first step in a modeling development procedure. Figure 2.1 shows a nitrogen cycle in a lake and it represents a conceptual diagram and the ecological network for a model of the nitrogen cycle. The complexity of the ecological network in Figure 2.1 cannot be used as ecological indicator because the real network is simplified too much in the figure; but if observations of the real network make it possible to draw close to the real network, a similar figure is obtained; but much more complicated. The complexity of the network in this figure could be used as an indicator for the function of the real ecosystem — even if the network was still a simplification of the real ecosystem.

Gardner and Ashby examined the influence on stability of connectivity (defined as the number of food links in the food web as a fraction of the number of topologically possible links) of large dynamic systems.[134] They suggest that all large complex dynamic systems may show the property of being stable up to a critical level of connectivity and then as the connectivity increases further, the system suddenly goes unstable. A connectivity of about 0.3 to 0.5 seems to give the highest stability.

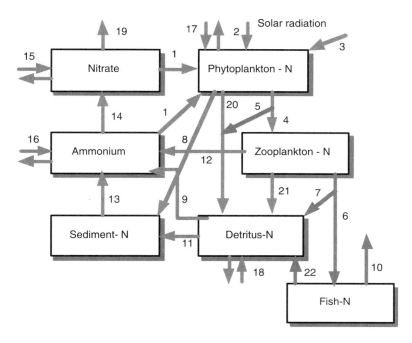

Figure 2.1   A conceptual diagram of the nitrogen cycle in a lake. The figure gives an illustration of the ecological network; but the real network is much more complex and the figure can therefore hardly be applied as an ecological indicator.

O'Neill examined the role on heterotrophs on the resistance and resilience and found that only small changes in heterotroph biomass could re-establish system equilibrium and counteract perturbations.[135] He suggests that the many regulation mechanisms and spatial heterogeneity should be accounted for when the stability concepts are applied to explain ecosystem responses.

These observations explain why it has been very difficult to find a relationship between ecosystem stability in its broadest sense and species diversity. Rosenzweig draws nearly the same conclusions.[136]

It can be observed that increased phosphorus loading gives decreased diversity,[137] but very eutrophic lakes are very stable. Figure 2.2 gives the result of a statistical analysis from a number of Swedish lakes. The relationship shows a correlation between number of species and the eutrophication, measured as chlorophyll-a in μg/l. A similar relationship is obtained between the diversity of the benthic fauna and the phosphorus concentration relative to the depth of the lakes.

Therefore it seems appropriate to introduce another but similar concept, named buffer capacity, β. It is defined as follows:[138,139]

$$\beta = 1/(\partial(\text{state variable})/\partial(\text{forcing function}))$$

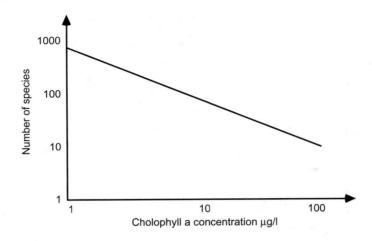

Figure 2.2 Weiderholm obtained the relationship shown for a number of Swedish lakes between the number of species and eutrophication, expressed as chlorophyll-a in μg/l.

Forcing functions are the external variables that drive the system such as discharge of waste water, precipitation, wind and so on, while state variables are the internal variables that determine the system, for instance the concentration of soluble phosphorus, the concentration of zooplankton and so on.

As demonstrated, the concept of buffer capacity has a definition that allows us to quantify, for instance, modeling. Furthermore, it is applicable to real ecosystems as it acknowledges that some changes will always take place in the ecosystem as response to changed forcing functions. The question is how large these changes are relative to changes in the conditions (the external variables or forcing functions).

The concept should be considered multidimensionally, as we may consider all combinations of state variables and forcing functions. It implies that even for one type of change there are many buffer capacities corresponding to each of the state variables. Ecological stability is defined as the ability of the system to resist changes in the presence of perturbations. It is a definition very close to buffer capacity, but it is lacking the multidimensionality of ecological buffer capacity.

The relation between forcing functions (impacts on the system) and state variables indicating the conditions of the system are rarely linear and buffer capacities are therefore not constant. In environmental management, it may therefore be important to reveal the relationships between forcing functions and state variables to observe under which conditions buffer capacities are small or large (compare with Figure 2.3).

Model studies have revealed that in lakes with a high eutrophication level, a high buffer capacity to nutrient inputs is obtained by a relatively small diversity.[139–141] The low diversity in eutrophic lakes is consistent with the

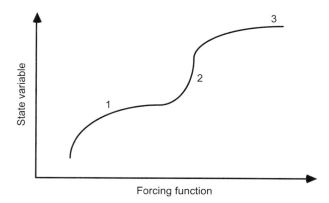

Figure 2.3   The relationship between state variables and forcing functions is shown. At point 1 and 3 the buffer capacity is high; at point 2 it is low.

above-mentioned results by Ahl and Weiderholm.[137] High nutrient concentrations equal a large phytoplankton species. The specific surface does not need to be large, because there are plenty of nutrients. The selection or competition is not on the uptake of nutrients but rather on escaping the grazing by zooplankton and here greater size is an advantage. In other words, the spectrum of selection becomes more narrow, which means reduced diversity. It demonstrates that a high buffer capacity may be accompanied by low diversity.

If a toxic substance is discharged into an ecosystem, the diversity will be reduced. The species most susceptible to the toxic substance will be extinguished, while other species — the survivors — will metabolize, transform, isolate, excrete, etc., the toxic substance and thereby decrease its concentration. A reduced diversity is observed, but simultaneously a high buffer capacity to the input of toxic compounds is maintained, which means that only small changes caused by the toxic substance will be observed. Model studies of toxic substance discharge into a lake[140,141] demonstrate the same inverse relationship between the buffer capacity to the considered toxic substance and diversity.

Ecosystem stability is therefore a very complex concept and it seems impossible to find a simple relationship between ecosystem stability and ecosystem properties.[142] Buffer capacity seems to be an applicable stability concept, as it is based on: (1) an acceptance of the ecological complexity — it is a multidimensional concept; and (2) reality — that is, that an ecosystem will never return to exactly the same situation again.

Another consequence of the complexity of ecosystems mentioned above should be considered here. For mathematical ease, the emphasis has been on equilibrium models, particularly with regard to population dynamics. The dynamic equilibrium conditions (steady state, not thermodynamic equilibrium) may be used as an attractor (in the mathematical sense, the ecological attractor is the thermodynamic equilibrium) for the system, but equilibrium will never be attained. Before the equilibrium is reached, the conditions, determined by the external factors and all ecosystem components, have changed and a new

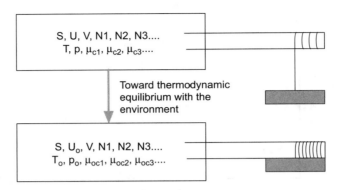

Figure 2.4   The definition of "technological" exergy is illustrated.

dynamic equilibrium (and thereby a new attractor) is effective. Before this attractor point has been reached, new conditions will again emerge and so on. A model based upon the equilibrium state will therefore give a wrong picture of ecosystem reactions. The reactions are determined by the present values of the state variables and they are different from those in the equilibrium state. We know from many modeling exercises that the model is sensitive to the initial values of the state variables. These initial values are a part of the conditions for further reactions and development. Consequently, the steady-state models may give results other than the dynamic models and it is therefore recommended to be very careful when drawing conclusions on the basis of equilibrium models. We must accept the complication that ecosystems are dynamic systems and will never attain equilibrium. We therefore need to apply dynamic models as widely as possible and it can easily be shown that dynamic models give results other than static ones.

Exergy is strictly defined as the amount of work the system can perform when it is brought into thermodynamic equilibrium with its environment. It is therefore dependent on both the environment and the system and not just on the system (see Figure 2.4). Exergy is therefore not a state variable, as free energy and entropy are.

If we choose the same ecosystem as a homogeneous "inorganic soup" and at same temperature and pressure as reference state (the environment), exergy will measure the thermodynamic distance from the "inorganic soup" in energy terms. This form for exergy is not strictly in accordance with the exergy introduced to calculate the efficiency of technological processes, but with the same system as the thermodynamic equilibrium at the same temperature and pressure as the reference state, we can calculate the exergy content of the system as coming entirely from biochemical energy and from the information embodied in the organisms (see Figure 2.5). The exergy of the system measures the contrast — the difference in work capacity — against the surrounding environment. To distinguish this exergy from technological exergy, we can call this exergy "eco-exergy." Wherever the expression exergy is used in this volume, it is assumed that it is eco-exergy.

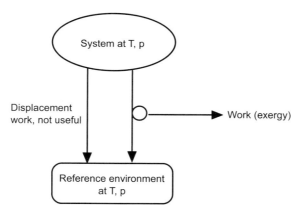

Figure 2.5 The definition of eco-exergy is illustrated. Eco-exergy is the amount of work that a system can perform when it is brought into equilibrium with the same system but with all the chemical compounds in form of inorganic decomposition products at the highest possible oxidation state. The reference system is an inorganic soup without life and without gradients. The reference state therefore has no eco-exergy.

If the system is in equilibrium with the surrounding environment the exergy is zero. The only way to move systems away from thermodynamic equilibrium is to perform work on them, and as the available work in a system is a measure of the ability, we have to distinguish between the system and its environment or thermodynamic equilibrium (i.e., the inorganic soup).

Survival implies maintenance of the biomass, and growth means increase of biomass. It costs exergy to construct biomass and obtain and store information. Survival and growth can therefore be measured by use of the thermodynamic concept exergy. Darwin's theory may therefore be reformulated in thermodynamic terms and expanded to the system level. The prevailing conditions of an ecosystem steadily change and the system will continuously select the species that can contribute most to the maintenance or even growth of the exergy of the system.

Notice that the thermodynamic translation of Darwin's theory requires that populations have the properties of reproduction, inheritance, and variation. The selection of the species that contribute most to the exergy of the system under the prevailing conditions requires that there are enough individuals with different properties to allow a selection can take place. It means that the reproduction and the variation must be high and that once a change has taken place due to a combination of properties that give better fitness, it can be conveyed to the next generation.

As proposed above, if we presume a reference environment that represents the system (ecosystem) at thermodynamic equilibrium, we can calculate the approximate exergy content of the system as coming entirely from the chemical energy: $\sum(\mu_c - \mu_{co})$ Ni where μ represents the echemical potential, respectively in the system (index c) and at thermodynamic equilibrium (index co) and Ni is the number of moles. Only what is called ca (chemical exergy) is therefore included in the computation of exergy. The physical exergy is omitted in these calculations as there are no temperature and pressure differences between the system and the reference system. We can determine with these calculations the exergy of the system compared with the same system at the same temperature and pressure but in the form of an inorganic soup without any life, biological structure, information, or organic molecules. As $(\mu_c - \mu_{co})$ can be found from the definition of the chemical potential replacing activities by concentrations, we get the following expressions for the eco-exergy:

$$\mathrm{Ex} = RT \sum_{i=0}^{i=n} c_i \ln c_i/c_{ieq} \quad [\mathrm{ML}^2\,\mathrm{T}^{-2}] \tag{2.1}$$

where R is the gas constant, T is the temperature of the environment, while c_i is the concentration of the ith component expressed in a suitable unit; for example, for phytoplankton in a lake c_i could be expressed as mg/l or as mg/l of a focal nutrient. c_{ieq} is the concentration of the ith component at thermodynamic equilibrium and n is the number of components. c_{ieq} is of course a very small concentration (except for $i = 0$, which is considered to cover the inorganic compounds), but is not zero, corresponding to a very low probability of forming complex organic compounds spontaneously in an inorganic soup at thermodynamic equilibrium.

The problem related to the assessment of c_{ieq} has been discussed and a possible solution proposed by Jørgensen et al.,[143] but the most essential arguments should be repeated here. Dead organic matter — detritus — which is given the index 1, can be found from classical thermodynamics:

$$\mu_1 = \mu_{1eq} + RT \ln c_1/c_{1eq} \quad [\mathrm{ML}^2\,\mathrm{T}^{-2}\,\mathrm{moles}^{-1}] \tag{2.2}$$

where μ indicates the chemical potential. The difference $\mu_1 - \mu_{1eq}$ is known for organic matter; for example, detritus, which is a mixture of carbohydrates, fats, and proteins. We find that detritus has approximately 18.7 kJ/g corresponding to the free energy of the mixture of carbohydrates, fats and proteins.

Generally, c_{ieq} can be calculated from the definition of the probability P_{ieq} to find component i at thermodynamic equilibrium:

$$P_{ieq} = c_{ieq} \Big/ \sum_{i=0}^{N} c_{ieq} \quad [-] \tag{2.3}$$

If we can find the probability, P_i, to produce the considered component i at thermodynamic equilibrium, we have determined the ratio of c_{ieq} to the total concentration. As the inorganic component, c_0, is very dominant in the thermodynamic equilibrium, Equation 2.3 may be rewritten as:

$$P_{ieq} \approx c_{ieq}/c_{0eq} \quad [-] \tag{2.4}$$

By a combination of equations, we get:

$$P_{1eq} = [c_1/c_{0eq}] \exp[-(\mu_1 - \mu_{1eq})/RT] \quad [-] \tag{2.5}$$

For the biological components, $2, 3, 4, \ldots N$, the probability, P_{ieq}, consists of the probability of producing the organic matter (detritus) — that is, P_{1eq}, and the probability, $P_{i,a}$, to obtain the information embodied in the genes, which determine the amino acid sequence. Living organisms use 20 different amino acids and each gene determines the sequence of about 700 amino acids. $P_{i,a}$, can be found from the number of permutations among which the characteristic amino acid sequence for the considered organism has been selected. It means that we have the following two equations available to calculate P_i:

$$P_{ieq} = P_{1eq} \times P_{i,a} \tag{2.6a}$$

($i \geq 2$; 0 covers inorganic compounds and 1 detritus) and

$$P_{i,a} = 20^{-700g} \quad [-] \tag{2.6b}$$

where g is the number of genes.
   Equation 2.4 is reformulated to:

$$c_{ieq} \approx P_{ieq} \times c_{0eq} \quad [\text{Moles L}^{-3}] \tag{2.7}$$

Equation 2.7 and Equation 2.2 are combined:

$$\text{Ex} \approx RT \sum_{i=0}^{N} [c_i \ln(c_i/(P_{ieq}c_{0eq}))] \quad [\text{ML}^2 \, \text{T}^{-2}] \tag{2.8}$$

This equation may be simplified by the use of the following approximations (based upon $P_{ieq} \ll c_i$, $P_{ieq} \ll P_0$ and $1/P_{ieq} \gg c_i$, $1/P_{ieq} \gg c_{0eq}/c_i$): $c_i/c_{0eq} \approx 1$, $c_i \approx 0$, $P_i c_{0eq} \approx 0$ and the inorganic component can be omitted. The significant contribution is coming from $1/P_{ieq}$, see Equation 2.8. We obtain:

$$\text{Ex} = -RT \sum_{i=1}^{N} c_i \ln(P_{ieq}) \quad [\text{ML}^2 \, \text{T}^{-2}] \tag{2.9}$$

where the sum starts from 1, because $P_{0,eq} \approx 1$.
   Expressing P_{ieq} as in Equation 2.6 and P_{1eq} as in Equation 2.5, we obtain the following expression for the calculation of an exergy index:

$$\text{Ex}/RT = \sum_{i=1}^{N} \left[c_i \cdot \ln\left(c_1/(c_{0eq})\right) - (\mu_1 - \mu_{1eq}) \sum_{i=2}^{N} c_i/RT - \sum_{i=1}^{N} c_i \ln P_{i,a} \right]$$
$$[\text{Moles L}^{-3}]$$

As the first sum is minor compared with the following two sums (use for instance $c_i/c_{0eq} \approx 1$), we can write:

$$Ex/RT = (\mu_1 - \mu_{1eq}) \sum_{i=1}^{N} c_i/RT - \sum_{i=1}^{N} c_i \ln P_{i,a} \quad [\text{Moles L}^{-3}] \tag{2.10}$$

This equation can now be applied to calculate contributions to the exergy index by important ecosystem components. If we consider detritus only, we know that the free energy of organic matter released is about 18.7 kJ/g. R is 8.4 J/mole, and the average molecular weight of detritus is assumed to be 100,000. We get the following contribution of exergy by detritus per liter of water, when we use the unit g detritus exergy equivalent/l:

$$Ex_1 = 18.7 \, c_i \, \text{kJ/l} \quad \text{or} \quad Ex_1/RT = 7.34 \times 105 \, c_i \quad [\text{ML}^{-3}] \tag{2.11}$$

A typical single-celled alga has on average 850 genes. Previously we have deliberately used the number of genes and not the amount of DNA per cell, which would include unstructured and nonsense DNA. In addition, a clear correlation between the number of genes and the complexity has been shown. However, recently it has been discussed that the nonsense genes are playing an important role; for example, they may be considered as spare parts, which are able to repair genes when they are damaged or be exposed to mutations. If it is assumed that only the informative genes contributes to the embodied information in organisms, an alga has 850 information genes in total — that is, they determine the sequence of $850 \times 700 = 595,000$ amino acids, the contribution of exergy per liter of water, using g detritus equivalent/l as concentration unit would be:

$$Exalgae/RT = 7.34 \times 10^5 \, c_i - c_i \ln 20^{-595000} = 25.2 \times 10^5 \, c_i \quad [\text{g/l}] \tag{2.12}$$

The contribution to exergy from a simple prokaryotic cell can be calculated similarly as:

$$Exprokar/RT = 7.34 \times 10^5 \, c_i + c_i \ln 20329000 = 17.2 \times 10^5 \, c_i \quad [\text{g/l}] \tag{2.13}$$

Organisms with more than one cell will have DNA in all cells determined by the first cell. The number of possible microstates becomes therefore proportional to the number of cells. Zooplankton has approximately 100,000 cells and 15,000 genes per cell, each determining the sequence of approximately 700 amino acids. In P_{zoo} (the probability to find 200 plankton at thermodynamic equilibrium) can therefore be found as:

$$- \ln P_{zoo} = -\ln (20^{-15000 \times 700} \times 10^{-5}) \approx 315 \times 105^5 \tag{2.14}$$

It can be seen that the contribution from the numbers of cells is insignificant. Similarly, P_{fish} and the P values for other organisms can be found.

The contributions from phytoplankton, zooplankton and fish to the exergy of the entire ecosystem are significant and far more than corresponding to the biomass. Notice that the unit of Ex/RT is g/l. Exergy can always be expressed in joules per liter, provided that the right units for R and T are used. Equation 2.12 to Equation 2.14 can be rewritten by converting g/l to g detritus/l by dividing by (7.34×10^5).

The exergy index can as seen be found as the concentrations of the various components, c_i, multiplied by weighting factors, β_i, reflecting the exergy that the various components possess due to their chemical energy and to the information embodied in DNA:

$$\text{Ex} = \sum_{i=n}^{i=0} \beta_i c_i \tag{2.15}$$

β_i values based on exergy detritus equivalents have been found for various species. The unit exergy detritus equivalents expressed in g/l can be converted to kJ/l by multiplication by 18.7 (corresponding to the approximate average energy content of 1 g detritus). Table 2.1 shows the number of information genes and the corresponding β-values calculated from the above equations.

The index 0 covers the inorganic components, which of course should be included in the calculations of exergy in principle, but in most cases they can be neglected as the contributions from detritus and to an even bigger extent from the biological components are much higher due to an extremely low concentration of these components in the reference system (the ecosystem converted to an inorganic dead system). The calculation of exergy index accounts by use of this equation for the chemical energy in the organic matter as well as for the information embodied in the living organisms. It is measured by the extremely small probability to form the living components; for instance, algae, zooplankton, fish, mammals etc., spontaneously from inorganic matter. The weighting factors may also be considered quality factors reflecting how developed the various groups are and to which extent they contribute to the exergy due to their content of information which is reflected in the computation. Boltzmann gave the following relationship for the work, W, that is embodied in the thermodynamic information:

$$W = RT \ln N \quad [\text{ML}^2 \text{T}^{-2}] \tag{2.16}$$

where N is the number of possible states, among which the information has been selected. N is for species considered to be the inverse of the probability to obtain the valid amino acid sequence spontaneously.

It is furthermore consistent with the following reformulation: "Information appears in nature when a source of energy (exergy) becomes available but the corresponding (entire) entropy production is not emitted immediately, but is held back for some time (as exergy)."

The total eco-exergy of an ecosystem cannot be calculated exactly, as we cannot measure the concentrations of all the components or determine all possible contributions to exergy in an ecosystem. If we calculate the exergy of a fox for instance, the calculations above will only give the contributions coming from the biomass and the information embodied in the genes, but what is the contribution from the blood pressure, the sexual hormones and so on? These properties are at least partially covered by the genes but is that the entire story? We can calculate the contributions from the dominant components, for instance by the use of a model or measurements, that covers the most essential components for a focal problem.

Exergy calculated by use of the above equations has some shortcomings. It is therefore proposed to consider the exergy found by these calculations as a relative exergy index:

1. We account only for the contributions from the organisms' biomass and information in the genes. Although these contributions most probably are the most important ones, it cannot be completely excluded that other important contributions are omitted.
2. We don't account for the information embodied in the network — the relations between organisms. The information in the model network that we use to describe ecosystems is negligible compared with the information in the genes, but we cannot exclude that the real, much more complex network may contribute considerably to the total exergy of a natural ecosystem.
3. We have made approximations in our thermodynamic calculations. They are all indicated in the calculations and are in most cases negligible.
4. We can never know all the components in a natural (complex) ecosystem. Therefore, we will only be able to utilize these calculations to determine exergy indices of our simplified images of ecosystems, for instance of models.
5. The exergy indices are, however, useful, as they have been successfully used as goal function (orientor) to develop structural dynamic models. The difference in exergy by comparison of two different possible structures (species composition) is decisive here. Moreover, exergy computations give always only relative values, as the exergy is calculated relatively to the reference system.

As already stressed, the presented calculations do not include the information embodied in the structure of the ecosystem — that is, the relationships between the various components, which is represented by the network. The information of the network encompasses the information of the components and the relationships of the components. The latter contribution has been calculated by Ulanowicz[144–146] as a part of the concept of ascendancy. In principle, the information embodied in the network should be included in the calculation of the exergy index of structural dynamic models, as the network is dynamically also changed. It may, however, often be omitted in most dynamic model calculations because the contributions from the network relationships of models (not from the components of the network, of course) are minor compared with the contributions from the components. This is due to the extreme simplifications made in the models compared with the networks in real ecosystems. It can therefore not be excluded that networks of real ecosystems may contribute considerably to the total exergy of the ecosystems, but for the type of models that we are using at present, we can probably omit the exergy of the information embodied in the network.

Specific exergy is defined as the exergy or rather exergy index divided by the biomass. Specific exergy expresses with other words the dominance of the higher organisms as they per unit of biomass carry more information — that is, have higher β values. A very eutrophic ecosystem has a very high exergy due to the high concentration of biomass, but the specific exergy is low as the biomass is dominated by algae with low β values.

The combination of the exergy index and the specific exergy index gives usually a more satisfactory description of the health of an ecosystem than the exergy index alone, because it considers the diversity and the life conditions for higher organisms. The combination of exergy, specific exergy and buffer capacities, defined as the change in a forcing function relatively to the corresponding change in a state variable, has been used as ecological indicators for lakes.

H.T. Odum defines the maximum power principle as a maximization of useful power.[147] It implies that the contributions to the total power that are useful. It means, that non-useful power is not included in the summation. The difference between useful and non-useful power is perhaps the key to understand Odum's principle and to utilize it to interpret ecosystem properties.

According to Odum it is the transformation of energy into work (consistent with the term useful power), that determines success and fitness. Many ecologists have incorrectly assumed that natural selection tends to increase efficiency. If this were true, endothermy could never have evolved. Endothermic birds and mammals are extremely inefficient compared with reptiles and amphibians. They expend energy at high rates in order to maintain a high, constant body temperature, which gives high levels of activity independent of environmental temperature. Fitness may be defined as reproductive power, dW/dt, the rate at which energy can be transformed into work to produce offspring. This interpretation of the maximum power principle is consistent with the maximum exergy.

In a book named *Maximum Power — The Ideas and Applications of H.T. Odum*, by Hall, was presented a clear interpretation of the maximum power principle, as had been applied in ecology by Odum. The principle claims that power or output of useful work is maximized — not the efficiency and not the rate but the trade-off between a high rate and high efficiency yielding most useful energy is equal to useful work. As illustrated in Figure 2.6, Hall is using an interesting semi-natural experiment to illustrate the application of the principle in ecology. Streams were stocked with different levels of predatory cutthroat trout. When predator density was low, there was considerable invertebrate food per predator, and the fish used relatively little maintenance of food-searching energy per unit of food obtained. With a higher fish-stocking

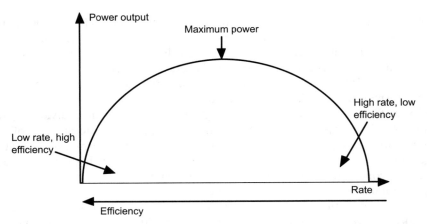

Figure 2.6   The maximum power principle claims that the development of an ecosystem is a trade-off (a compromise) between the rate and the efficiency — that is, the maximum power output per unit of time.

rate, food became less available per fish, and each fish had to use more energy searching for it. Maximum production occurred at intermediate fish-stocking rates, which means intermediate rates at which the fish utilized their food.

Hall mentions another example. Deciduous forests in moist and wet climates tend to have a leaf area index of about 6. Such an index is predicted from the maximum power hypothesis applied to the net energy derived from photosynthesis. Higher leaf area index values produce more photosynthate, but do so less efficiently because of the respirational demand of the additional leaf. Lower leaf area indices are more efficient per leaf, but draw less power than the observed intermediate values of roughly 6.

According to Andresen the same concept applies for regular fossil fuel power generation. The upper limit of efficiency for any thermal machine such as a turbine is determined by the Carnot efficiency. A steam turbine could run at 80% efficiency, but it would need to operate at a nearly infinitely slow rate. Obviously we are not interested in a machine that generates revenues infinitely slowly, no matter how efficiently. Actual operating efficiencies for modern steam powered generator are therefore closer to 40%, roughly half the Carnot efficiency. The examples show that the maximum power principle is embedded in the irreversibility of the world. The highest process efficiency can be obtained by endoreversible conditions, meaning that all irreversibilities are located in the coupling of the system to its surroundings — there are no internal irreversibilities. Such systems will, however, operate too slowly. Power is zero for any endoreversible system. If we want to increase the process rate, we need to increase the irreversibility and thereby decrease the efficiency. The maximum power is the compromise between endoreversible processes and very fast, completely irreversible processes.

Emergy was introduced by H.T. Odum[147] and attempts to account for the energy required in formation of organisms in different trophic levels. The idea

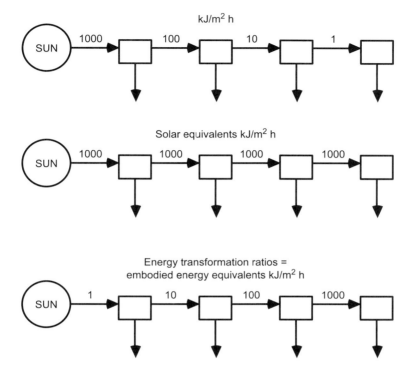

Figure 2.7   Energy flow, solar equivalents and energy transformation ratios equal the embodied energy equivalents in a food chain.

is to correct energy flows for their quality. Energies of different types are converted into equivalents of the same type by multiplying by the energy transformation ratio. For example fish, zooplankton, and phytoplankton can be compared by multiplying their actual energy content by their solar energy transformation ratios. The more transformation steps there are between two kinds of energy, the greater the quality and the greater the solar energy required to produce a unit of energy (J) of that type. When one calculates the energy of one type, that generates a flow of another. This is sometimes referred to as the embodied energy of that type. Figure 2.7 presents the concept of embodied energy in a hierarchical chain of energy transformation and Table 2.4 gives embodied energy equivalents for various types of energy.

Odum reasons that surviving systems develop designs that receive as much energy amplifier action as possible. The energy amplifier ratio is defined in Figure 2.8 as the ratio of output B to control flow C. Odum suggests that in surviving systems the amplifier effects are proportional to embodied energy, but full empirical testing of this theory still needs to be carried out in the future.

One of the properties of high-quality energies is their flexibility. Whereas low-quality products tend to be special, requiring special uses, the

Table 2.4 Embodied energy equivalents for various types of energy

Type of energy	Embodied energy equivalents
Solar energy	1.0
Winds	315
Gross photosynthesis	920
Coal	6800
Tide	11,560
Electricity	27,200

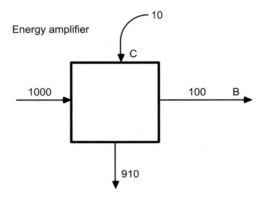

Figure 2.8 The energy amplifier ratio, R, is defined as the ratio of output B to control flow C. It means that $R = 10$ in this case.

higher-quality part of a web is of a form that can be fed back as an amplifier to many different units throughout the web. For example, the biochemistry at the bottom of the food chain in algae and microbes is diverse and specialized, whereas the biochemistry of top animal consumer units tends to be similar and general, with services, recycles, and chemical compositions usable throughout.

Hannon[149] and Hannon and Ruth[150] applied energy intensity coefficients as the ratios of assigned embodied energy to actual energy to compare systems with different efficiencies. The difference between embodied energy flows and power (see Equation 2.17) simply seems to be a conversion to solar energy equivalents of the free energy ΔF. The increase in biomass in Equation 2.17 is a conversion to the free energy flow and the definition of embodied energy is a further conversion to solar energy equivalents.

Embodied energy is (as seen from these definitions) determined by the bio-geo-chemical energy flow into an ecosystem component, measured in solar energy equivalents. The stored emergy, Em, per unit of area or volume to be distinguished from the emergy flows can be found from:

$$Em = \sum_{i=1}^{i=n} \Omega_i \times c_i \tag{2.17}$$

where Ω_i is the quality factor which is the conversion to solar equivalents, as illustrated in Table 2.2 and Figure 2.7, and c_i is the concentration expressed per unit of area or volume. The calculations reduce the difference between stored emergy (embodied energy) and stored exergy, which can also be determined with good approximation as the sum of concentrations multiplied by a quality factor (see Equation 2.15), to a difference between the applied quality factors. Emergy uses as a quality factor the cost in form of solar energy, while the exergy quality accounts for the information embodied in the biomass. Emergy gives the costs while exergy gives the result. The ratio emergy paid to resulting exergy (see chapter 10).

Emergy calculates thereby how much solar energy (which is our ultimate energy resource) is required to obtain 1 unit of biomass of various organisms, while exergy accounts for how much "first class" energy (i.e., energy that can do work) the organisms, as a result of the complex interactions in an eco-system, possess. Both concepts attempt to account for the quality of the energy: emergy by looking into the energy flows in the ecological network to express the energy costs in solar equivalents; and exergy by considering the amount of information that the components have embodied.

The differences between the two concepts may be summarized as follows:

1. Emergy has no clear reference state, which is not needed as it is a measure of energy flows, while eco-exergy is defined relatively to the same system at thermodynamic equilibrium.
2. The quality factor of exergy is based on the content of information, while the quality factor for emergy is based on the cost in solar equivalents.
3. Exergy is better anchored in thermodynamics and has a wider theoretical basis.
4. The quality factor, Ω, may be different from ecosystem to ecosystem and in principle it is necessary to assess in each case the quality factor based on an energy flow analysis, which is sometimes cumbersome to make. The quality factors listed in Table 2.2 may be used generally as good approximations. The quality factors used for computation of exergy, β, require a knowledge to the non-nonsense genes of various organisms, which sometimes is surprisingly difficult to assess (see Appendix A).

In his book *Environmental Accounting — Emergy and Environmental Decision Making*, Odum used calculations of emergy to estimate the sustainability of the economy of various countries. As emergy is based on the cost in solar equivalents, which is the only long-term available energy, it seems to be a sound first estimation of sustainability, although it sometimes is an extremely difficult concept to quantify.

The diversity index (DI) for an ecosystem is usually represented as:

$$\text{DI} = -\sum_{i=1}^{s} \left(P_i \times \log_2 P_i \right) \tag{2.18}$$

which originates from Shannon's theory deriving the average entropy of discrete information. Generally, many kinds of similar indices are proposed and used, but Equation 2.18 is comparatively sound in its theoretical basis of statistical mechanics[37]. P_i in Equation 5-4 originally signified the probability of occurrence of the ith information, which was later replaced by n_i/N in an ecosystem by Margalef, where N is the total number of living elements in the ecosystem and n_i the number of living member of the ith species — that is, $P_i = n_i/N$. Therefore the diversity index is denoted by the relationship:

$$DI = \sum_{i=1}^{s} \left(n_i/N\right) \times \log_2\left(n_i/N\right) \tag{2.19}$$

where N is the total number of living elements; n_i is the living number of the ith species.

2.9 AN OVERVIEW OF APPLICABLE ECOLOGICAL INDICATORS FOR EHA

Table 2.5 gives an overview of the classes of ecological indicators (see the eight levels in section 2 of this chapter) applied in the chapters 3 to 15. It is indicated in the table which ecosystems the various chapters are considering in the presentation of proposed ecological indicators. It is of course not possible to present all applicable indicators in 13 case studies of the use of ecological indicators for EHA. As mentioned in section 2, level 1 indicators have been widely used for EHA of rivers and may also be used in most other ecosystems. Similarly, concentrations of chemical compounds are obvious to use for all unhealthy conditions caused by toxic substances. The experience gained by the use of level 6 to 8 indicators is usually of more general value in the EHA, because the higher level indicators give an overall (holistic) picture of how far

Table 2.5 Overview of applied ecological indicators in chapters 3 to 16

Chapter	Ecosystem	Indicator level							
		1	2	3	4	5	6	7	8
3	Coastal, estuary							+	+
4	Lake	+	+	+	+	+	+		
5	Lake				+	+	+	+	+
6	Coastal			+	+	+	+	+	+
7	Coastal				+	+	+	+	+
8	Marine				+	+	+		
9	Wetland			+	+	+	+	+	
10	Pond, lagoon, lake, basin							+	+
11	Agroecosystem								+
12	Landscape			+	+	+	+	+	+
13	Landscape						+		
14	Regional								+
15	Regional			+	+				
16	River and estuary	+	+	+	+	+	+		

a focal ecosystem on the system level is from healthy conditions. The overview that is a result of this volume is therefore to a higher extent giving information on the applicability of level 6 to 8 indicators. These indicators should in practical EHA be supplemented with level 1 to 5 indicators, which are more specific. The selection of level 1 to 5 indicators is furthermore obvious in most cases; for instance the use of PCB and zebra mussels in the EHA of the Great North American Lakes (see chapter 4).

2.10 EHA: PROCEDURES

2.10.1 Direct Measurement Method (DMM)

The procedures established for the direct measurement method (DMM) are as follows:

1. Identify the necessary indicators to be applied in the assessment process: use Table 2.3, section 2.2.
2. Measure directly or calculate indirectly the selected indicators.
3. Assess ecosystem health based on the resulting indicator values.

2.10.2 Ecological Model Method (EMM)

The procedures established for the ecological modeling method (EMM) for lake ecological health assessment are shown in Figure 2.9.

Five steps are necessary when assessing lake ecosystem health using EMM procedure:

1. Determine the model structure and complexity according to the ecosystem structure.
2. Establish an ecological model through designing a conceptual diagram, developing model equations, and estimating model parameters.
3. Calibrate the model as necessary in order to assess its suitability in application to ecosystem health assessment process.
4. Calculate ecosystem health indicators.
5. Assess ecosystem health based on the values of the indicators.

2.10.3 Ecosystem Health Index Method (EHIM)

In order to assess quantitatively the state of ecosystem health, an ecosystem health index (EHI) in a scale of 0 to 100 was developed. It was assumed that, when EHI is zero, the healthy state is worst; when EHI is 100, the healthy state is best. In order to facilitate the description of healthy states, EHI is equally divided into five segments or ranges as: 0–20%, 20–40%, 40–60%, 60–80%, 80–100%, which correspond with the five health states, "worst," "bad," "middle," "good" and "best," respectively.

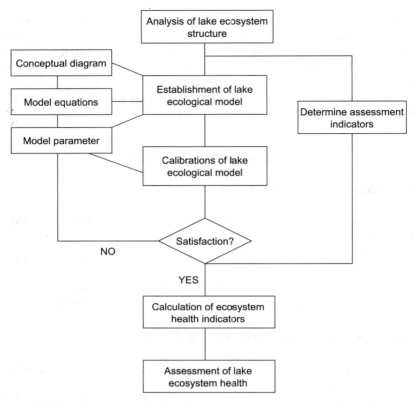

Figure 2.9 The procedure of ecological model method (EMM) for ecological health assessment. (Modified from Xu et al. *Wat. Res.* 35(1), 3160, 2001. With permission.)

EHI can be calculated by the following equation:

$$\text{EHI} = \sum_{i=1}^{n} \omega_i \times \text{EHI}_i \qquad (2.20)$$

where EHI is a synthetic ecosystem health index, EHI_i is the ith ecosystem health index for the ith indicator, ω_i is the weighting factor for the ith indicator. It can be seen from Equation 2.20 that the synthetic EHI depends on sub-EHIs and weighting factors for each indicators. The procedure established for lake ecosystem health assessment using EHI method is shown in Figure 2.10. Five steps are necessary for EHI method:

1. Select basic and additional indicators
2. Calculate sub-EHIs for all selected indicators
3. Determine weighting factors for all selected indicators
4. Calculate synthetic EHI using sub-EHIs and weighting factors for all selected indicators
5. Assess ecosystem health based on synthetic EHI values.

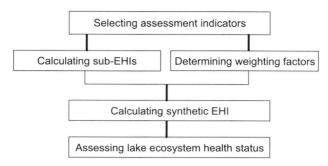

Figure 2.10   The procedure of EHI method for lake ecosystem health assessment.

2.11 AN INTEGRATED, CONSISTENT ECOSYSTEM THEORY THAT CAN BE APPLIED AS THE THEORETICAL BASIS FOR EHA

Several ecosystem theories have been presented in the scientific literature during the last two or three decades. At first glance they look very different and seem to be inconsistent, but a further examination reveals that they are not so different and that it should be possible to unite them in a consistent pattern, which is the idea behind this volume. This has been accepted among system ecologists since 1998/1999, but as a result of two meetings in 2000 (one in Italy, Porto Venere late May and one in Copenhagen, early June in conjunction with an American Society for Limnology and Oceanography (ASLO) meeting), it can now be concluded that a consistent pattern of ecosystem theories has been formed. Several system ecologists agreed on the pattern presented below as a working basis for further development in system ecology. This is of utmost importance for progress in system ecology, because with a theory in hand it will be possible to explain many rules that are published in ecology and applied ecology which again explain many ecological observations. We should, in other words, be able attain the same theoretical basis that characterizes physics: a few basic laws, which can be used to deduce rules that explain observations. It has therefore also been agreed that one of the important goals in system ecology would be to demonstrate (prove) the links between ecological rules and ecological laws.

Ten to fifteen years ago the presented theories seem very inconsistent and chaotic. How could E.P. Odum's attributes,[7] H.T. Odum's maximum power,[147] Ulanowicz ascendancy,[144] Patten's indirect effect,[151] Kay and Schneider's maximum exergy degradation,[152] Jørgensen's maximum exergy principle,[138,140,153], and Prigogine[154] and Mauersberger's minimum entropy dissipation[155–156] be valid at the same time? Everybody insisted that his version of a law for ecosystem development was right, and all the other versions were wrong. New results and an open discussion among the contributing scientists have led to a formation of a pattern whereby all the theories contribute to the total picture of ecosystem development.

The first contribution to a clear pattern of the various ecosystem theories came from the network approach used often by Patten (Patten and Fath, personal communication). Patten and Fath have shown by a mathematical analysis of networks in steady state (representing for instance an average annual situation in an ecosystem with close to balanced inputs and outputs for all components in the network) that the sum of through flows in a network (which is maximum power) is determined by the input and the cycling within the network. The input (the solar radiation) again is determined by the structure of the system (the stored exergy, the biomass). Furthermore, the more structure the more maintenance is needed and therefore more exergy must be dissipated, the greater the inputs are. Cycling on the other hand means that the same energy (exergy) is better utilized in the system, and therefore more biomass (exergy) can be formed without an increase in inputs. It has been shown previously that more cycling means increased ratio of indirect to direct effects, while increased input has no effect on the ratio of indirect to direct effects. Fath and Patten used these results to determine the development of various variables used as goal functions (exergy, power, entropy etc.). An ecosystem is of course not setting goals, but a goal function is used to describe the direction of development an ecosystem will take in an ecological model. Their results can be summarized as follows:

1. Increased inputs (more solar radiation is captured) imply more biomass, more exergy stored, more exergy degraded, therefore also higher entropy dissipation, more through-flow (power), increased ascendency, but no change in the ratio of indirect to direct effects or in the retention time for the energy in the system equal to the total exergy/input exergy per unit of time.
2. Increased cycling implies more biomass, more exergy stored, more through-flow, increased ascendency, increased ratio indirect to direct effect, increased retention but no change in exergy degradation.

Almost simultaneously, Jørgensen et al. published a paper which claimed that ecosystems show three growth forms:[157]

1. Growth of physical structure (biomass) which is able to capture more of the incoming energy in form of solar radiation, but also requires more energy for maintenance (respiration and evaporation).
2. Growth of network, which means more cycling of energy and matter.
3. Growth of information (more develop plants and animals with more genes), from r strategists to K strategists, which waste less energy but also usually carry more information.

These three growth forms may be considered an integration of E.P. Odum's attributes which describe changes in ecosystem associated with development from the early stage to the mature stage. Eight of the most applied attributes associated to the three growth forms should be mentioned (the complete list of attributes is given in Table 2.1):

1. Ecosystem biomass (physical structure) increases
2. More feed back loops (including recycling of energy and matter) are built

3. Respiration increases
4. Respiration relative to biomass decreases
5. Bigger animals and plants (trees) become more dominant
6. The specific entropy production (relative to biomass) decreases
7. The total entropy production will first increases and then stabilize on approximately the same level
8. The amount of information increases (more species, species with more genes, the biochemistry becomes more diverse).

Growth form 1 covers attributes 1, 3, and 7. Growth form 2 covers 2 and 6, and growth form 3 covers the attributes 4, 5, 7, and 8.

In the same paper, Figure 2.11 was presented to illustrate the concomitant development of ecosystems, exergy captured (most of that being degraded) and exergy stored (biomass, structure, information). The points in the figures correspond to different "ecosystems": an asphalt road, bare soil, a desert, grassland, young spruce plantation, older spruce plantation, old temperate forest and rain forest. Debeljak has shown that he gets the same shape as in Figure 2.11 when he determines exergy captured and exergy stored in managed forest and virgin forest on different stages of development[158] (see Figure 2.12).

Holling has suggested how ecosystem progress through the sequential phases of renewal (mainly growth form 1), exploitation (mainly growth form 2), conservation (dominant growth form 3) and creative destruction (see Figure 2.13).[159] The latter phase fits also into the three growth forms but will

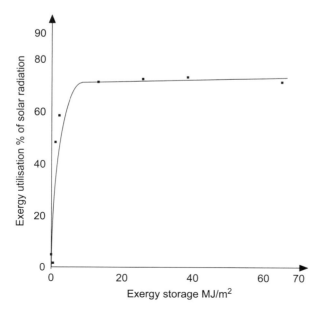

Figure 2.11 Exergy storage vs. exergy utilzation (percentage of solar radiation) for various ecosystems.

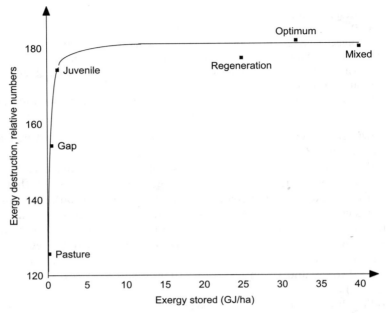

Figure 2.12   The plot shows the result by Debeljak. He examined managed and virgin forest in different stages. Gap has no trees, while the virgin forest changes from optimum to mixed to regeneration and back to optimum, although the virgin forest can be destroyed by catastrophic events as fire or storms. The juvenile stage is a development between the gap and the optimum. Pasture is included for comparison.

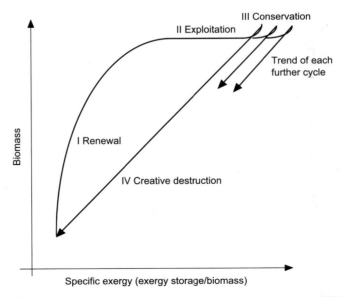

Figure 2.13   Holling's four stages are expressed in terms of biomass and specific exergy.

require a further explanation. The creative destruction phase is either a result of external or internal factors. In the first case (for instance hurricanes and volcanic activity), further explanation is not needed as an ecosystem has to use the growth forms under the prevailing conditions, which are determined by the external factors.

If the destructive phase is a result of internal factors, the question is "why would a system be self-destructive?" A possible explanation is that a result of the conservation phase is that almost all nutrients will be contained in organisms which implies that there are no nutrients available to test new and possibly better solutions to move further away from thermodynamic equilibrium or, expressed in Darwinian terms, to increase the probability of survival. This is also implicitly indicated by Holling, as he talks about creative destruction. Therefore when new solutions are available, it would in the long run be beneficial for the ecosystem to decompose the organic nutrients into inorganic components which can be utilized to test the new solutions. The creative destruction phase can be considered a method to utilize the three other phases and the three growth forms more effectively in the long run.

Five hypothesis have been proposed to describe ecosystem growth and development, namely:

1. The entropy production tends to be minimum (this was proposed by Prigogine in 1947 and 1980, for linear systems at steady nonequilibrium state, not for far from equilibrium systems). It is applied by Mauersberger to derive expressions for bioprocesses at a stable stationary state (see chapter 5).[155,156]

2. Natural selection tends to make the energy flux through the system a maximum, so far as compatible with the constraints to which the system is subject.[147] This is also called the maximum power principle (see Section 2.3).

3. Ecosystems will organize themselves to maximize the degradation of exergy.[160]

4. A system that receives a through flow of exergy will have a propensity to move away from thermodynamic equilibrium, and if more combinations of components and processes are offered to utilize the exergy flow, the system has the propensity to select the organization that gives the system as much stored exergy as possible. See section 3 of this chapter and references 140 and 153.

5. Ecosystems will have a propensity to developed toward a maximization of the ascendancy.[144]

The usual description of ecosystem development illustrated for instance by the recovery of Yellowstone Park after fire, an island born after a volcanic eruption, reclaimed land etc. is well covered by E.P. Odum:[7] at first the biomass increases rapidly which implies that the percentage of captured incoming solar radiation increases but also the energy needed for the maintenance. Growth form 1 is dominant in this first phase, where stored exergy increases (more biomass, more physical structure to capture more solar radiation), but also the through-flow (of useful energy), exergy

Table 2.6   Accordance between growth forms and the proposed descriptors

	Hypothesis		
	Growth form 1	Growth form 2	Growth form 3
Exergy storage	Up	Up	Up
Power/through flow	Up	Up	Up
Ascendency	Up	Up	Up
Exergy dissipation	Up	Equal	Equal
Retention time	Equal	Up	Up
Entropy production	Up	Equal	Equal
Exergy/biomass = specific exergy	Equal	Up	Up
Entropy/biomass = spec. entropy prod.	Equal	Down	Down
Ratio indirect/direct effects	Equal	Up	Up

dissipation and the entropy production increases due to increased need of energy for maintenance. Growth forms 2 and 3 become dominant later, although an overlap of the three growth forms does take place. When the percentage of solar radiation captured reaches about 80%, it is not possible to increase the amount of captured solar radiation further (due in principle to the second law of thermodynamics). Therefore, further growth of the physical structure (biomass) does not improve the energy balance of the ecosystem. In addition, all or almost all the essential elements are in form of dead or living organic matter and not as inorganic compounds ready to be used for growth. Growth form 1 will therefore not proceed, but growth forms 2 and 3 can still operate. The ecosystem can still improve the ecological network and can still change r strategists with K strategists, small animals and plants with bigger ones and less developed with more developed with more information genes. A graphic representation of this description of ecosystem development is presented in Figure 2.11 and Figure 2.12. The accordance with the five descriptors, specific entropy production, and the three growth forms based on this description of ecosystem development is shown in Table 2.6.

The presented integrated ecosystem theory can be applied in EHA in two ways:

1. The widely applied E.P. Odum's attributes are as demonstrated covered by the use of several of the presented holistic indicators; for instance, exergy, emergy, ascendency, specific exergy, and entropy production/ biomass. The application of the holistic indicators thereby gets wider perspectives.
2. The development of the three growth forms may be used to explain the thermodynamic holistic indicators.

It is mandatory to understand the development of ecosystems and their reactions to stress when the results of an EHA are interpreted in environmental management context. Therefore it is important not to consider the indicators just as classification numbers but to attempt to understand "the story" behind the indicators to be able to answer the questions: why and where is the ecosystem unhealthy? How did it happen? When will the ecosystem become healthy again? What should we do to recover the ecosystem? This will often require a profound knowledge of ecology and ecosystem theory.

REFERENCES

1. Von Bertalanffy, L. *Problems of Life*. Wiley, New York, 1952.
2. Costanza, R., Norton, B.G., and Haskell, B.D., Eds. *Ecosystem Health: New Goals for Environmental Management*. Island Press, Washington, D.C., 1992.
3. Kay, J.J. and Schneider, E.D. "Thermodynamics and measurements of ecosystem integrity," in *Ecological Indicators*, McKenzie, D., Ed. Elsevier, Amsterdam, 1991, pp. 159–182.
4. Akbari, H. and Sezgen, O. Performance evaluation of thermal-energy storage-systems. *Energ. Build.* 22, 15–24, 1995.
5. O'Connor, J.S. and Dewling, R.T. Indices of marine degradation: their utility. *Environ. Manage.* 10, 335–343, 1986.
6. Hynes, H.B.N. *Ecology of Running Water*. Liverpool University Press, Liverpool, 1971.
7. Reference removed.
8. Reference removed.
9. Bellan, G. Pollution et peuplements bentiques sur substrat meuble dans la region de Marseille. 1. Le secteur de Cortiou. *Rev. Intern. Oceanogr. Med.* 6, 53–87, 1967.
10. Reference removed.
11. Bellan, G. Annélides polychétes des substrats solids de troits mileux pollués sur les côrtes de Provence (France): Cortiou, Golfe de Fos, Vieux Port de Marseille. *Téthys* 9, 260–278, 1980.
12. Bellan-Santini, D. Relationship between populations of amphipods and pollution. *Mar. Pollut. Bull.* 11, 224–227, 1980.
13. Warwick, R.M. Environmental impact studies on marine communities: pragmatical considerations. *Aust. J. Ecol.* 18, 63–80, 1993.
14. Reference removed.
15. Reference removed.
16. Roberts, R.D., Gregory, M.G., and Foster, B.A. Developing an efficient macrofauna monitoring index from an impact study — a dredge spoil example. *Mar. Pollut. Bul.* 36, 231–235, 1998.
17. Smith, R.W., Bergen, M., Weisberg, S.B., Cadien, D., Dalkey, A., Montagne, D., Stull., J.K., and Velarde, R.G. Benthic response index for assessing infaunal communities on the Mainland shelf of southern California. *Ecol. Appl.* 11, 1073–1087, 2001.
18. Pearson, T.H. and Rosenberg, R. Macrobenthic succession in relation to organic enrichment and pollution of the marine environment. *Oceanogr. Mar. Biol. Ann. Rev.* 16, 229–331, 1978.
19. Niell, F.X. and Pazó, J.P. Incidencia de vertidos industriales en la estructura de poblaciones intermareales. II. Distribución de la biomasa y de la diversidad específica de comunidades de macrófitos de facies rocosa. *Inv. Pesq.* 42, 231–239, 1978.

20. Pérez-Ruzafa, A. and Marcos, C. "Contaminación marina: Enfoques y herramientas para abordar los problemas ambientales del medio marino," in *Perspectivas y Herramientas en el Estudio de la Contaminación Marina*, Pérez-Ruzafa, A., Marcos, C., Salas, F., and Zamora, S., Eds. Servicio de Publicaciones, Universidad de Murcia, Murcia, 2003, pp. 11–33.

21. Pergent, G., Pergent-Martini, C., and Boudouresque, C.F. Utilisation de l'herbier à *Posidonie oceanica* comme indicateur biologique de la qualité du milieu littoral en Méditerranée: état des connaissances. *Mésogée.* 54, 3–27, 1995.

22. Pergent, G., Mendez, S., Pergent-Martini, C., and Pasqualini, V. Preliminary data on impact of fish farming facilities on *Posidonia oceanica* meadows in the Mediterranean. *Oceanolog. Acta* 22, 95–107, 1999.

23. Moreno, D., Aguilera, P.A., and Castro, H. Assessment of the conservation status of seagrass (*Posidonia oceanica*) meadows: implications for monitoring strategy and the decision-making process. *Biol. Conserv.* 102, 325–332, 2001.

24. Ros, J.D., Cardell, M.J., Alva, V., Palacin, C., and Llobet, I. Comunidades sobre fondos blandos afectados por un aporte masivo de lodos y aguas residuales (litoral frente a Barcelona, Mediterráneo occidental): resultados preliminares. *Bentos* 6, 407–423, 1990.

25. Goff, F.G. and Cottam, G. Gradient analysis: the use of species and synthetic indices. *Ecology* 48, 793–806, 1967.

26. Gauch, H.G. *Multivariate Analysis in Community Ecology. Cambridge Studies in Ecology.* Cambridge University Press, New York, 1982.

27. Philips, D.H.J. The use of biological indicator organisms to monitor trace metal pollution in marine and estuarine environments — a review. *Environ. Pollut.* 13, 281–317, 1977.

28. De Wolf, P. Mercury content of mussel from West European Coasts. *Mar. Pollut. Bull.* 6, 61–63, 1975.

29. Goldberg, E., Bowen, V.T., Farrington, J.W., Harvey, G., Martin, P.L., Parker, P.L., Risebrough, R.W., Robertson, W., Schnider, E., and Gamble, E. The Mussel Watch. *Environ. Conserv.* 5, 101–125, 1978.

30. Dabbas, M., Hubbard, F., and McManus, J. The shell of *Mytilus* as an indicator of zonal variatons of water quality within an estuary. *Estuar. Coast. Shelf. Sci.* 18, 263–270, 1984.

31. Cossa, F., Picard, M., and Gouygou, J.P. Polynuclear aromatic hydrocarbons in mussels from the Estuary and northwestern Gulf of St. Lawrence, Canada. *Bull. Environ. Contam. Toxicol.* 31, 41–47, 1983.

32. Miller, B.S. Trace metals in the common mussel *Mytilus edulis* (L.) in the Clyde Estuary. *Proc. R. Soc. Edinb.* 90, 379–391, 1986.

33. Renberg, L., Tarkpea, M., and Sundstroem, G. The use of the bivalve *Mytilus edulis* as a test organism for bioconcentration studies: II. The bioconcentration of two super (14)C-labeled chlorinated paraffins. *Ecotoxicol. Environ. Saf.* 11, 361–372, 1986.

34. Carrell, B., Forberg, S., Grundelius, E., Henrikson, L., Johnels, A., Lindh, U., Mutvei, H., Olsson, M., Svaerdstroem, K., and Westermark, T. Can mussel shells reveal environmental history? *Ambio* 16, 2–10, 1987.

35. Lauenstein, G., Robertson, A., and O'Connor, T. Comparison of trace metal data in mussels and oysters from a Mussel Watch programme of the 1970s with those from a 1980s programme. *Mar. Pollut. Bull.* 21, 440–447, 1990.

36. Viarengo, A. and Canesi, L. Mussels as biological indicators of pollution. *Aquaculture* 94, 225–243, 1991.

37. Regoli, F. and Orlando, E. "*Mytilus galloprovincialis* as a bioindicator of lead pollution: biological variables and cellular responses," in *Proceedings of the Second European Conference on Ecotoxicology*, Sloof, W. and De Kruijf, H., Eds. 1993, pp. 1–2.
38. Riisgard, H.U., Kierboe, T., Molenberg, F., Drabaek, I., and Madsen, P. Accumulation, elimination and chemical speciation of mercury in the bivalves *Mytilus edulis* and *Macoma balthica*. *Mar. Biol.* 86, 55–62, 1985.
39. Mohlenberg, F. and Riisgard, H.U. Partioning of inorganic and organic mercury in cockles *Cardium edule* (L.) and *C. glaucum* (Bruguiere) from a chronically polluted area: influence of size and age. *Environ. Pollut.* 55, 137–148, 1988.
40. Brock, V. Effects of mercury on the biosynthesis of porphyrins in bivalve molluscs (*Cerastoderma edule* and *C. lamarcki*). *J. Exp. Mar. Bio. Ecol.* 164, 17–29, 1992.
41. Mo, C. and Neilson, B. Variability in measurements of zinc in oysters, *C. virginica*. *Mar. Pollut. Bull.* 22, 522–525, 1991.
42. Marina, M. and Enzo, O. Variability of zinc and manganeso concentrations in relation to sex and season in the bivalve *Donax trunculus*. *Mar. Pollut. Bull.* 14, 342–346, 1983.
43. Romeo, M. and Gnassia-Barelli, M. *Donax trunculus* and *Venus verrucosa* as bioindicators of trace metal concentrations in Mauritanian coastal waters. *Mar. Biol.* 99, 223–227, 1988.
44. Picard-Berube, M. and Cossa, D. Teneurs en benzo 3,4 pyréne chz *Mytilus edulis*. L. de léstuarie et du Golfe du Saint-Laurent. *Mar. Environ. Res.* 10, 63–71, 1983.
45. Leonzio, C., Bacci, E., Focardi, S., and Renzoni, A. Heavy metals in organisms from the northern Tyrhenian sea. *Sci. Total Environ.* 20, 131–146, 1981.
46. Golovenko, V.K., Shchepinsky, A.A., and Shevchenko, V.A. Accumulation of DDT and its metabolism in Black Sea mussels. *Izv. Akad. Nauk. SSR. Ser. Biol.* 4, 453–550, 1981.
47. Cooper, R.J., Langlois, D., and Olley, J. Heavy metals in Tasmanian Shellfish. I. Monitoring heavy metals contamination in the Dermwert Estuary: use of oysters and mussels. *Journal of Applied Toxicology* 2(2), 99–109, 1982.
48. Ritz, D.A., Swain, R., and Elliot, N.G. Use of the mussel *Mytilus edulis planulatus* (Lamarck) in monitoring heavy metal levels in seawater. *Aust. J. Mar. Freshwater Res.* 33, 491–506, 1982.
49. Richardson, B.J. and Waid., J.S. Polychlorinated biphenyls (PCBs) in Shellfish from Australian coastal waters. *Ecol. Bull. (Stockholm)* 35, 511–517, 1983.
50. Albrecht, W.N., Woodhouse, C.A., and Miller, J.N. Nearshore dredge-spoil dumping and cadmium, copper and zinc levels in a dermestid shrimp. *Bull. Environ. Contamin. Toxicol.* 26, 219–223, 1981.
51. Reish, D.J. Effects of metals and organic compounds on survival and bioaccumulation in two species of marine gammaridean amphipod, together with a summary of toxicological research on this group. *J. Nat. Hist.* 27, 781–794, 1993.
52. Langston, W.J., Burt, G.R., and Zhou, M.J.A.F. Tin and organotin in water, sediments and benthic organisms of Poole Harbour. *Mar. Pollut. Bull.* 18, 634–639, 1987.
53. McElroy, A.E. Trophic transfer of PAH and metabolites (fish, worm). *Responses of Marine Organisms to Pollutants* 8, 265–269, 1988.
54. Reish, D.J. and Gerlinger, T.V. The effects of cadmium, lead, and zinc on survival and reproduction in the polychaetous annelid *Neanthes arenaceodentata*

(F. Nereididae). *Proceedings of the First International Polychaete Conference,* 1984, pp. 383–389.

55. Gibbs, P.E., Langston, W.J., Burt, G.R., and Pascoe, P.L. *Tharyx marioni* (Polychaeta): a remarkable accumulator of arsenic. *J. Mar. Biol. Assoc. U.K.* 63, 313–325, 1983.

56. Bryan, G. W. and Gibbs, P.E. "Polychaetes as indicators of heavy-metal availability in marine deposits," in *Oceanic Processes in Marine Pollution, vol 1, Biological Processes and Wastes in the Ocean,* Capuzzo, J.M. and Kester, D.R., Eds. Krieger, Melbourne, 1987, pp. 37–49.

57. Eadie, J.B., Faust, W., Gardner, W.S., and Nalepa, T. Polycyclic aromatic hydrocarbons in sediments and associated benthos in Lake Erie. *Chemosphere* 11, 185–191, 1982.

58. Gosset, R.W., Brown, D.A., and Young, D.R. Predicting the bioaccumulation of organic compounds in marine organisms using octanol/water partition coefficients. *Mar. Pollut. Bull.* 14, 387–392, 1983.

59. Varanasi, U., Chan, S., and Clark, R. *National Benthic Surveillance Project: Pacific Coast.* National Ocean Service, Washington, 1989.

60. Malins, D.C., McCain, B.B., Myers, M.S., Brown, D.W., Sparks, A.K., Morado, J.F., and Hodgins, H.O. Toxic chemicals and abnormalities in fish and shellfish from urban bays of Puget Sound. *Responses of Marine Organisms to Pollutants* 14, 527–528, 1984.

61. Couch, J.A. and Harshbarger, J.C. Effects of carcinogenic agents on aquatic animals: an environmental and experimental overview. *Environ. Carcigneosis Revs.* 3C, 63–105, 1985.

62. Myers, M.S., Rhodes, L.D., and McCain, B.B. Pathologic anatomy and patterns of occurrence of hepatic neoplasm, putative preneoplastic lesions and other idiopathic hepatic conditions in English sole (*Parophrys vetulus*) from Puget Sound, Washington, U.S.A. *J. Natl. Cancer Inst.* 78, 333–363, 1987.

63. Levine, H.G. "The use of seaweeds for monitoring coastal waters," in *Algae as Ecological Indicators,* Shubert, L.E., Ed. Academic Press, Ann Arbor, 1984.

64. Maeda, S. and Sakaguchi, T. "Accumulation and detoxification of toxic metal elements by algae," in *Introduction to Applied Phycology,* Akatsuka I., Ed. SPB Academic Publishing, Tokyo, 1990.

65. Newmann, G., Notter, M., and Dahlgaard, H. Bladder-wrack (*Fucus vesiculosus* L) as an indicator for radionuclides in the environment of Swedish nuclear power plants. *Swedish Environmental Protection Agency* 3931, 1–35, 1991.

66. Storelli, M.M. and Marcotrigiano, G.O. Persistent organochlorine residues and toxic evaluation of polychlorinated biphenyls in sharks from the Mediterranean Sea (Italy). *Mar. Pollut. Bull.* 42, 1323–1329, 2001.

67. Word, J.Q. The infaunal trophic index, in Southern California Coastal Water Research Project Annual Report, El Segundo, CA, 1979, pp. 19–39.

68. Rafaelli, D.G. and Mason, C.F. Pollution monitoring with meiofauna using the ratio of nematodes to copepods. *Mar. Poll. Bull.* 12, 158–163, 1981.

69. Gee, J.M., Warwick, R.M., Schaanning, M., Berge, J.A., and Ambrose, W.G. Effects of organic enrichment on meiofaunal abundance and community structure in sublittoral soft sediments. *J. Exp. Mar. Biol. Ecol.* 91, 247–262, 1985.

70. Lambshead, P.J.D. and Platt, H.M. "Structural patterns of marine benthic assemblages and their relationship with empirical statistical models," in *Proceedings of the nineteenth European marine biology symposium,* Gibbs, P.E., Ed. Plymouth, 1985, pp. 16–21.

71. Fauchald, K. and Jumars, P. The diet of the worms: a study of polychaete feeding guilds. *Oceanogr. Mar. Biol. Ann. Rev.* 17, 193–284, 1979.
72. Taghon, G.L., Nowell, A.R.M., and Jumars, P. Induction of suspension feeding in spionid polychaetes by high particulate fluxes. *Science* 210, 562–564, 1980.
73. Dauer, D.M., Maybury, C.A., and Ewing, R.M. Feeding behavior and general ecology of several spionid polychaetes from the Chesapeake Bay. *J. Exp. Mar. Biol. Ecol.* 54, 21–38, 1981.
74. Gambi, M.C. and Giangrande, A. Characterization and distribution of polychaete trophic groups in the soft-bottoms of the Gulf of Salerno. *Oebalia* 11, 223–240, 1985.
75. Maurer, D., Leathem, W., and Menzie, C. The impact of drilling fluid and well cuttings on polychaete feeding guilds from the U.S. northeastern continental shelf. *Mar. Pollut. Bull.* 12, 342–347, 1981.
76. Pires, S. and Muniz, P. Trophic structure of polychaetes in the Sao Sebastiao Channel (southeastern Brazil). *Mar. Biol.* 5, 517–528, 1999.
77. Cormaci, M. and Furnari, G. Phytobenthic communities as monitor of the environmental conditions of the Brindisi coast-line. *Oebalia* XVII-I suppl., 177–198, 1991.
78. Verlaque, M. Impact du rejet thermique de Martigues-Ponteau sur le macro-phytobenthos. *Tethys* 8, 19–46, 1977.
79. Belsher, T. and Boudouresque, C.F. L'impact de la pollution sur la fraction algale des peuplements benthiques de Méditerranée. *Atti Tavola Rotonda Internazionale, Livorno* 215–260, 1976.
80. Magurran, A.E. *Diversidad Ecológica y su Medición.* Vedrá, Barcelona, 1989.
81. Gray, J.S. Pollution-induced changes in populations. *Phil. Trans. R. Soc. London* 286, 545–561, 1979.
82. Shaw, K.M., Lambshead, P.J.D., and Platt, H.M. Detection of pollution-induced disturbance in marine benthic assemblages with special reference to nematodes. *Mar. Ecol. Progr. Ser.* 11, 195–202, 1983.
83. Hughes, R.G. A model of the structure and dynamics of benthic marine invertebrate communities. *Mar. Ecol. Prog. Ser.* 15, 1–11, 1984.
84. Warwick, R.M. and Clarke, K.R. New 'biodiversity' measures reveal a decrease in taxonomic distinctness with increasing stress. *Mar. Ecol. Prog. Ser.* 129, 301–305, 1995.
85. Warwick, R.M. and Clarke, K.R. Taxonomic distinctness and environmental assessment. *J. Appl. Ecol.* 35, 532–543, 1998.
86. Clarke, K.R. and Warwick, R.M. The taxonomic distinctness measures of biodiversity: weighting of steps lengths between hierarchical levels. *Mar. Ecol. Prog. Ser.* 184, 21–29, 1999.
87. Shannon, C.E. and Wiener, W. *The Mathematical Theory of Communication.* University of Illinois Press, Chicago, Illinois, 1963.
88. Margalef, R. *Ecología.* Omega, Barcelona, 1978.
89. Stirn, J., Avcin, A.J., Kerzan, I., Marcotte, B.M., Meith-Avcin, N., Vriser, B., and Vukovic, S. "Selected biological methods for assesment of pollution," in *Marine Pollution and Waste Disposal*, Pearson, E.A.W. and Defraja, E., Eds. Pergamon Press, Oxford, 1971, pp. 307–328.
90. Anger, K. On the influence of sewage pollution on inshore benthic communities in the south of Kiel Bay. 2. Quantitative studies on community structure. *Helgoländer. Wiss. Meeresunters.* 27, 408–438, 1975.

91. Hong, J. Impact of the pollution on the benthic community. *Bull. Korean Fish. Soc.* 16, 273–290, 1983.
92. Zabala, K., Romero, A., and Ibáñez, M. La contaminación marina en Guipuzcoa: I. Estudio de los indicadores biológicos de contaminación en los sedimentos de la Ría de Pasajes. *Lurralde* 3, 177–189, 1983.
93. Encalada, R.R. and Millan, E. Impacto de las aguas residuales industriales y domesticas sobre las comunidades bentónicas de la Bahía de Todos Santos, Baja California, Mexico. *Ciencias Marinas* 16, 121–139, 1990.
94. Calderón-Aguilera, L.E. Análisis de la infauna béntica de Bahía de San Quintín, Baja California, con énfasis en su utilidad en la evaluación de impacto ambiental. *Ciencias Marinas* 18, 27–46, 1992.
95. Pocklington, P., Scott, D.B., and Schafer, C.T. Polychaete response to different aquaculture activities. *Proceedings of the 4th International Polychaete Conference*, Quingdao, China, 1994, 162, pp. 511–520.
96. Engle, V., Summers, J.K., and Gaston, G.R. A benthic index of environmental condition of Gulf of Mexico. *Estuaries* 17, 372–384, 1994.
97. Mendez-Ubach, N. Polychaetes inhabiting soft bottoms subjected to organic enrichment in the Topolobampo lagoon complex, Sinaloa, México. *Océanides* 12, 79–88, 1997.
98. Yokoyama, H. Effects of fish farming on macroinvertebrates. Comparison of three localities suffering from hypoxia. *UJNR Tech. Rep.* 24, 17–24, 1997.
99. Molvær, J., Knutzen, J., Magnusson, J., Rygg, B., Skei, J., and Sørensen, P. *Classification of Environmental Quality in Fjords and Coastal Waters.* SFT guidelines. 97:03, 36, Oslo, 1997.
100. Leppäkoski, E. Assessment of degree of pollution on the basis of macrozoobenthos in marine and brakish water environments. *Acta. Acad. Aba.* 35, 1–98, 1975.
101. Rygg, B. Distribution of species along a pollution gradient induced diversity gradients in benthic communities in Norwegian Fjords. *Mar. Pollut. Bull.* 16, 469–473, 1985.
102. Gray, J.S. and Mirza, F.B. A possible method for the detection of pollution induced disturbance on marine benthic communities. *Mar. Pollut. Bull.* 10, 142–146, 1979.
103. Ros, J.D. and Cardell, M.J. La diversidad específica y otros descriptores de contaminación orgánica en comunidades bentónicas marinas. *Actas del Symposium sobre Diversidad Biológica.* Centro de Estudios Ramón Areces, Madrid, 219–223, 1991.
104. Salas, F. *Valoración de los Indicadores Biológicos de Contaminación Orgánica en la Gestión del Medio Marino.* Tesis de Licenciatura, Universidad de Murcia, 1996.
105. Simpson, E.H. Measurement of diversity. *Nature* 163, 688, 1949.
106. Lambshead, P.J.D., Platt, H.M., and Shaw, K.M. The detection of differences among assemblages of marine benthic species based on an assessment of dominance and diversity. *J. Nat. Hist.* 17, 847–859, 1983.
107. Somerfield, P.J. and Clarke, K.R. A comparison of some methods commonly used for the collection of sublittoral sediments and their associated fauna. *Mar. Environ. Res.* 43, 145–156, 1997.
108. Warwick, R.M. A new method for detecting pollution effects on marine macrobenthic communities. *Mar. Biol.* 92, 557–562, 1986.
109. Beukema, J.J. An evaluation of the ABC method (abundance/biomass comparison) as applied to macrozoobenthic communities living on tidal flats in the Dutch Wadden Sea. *Mar. Biol.* 99, 425–433, 1988.

110. Clarke, K.R. Comparison of dominance curves. *J. Exp. Mar. Biol. Ecol.* 138, 130–143, 1990.

111. McManus, J.W. and Pauly, D. Measuring ecological stress: variations on a theme by R.M. Warwick. *Mar. Biol.* 106, 305–308, 1990.

112. Meire, P.M. and Dereu, J. Use of the abundance/biomass comparison method for detecting environmental stress: some considerations based on intertidal macro-zoobenthos and bird communities. *J. Appl. Ecol.* 27, 210–223, 1990.

113. Anderlini, V.C. and Wear, R.G. The effects of sewage and natural seasonal disturbance on benthic macrofaunal communites in Fitzray Bay, Wellington, New Zealand. *Mar. Pollut. Bull.* 24, 21–26, 1992.

114. Agard, J.B.R., Gobin, J., and Warwick, R.M. Analysis of marine macrobenthic community structure in relation to pollution, natural oil seepage and seasonal disturbance in a tropical environment (Trinidad, West Indies). *Mar. Ecol. Progr. Ser.* 92, 233–243, 1993.

115. Gray, J.S., Aschan, M., Carr, M.R., Clarke, K.R., Green, R.H., Pearson, T.H., Rosenberg, R., Warwick, R.M., and Bayne, B.L. Analysis of community attributes of the benthic macrofauna of Frierfjord/Langesundfjord and in a mesocosm experiment. *Marine Ecology-Progress Series* 46, 1–3, 151–165, 1988.

116. Ritz, D.A., Lewis, M.E., and Shen, M. Response to organic enrichment of infaunal macrobenthic communities under salmonid seacages. *Mar. Biol.* 103, 211–214, 1989.

117. Reizopoulou, S., Thessalou-Legaki, M., and Nicolaidou, A. Assessment of disturbance in Mediterranean lagoons: an evaluation of methods. *Mar. Biol.* 125, 189–197, 1996.

118. Salas, F. Valoración y aplicabilidad de los índices y bioindicadores de contaminación orgánica en la gestión del medio marino. Doctoral thesis, Universidad de Murcia, 2002.

119. Ibanez, F. and Dauin, J.C. Long-term changes (1977–1987) in a muddy fine sand *Abra alba-Melina palmata* community from the Western Channel: multivariate time series analysis. *Mar. Ecol. Progr. Ser.* 19, 65–81, 1988.

120. Weston, D.P. Quantitative examination of macrobenthic community changes along an organic enrichment gradient. *Mar. Ecol. Progr. Ser.* 61, 233–244, 1990.

121. Craeymeersch, J.A. Applicability of the abundance/biomass comparison method to detect pollution effects on intertidal macrobenthic communities. *Hydrobiol. Bull.* 24, 133–140, 1991.

122. Salas, F., Neto, J.M., Borja, A., and Marques, J.C. Evaluation of the applicability of a marine biotic index to characterise the status of estuarine ecosystems: the case of Mondego estuary (Portugal). *Ecological Indicators* 4, 215–255, 2004.

123. Abou-Aisha, K.H., Kobbia, I.A., Abyad, M.S., Shabana, E.F., and Schanz, F. Impact of phosphorous loadings on macro-algal communities in the Red sea coast of Egypt. *Wat. Air Soil Pollut.* 83, 285–297, 1995.

124. Satsmadjis, J. Analysis of benthic fauna and measurement of pollution. *Revue Internationale Oceanographie Medicale* 66–67, 103–107, 1982.

125. Wollenweider, R.A., Giovanardi, F., Montanari, G., and Rinaldi, A. Characterisation of the trophic conditions of marine coastal waters with special reference to the NW Adriatic Sea: proposal for a trophic scale, turbidity and generalised water quality index. *Environmetrics* 9, 329–357, 1998.

126. Nelson, W.G. Prospects for development of an index of biotic integrity for evaluating habitat degradation in coastal systems. *Chem. Ecol.* 4, 197–210, 1990.

127. Weisberg, S.B., Ranasinghe, J.A., Dauer, D.M., Schaffner, L.C., Díaz, R.J., and Frithsen, J.B. An estuarine benthic index of biotic integrity (B-IBI) for the Chesapeake Bay. *Estuaries* 20, 149–158, 1997.
128. Cooper, J.A.G., Harrison, T.D., Ramm, A.E.L., and Singh, R.A. *Refinement, Enhancement and Application of the Estuarine Health Index to Natal's Estuaries, Tugela–Mtamvuna.* Unpublished Technical Report. CSIR, Durban, 1993.
129. Deegan, L.A., Finn, J.T., Ayvazian, S.G., and Ryder, C. *Feasibility and Application of the Index of Biotic Integrity to Massachusetts Estuaries (EBI).* Final report to Massachusetts Executive Office of Environmental Affairs. Department of Environmental Protection. North Grafton, MA, 1993.
130. Vaas, P.A. and Jordan, S.J. "Long term trends in abundance indices for 19 species of Chesapeake Bay fishes: reflection of trends in the Bay ecosystem," in *Proceedings of a Conference on New Perspectives in the Chesapeake System: a Research and Management Partnership*, Mihursky, J., and Chaney, A., Eds. Chesapeake Research Consortium Publ. 137, 1990.
131. Carmichael, J., Richardson, B., Roberts, M., and Jordan, S.J. *Fish Sampling in Eight Chesapeake Bay Tributaries.* Maryland Dept. of Natural Resources, CBRM-HI-92-2. Annapolis, 1992, 50 p.
132. Bettencourt, A., Bricker, S.B., Ferreira, J.G., Franco, A., Marques, J.C., Melo, J.J., Nobre, A., Ramos, L., Reis, C.S., Salas, F., Silva, M.C., Simas, T., and Wolff, W. *Typology and Reference Conditions for Portuguese Transitional and Coastal Waters*, Development of guidelines for the application of the European Union Water Framework Directive. Edição IMAR/INAG, 2004, 98p.
133. Maree, R.C., Whitfield, A.K. and Quinn, N.W. *Prioritisation of South African Estuaries Based on their Potential Importance to Estuarine-Associated Fish Species.* Unpublished report to Water Research Commission, 2000, 55 p.
134. Gardner, M.R. and Ashby, W.R. Connectance of large synamical (cybernetic) systems: critical values for stability. *Nature* 288, 784, 1970.
135. O'Neill, R.V. Ecosystem persistence and heterotrophic regulation. *Ecology* 57, 1244–1253, 1976.
136. Rosenzweig, M.L. Paradox of enrichment: destabilization of exploitation ecosystems in ecological time. *Science* 171, 385–387, 1971.
137. Ahl, T. and Weiderholm, T. Svenska vattenkvalitetskriterier. Eurofierande ämnen. SNV PM (Swed.), 918, 1977.
138. Jørgensen, S.E. Review and comparison of goal functions in system ecology. *Vie Mileu.* 44, 11–20, 1994.
139. Jørgensen, S.E. *Integration of Ecosystem Theories: A Pattern.* Kluwer, Dordrecht, The Netherlands, 2002, 428 p.
140. Jørgensen, S.E. and Mejer, H. A holistic approach to ecological modelling. *Ecol. Model.* 7, 169–189, 1979.
141. Jørgensen, S.E. and Mejer, H. "Exergy as a key function in ecological models," in Mitsch, W., Bosserman, R.W., and Klopatek, J.M., Eds. *Energy and Ecological Modelling. Developments in Environmental Modelling*, Vol 1. Elsevier. Amsterdam, 1981, pp. 587–590.
142. May, R.M. *Stability and Complexity in Model Ecosystems.* 3rd ed. Princeton University Press, Princeton, 1977.
143. Jørgensen, S.E., Nielsen, S.N., and Mejer, H. Energy, environ, exergy and ecological modelling. *Ecological Modelling* 77, 99–109, 1995.

144. Ulanowicz, R.E. *Growth and Development Ecosystems Phenomenology.* Springer-Verlag, New York, 1986.
145. Ulanowicz. R.E. *Ecology, the Ascendent Perspective.* Columbia University Press, New York, 1997, 201 p.
146. Ulanowicz, R.E. and Puccia, C.J. Mixed trophic impacts in ecosystems. *Coenoses* 5, 7–16, 1990.
147. Odum, H.T. *System Ecology.* Wiley Interscience, New York, 1983, 510 p.
148. Reference removed.
149. Hannon, B. The structure of ecosystems. *J. Theor. Biol.* 41, 535–546, 1973.
150. Hannon B. and Ruth, M. *Modeling Dynamic Biological Systems.* Springer-Verlag, Berlin, 1997, 396 p.
151. Patten, B.C. Energy, emergy and environs. *Ecol. Model.* 62, 29–70, 1992.
152. Kay, J. and Schneider, E.D. "Thermodynamics and measures of ecological integrity," in *Proceedings "Ecological Indicators,"* Elsevier, Amsterdam, 1992. pp. 159–182.
153. Jørgensen, S.E. "Exergy and buffering capacity in ecological system," in *Energetics and Systems,* Mitsch, W.J., Ragade, R.K., Bossermann, R.W., and Dillon, J., Eds. Ann Arbor Science, Ann Arbor, MI, 1982, 61 p.
154. Prigogine, I. Etude thermodynamique des phÇnomenäs irreversibles. Desoer, Liege, 1947.
155. Mauersberger, P. "General principles in deterministic water quality modeling." in *Mathematical Modeling of Water Quality: Streams, Lakes and Reservoirs (International Series on Applied Systems Analysis,* 12), Orlob, G.T., Ed. Wiley, New York, 1983, pp. 42–115.
156. Mauersberger, P. Optimal control of biological processes in aquatic ecosystem. *Gerlands Beitr. Geiophys.* 94, 141–147, 1985.
157. Jørgensen, S.E., Patten, B.C., and Straskraba, M. *Ecosystem Emerging IV: Growth. Ecological Modelling,* 126, 249–284, 2000.
158. Debeljak, M. Application of Exergy Degradation and Exergy Storage as Indicators for the Development of Managed and Virgin Forest. Ph.D thesis to be defended at the University of Ljubljana University, 2001.
159. Holling, C.S. "The resilience of terrestrial ecosystems: local surprise and global change," in *Sustainable Development of the Biosphere,* Clark, W.C. and Munn, R.E., Eds. Cambridge University Press, Cambridge, 1986, pp. 292–317.
160. Kay, J.J. *Self Organization in Living Systems [Thesis]. Systems Design Engineering,* University of Waterloo, Ontario, Canada, 1984.

Additional References

Buhr, K-J. Suspension-feeding and assimilation efficiency, in *Lanice conchilega. Marine Biology* 38, 373–383, 1976.
Boltzmann, L. The Second Law of Thermodynamics (Populare Schriften. Essay No. 3 (Address to Imperial Academy of Science in 1886)). Reprinted in English in: Theoretical Physics and Philosophical Problems, Selected Writings of L. Boltzmann. D. Riedel, Dordrecht, 1905.
Debeljak, M. Application of exergy degradation and exergy storage as indicators for the development of managed and virgin forest. Ph.D thesis to be defended at the Unviersity of Ljubljana University, 2002.

Fath, B. and Patten, B.C., Goal Functions and Network Theory. Presented in Porto Venere May 2000 at the Third Conference on Energy, 2000.

Hall, A.S. Maximum Power — The Ideas and Applications of H.T. Odum. University Press of Colorado, Denver, U.S., 1995, 390 p.

Margalef, R. Communication of structure in plankton population. *Limnol Oceanogr*. 6, 124–128, 1961.

Margalef, R. Teoná de los sistemas ecológicos. Publications de la Universitat de Barcelona. Barcelona, 1991, 290 p.

Odum, H.T. Environmental Accounting—Energy and Decision Making. Wiley, New York, 1996, 370 p.

Prigogine, I. L'Etude Thermodynamique des Processus Irreversibles. Desoer, Liege, 1947.

Prigogine, I. From being to becoming: Time and Complexity in the Physical Sciences. Freeman, San Franscisco, California, U.S., 1980, 220 p.

Weiderholm, T. Use of benthos in lake monitoring. J. Water Pollut. Control Fed. 52, 537, 1980.

APPENDIX A

β-values for Computation of Exergy (Eco-exergy)

Table 1 Exergy of living organisms (These values are according to Jørgensen, 2002)

Organisms	gi	$\beta = (\text{eichem} + \text{eibio})/\text{eichem}$	Exergy kJ/g
Detritus	0	1	18.7
Minimal cell 2	470	2.3	43.8
Bacteria	600	2.7	50.5
Algae	850	3.4	64.2
Yeast	2,000	5.8	108.5
Fungus	3,000	9.5	178
Sponges	9,000	26.7	499
Moulds	9,500	28.0	524
Plants, trees	10,000–30,000	29.6–86.8	554–1,623
Worms	10,500	30.0	561
Insects,	10,000–15,000	29.6–43.9	554–821
Jellyfish	10,000	29.6	554
Zooplankton	10,000–15,000	29.6–43.9	554–821
Fish	10,000–120,000	287–344	5,367–6,433
Birds	120,000	344	6,433
Amphibians	120,000	344	6,433
Reptiles	130,000	370	6,919
Mammals	140,000	402	7,517
Human	250,000	716	13,389

Recently, results of the genome mapping project (ten species included *homo sapiens*) have been applied to find new β-values that are higher than the previous ones (see Table 1). The number of genes are lower than previously expected but the number of amino acid codes per gene is considerably higher. Also, other measures of the complexity of organisms have been applied to expand the list of β-values; see Calculations of Exergy for Organisms By Sven Erik Jørgensen, Niels Ladegaad, Marko Debeljak and Joao Carlos Marques, in print (in Ecological Modelling). The new and higher values should be applied by computation of exergy (eco-exergy). The previously applied exergy calculations have only been applied relatively, and the results are therefore still valid, because the new higher values are relatively not very different from the previous values.

Table 2

New β-values	
Detritus	1.00
Virus	1.01
Minimal cell	5.8
Bacteria	8.5
Archaea	13.8
Protists Algae	20
Yeast	17.8
	33 Mesozoa, Placozoa
	39 Protozoa, amoebe
	43 Phasmida (stick insects)
Fungi, moulds	61
	76 Nemertina
	91 Cnidaria (corals, sea anemones, jelly fish)
Rhodophyta	92
	97 Gastroticha
Prolifera, sponges	98
	109 Brachiopoda
	120 Platyhelminthes (flatworms)
	133 Nematoda (round worms)
	133 Annelida (leeches)
	143 Gnathostomulida
Mustard weed	143
	165 Kinorhyncha
Seedless vascula plants (incl. ferns)	158
	163 Rotifera (wheel animals)
	164 Entoprocta
Moss	174
	167 Insecta (beetles, flies, bees, wasps, bugs, ants)
	191 Coleodiea (Sea squirt)
	221 Lipidoptera (buffer flies)
	232 Crustaceans
	246 Chordata
Rice	275
Gynosperms (inl. pinus)	314
	310 Mollusca, bivalvia, gastropodea
	322 Mosquito
Flowering plants	393
	499 Fish
	688 Amphibia
	833 Reptilia
	980 Aves (Birds)
	2,127 Mammalia
	2,138 Monkeys
	2,145 Anthropoid apes
	2,173 *Homo Sapiens*

CHAPTER 3

Application of Ecological Indicators to Assess Environmental Quality in Coastal Zones and Transitional Waters: Two Case Studies

J.C. Marques, F. Salas, J.M. Patrício, and M.A. Pardal

This chapter addresses the application of ecological indicators in assessing the biological integrity and environmental quality in coastal ecosystems and transitional waters. In this context, the question of what might be considered a good ecological indicator is approached, and the different types of data most often utilized to perform estimations are discussed. Moreover, we present a brief review on the application of ecological indicators in coastal and transitional waters ecosystems referring to: (1) indicators based on species presence vs. absence; (2) biodiversity as reflected in diversity measures; (3) indicators based on ecological strategies; (4) indicators based on species biomass and abundance; (5) indicators accounting for the whole environmental information; and (6) thermodynamically oriented and network analysis-based indicators. Algorithms are provided in an abridged way and the pros and cons regarding the application of each indicator are discussed. The question of how to choose the most adequate indicator for each particular case is discussed as a function of data requirements and data availability. Two case studies are used to illustrate whether a number of selected ecological indicators were satisfactory in describing the state of ecosystems, comparing their relative performances and discussing how their usage can be improved for environment health assessment. The possible relation between values of these indicators and the environmental quality status of ecosystems was analyzed. We reached the conclusion that to select an ecological indicator, we must account for its dependence on external factors beyond our control, such as the need for reference values that often do not exist, or particular characteristics regarding the habitat type. As a result, it is reasonable to say that no indicator will be valid in all situations, and that a single approach does not seem appropriate due to the complexity inherent in assessing the environmental quality status of a system. Therefore, as a principle, such evaluation should be always performed using several ecological indicators, which may provide complementary information.

3.1 INTRODUCTION

Ecological indicators are commonly used to supply synoptic information about the state of ecosystems. They usually address an ecosystem's structure and/or functioning accounting for a certain aspect or component; for example, nutrient concentrations, water flows, macroinvertebrates and/or vertebrates diversity, plants diversity, plants productivity, erosion symptoms, and sometimes ecological integrity at a systems level.

The main attribute of an ecological indicator is to combine numerous environmental factors in a single value, which might be useful in terms of management and for making ecological concepts compliant with the general public understanding. Moreover, ecological indicators may help in establishing a useful connection between empirical research and modeling since some of them are of use as orientors (also referred to in the literature as goal functions) in ecological models. Such application proceeds from the fact that conventional models of aquatic ecosystems are not effective in predicting the occurrence of qualitative changes in ecosystems; for example, shifts in species composition, which is due to the fact that measurements typically carried out — such as biomass and production — are not efficient at capturing such modifications (Nielsen, 1995). Nevertheless, it has been tried to incorporate this type of changes in structurally dynamic models (Jørgensen, 1992; Nielsen, 1992, 1994, 1995; Jørgensen et al., 2002), to improve their predictive capability, achieving a better understanding of ecosystem behavior, and consequently a better environmental management.

In structurally dynamic models, the simulated ecosystem behavior and development (Nielsen, 1995; Straškraba, 1983) is guided through an optimization process by changing the model parameters in accordance with a given ecological indicator, used as an orientor (goal function). In other words, this allows the introduction in models parameters that change as a function of changing forcing functions and conditions of state variables, optimizing the model outputs by a stepwise approach. In this case, the orientor is assumed to express a given macroscopic property of the ecosystem, resulting from the emergence of new characteristics arising from self-organization processes.

In general, the application of ecological indicators is not free from criticism. One such criticism is that aggregation results in oversimplification of the ecosystem under observation. Moreover, problems arise from the fact that indicators account not only for numerous specific system characteristics, but also other kinds of factors; for example, physical, biological, ecological, socio-economic etc. Indicators must therefore be utilized following the right criteria and in situations that are consistent with its intended use and scope; otherwise they may lead to confusing data interpretations.

This paper addresses the application of ecological indicators for assessing the biological integrity and environmental quality in coastal ecosystems and transitional waters. The possible characteristics of a good ecological indicator, or what kind of information regarding ecosystem responses can be obtained from the different types of biological data usually taken into account in evaluating the state of coastal areas, has already been discussed in chapter 2. Two cases studies are used to illustrate whether different types of indicators were satisfactory in describing the state of ecosystems, comparing their relative performances and discussing how can their usage be improved for environment health assessment.

3.2 BRIEF REVIEW ON THE APPLICATION OF ECOLOGICAL INDICATORS IN ECOSYSTEMS OF COASTAL AND TRANSITIONAL WATERS

Almost all coastal marine and transitional waters ecosystems all over the world have been under severe environmental stress following the settlement of human activities. Estuaries, for example, are the transition between marine, freshwater and land ecosystems, being characterized by distinctive biological communities with specific ecological and physiological adaptations. In fact, we may say that the estuarine habitat does not imply a simple overlap of marine and land factors, constituting instead an individualized whole with its own biogeochemical factors and cycles, which represents the environment for real estuarine species to evolve. In such ecosystems, besides resources available, fluctuating conditions, namely salinity and type of substrate, are a key issue regarding an organism's ecological distribution and adaptive strategies (see, for example, McLusky, 1989; Engle et al., 1994).

The most common types of problems in terms of pollution include illegal sewage discharges associated with nutrient enrichment; pollution due to toxic substances such as pesticides, heavy metals, and hydrocarbons; unlimited development; and habitat fragmentation or destruction.

In the case of transitional waters, limited water circulation and inappropriate water management tends to concentrate nutrients and pollutants, and to a certain extent we may say that sea pollution begins there (Perillo et al., 2001). Moreover, in estuaries, drainage of harbors and channels modifies geomorphology, water circulation, and other physicochemical features, and consequently the habitat's characteristics. In recent times, perhaps the most important problem is the excessive loading of nutrients mainly due to fertilizers used in agriculture, and untreated sewage water, which induces eutrophication processes. These problems can be observed all over the world.

Many ecological indicators used or tested in evaluating the status of these ecosystems can be found in the literature, resulting from just a few distinct theoretical approaches. A number of them focus on the presence or absence of given indicator species, while others take into account the different ecological strategies carried out by organisms, diversity, or the energy variation in the system through changes in the biomass of individuals. A last group of ecological indicators are thermodynamically oriented or based on network analysis, and look for capturing the information on the ecosystem from a more holistic perspective (Table 3.1).

3.2.1   Indicators Based on Species Presence vs. Absence

Determining the presence or absence of one species or group of species has been one of the most used approaches in detecting pollution effects. For instance, the Bellan, (based on polychaetes), or the Bellan–Santini (based on amphipods) indices attempt to characterize environmental conditions by analyzing the dominance of species that indicate some type of pollution in relation to the species considered to indicate an optimal environmental situation (Bellan, 1980; Bellan and Santini, 1980). Several authors do not advise the use of these indicators because often such indicator species may occur naturally in relative high densities. The point is that there is no reliable methodology to know at which level the indicator species can be well represented in a community that is not really affected by any kind of pollution, which leads to a significant exercise of subjectivity (Warwick, 1993). Despite these criticisms, even recently, the AMBI index (Borja et al., 2000), which is based on the Glemarec and Hily (1981) species classification regarding pollution; as well as the Bentix index (Simbora and Zenetos, 2002), have gone back to update such pollution detecting tools. Roberts et al. (1998) also proposed an index based on macrofauna species, which accounts for the ratio of each species abundance in control vs. samples proceeding from stressed areas. It is however semiquantitative as well as site- and pollution type-specific.

The AMBI index, for example, accounts for the presence of species indicating a type of pollution and of species indicating a reference situation assumed to be polluted. It has been considered useful in terms of the application of the European Water Framework Directive in coastal ecosystems and estuaries. In fact, although this index is very much based on the paradigm of Pearson and Rosenberg (1978), which emphasizes the influence of organic matter enrichment on benthic communities, it was shown to be useful in assessing other anthropogenic impacts, such as physical alterations in the habitat, heavy metal inputs, etc. in several European areas of the Atlantic (North Sea; Bay of Biscay; and southern Spain) and Mediterranean coasts (Spain and Greece) (Borja et al., 2003).

Table 3.1 Short review of environmental quality indicators regarding the benthic communities

Type of indicator	Requirements and applicability evaluation	Algorithm
Based on species presence vs. absence	List of species. Subjective in most of the cases. Only the use of AMBI and Bentix is recommended.	Bellan index (Bellan, 1980): $$IP = \sum \frac{\text{pollution species indicator}}{\text{no pollution species indicator}}$$ Pollution indicator species: *Platenereis dumerilii, Theosthema oerstedi, Cirratulus cirratus* and *Dodecaria concharum.* No-pollution indicator species: *Syllis gracillis, Typosyllis prolifera, Typosyllis* sp. and *Amphiglena mediterranea.* Bellan–Santini index (Bellan-Santini,1980): $$IP = \sum \frac{\text{pollution species indicator}}{\text{no pollution species indicator}}$$ Pollution indicator species: *Caprella acutrifans* and *Podocerus variegates* No-pollution indicator species: *Hyale* sp, *Elasmus pocillamunus* and *Caprella liparotensis* AMBI (Borja et al., 2000): $$AMBI = \frac{\{(0 \times \%GI) + (1.5 \times \%GII) + (3 \times \%GIII) + (4.5 \times \%GIV) + (6 \times \%GV)\}}{100}$$ GI: Species very sensitive to organic enrichment and present under unpolluted conditions GII: Species indifferent to enrichment GIII: Species tolerant to excess of organic matter enrichment GIV: Second-order opportunist species, mainly small sized Polychaetes GV: First-order opportunist species, essentially deposit-feeders Bentix (Simboura and Zenetos, 2002) : $$Bentix = \frac{\{(6 \times \%GI) + 2 \times (\%GII + \%GIII)\}}{100}$$ GI: Species very sensitive to pollution GII: Species tolerant to pollution GIII: Second-order and first-order opportunist species

(Continued)

Table 3.1 Continued

Type of indicator	Requirements and applicability evaluation	Algorithm
Based on ecological strategies	List of taxa (species or higher taxonomic groups) and knowledge on their life strategies, which can be in the literature. Subjective. Not recommended.	Nematodes/copepods ratio (Rafaelli and Mason, 1981): $I = \dfrac{\text{nematodes abundance}}{\text{copepodes abundance}}$ Polychaetes/amphipods ratio (Gómez Gesteira, 2000): $Log_{10}\left(\dfrac{\text{Polychaetes abundance}}{\text{Amphipodes abundance}} + 1\right)$ Infaunal index (Word, 1979): $ITI = 100 - 100/3 \times (0n_1 + 1n_2 + 2n_3 + 3n_4)/(n_1 + n_2 + n_3 + n_4)$ n_1 = number of individuals of suspensivores feeders n_2 = number of individuals of interface feeders n_3 = number of individuals of surface deposit feeders n_4 = number of individuals of subsurface deposit feeders
Diversity measures	Quantitative samples; adequate taxa identification; Data on species density (number of individuals and/or biomass). In the case of K-dominance curves, time series for the same local are desirable. Although not exempt from subjectivity, results might be useful.	Shannon–Wienner index (Shannon–Wienner, 1963): $H' = \sum p_i \log_2 p_i$ Where p_i is the proportion of abundance of species i in a community were species proportions are $p_1, p_2, p_3 \ldots p_n$. Margalef index: $D = (S-1)/\log_e N$ Where S is the number of species found and N is the total number of individuals

Berger-Parker index:

$D = (n_{max})/N$

Where n_{max} is the number of individuals of the dominant species and N is the total number of individuals

Simpson index:

$D = \sum n_i(n_i - 1)/N(N-1)$

Where n_i is the number of individuals of species i and N is the total number of individuals

Average taxonomic diversity index (Warwick and Clarke, 1995 1998):

$\Delta = [\sum\sum_{i<j} \omega_{ij} \times i \times j]/[N(N-1)/2]$

Where ω_{ij} is the taxonomic distance between every pair of individuals, the double summation is over all pairs of species i and j ($i, j = 1, 2, \ldots, S$; $i<j$), and $N = \sum_i \times_i$ is the total number of individuals in the sample

When the sample consists simply of a species list the index takes this form:

$\Delta^+ = [\sum\sum_{i<j} \omega_{ij} \times i \times j]/[S(S-1)/2]$

Where S is the number of the species in the sample

K-dominance curves (Lambshead et al., 1983):

Cumulative ranked abundance plotted against species rank, or log species rank

(Continued)

Table 3.1 Continued

Type of indicator	Requirements and applicability evaluation	Algorithm
Based on species biomass and abundance	Quantitative benthic samples; taxa identification; species density (number of individuals and/or biomass). Data along gradients in the same system are suitable. Results might be useful.	ABC curves (Warwick., 1986): K-dominance curves for species abundances and species biomasses on the same graph The ABC method derived the W statistic (Warwick and Clarke, 1994): $W = \sum (B_i - A_i)/50 \times (S-1)$ Where B_i is the biomass of species i, A_i the abundance of species i, and S is the number of species
Indicators accounting for the whole environmental information	Physical chemical parameters; Quantitative benthic samples; taxa identification; species density (number of individuals and/or biomass). Although it is a good idea to integrate the whole environmental information, they are difficult to apply as they need a large amount of data of different nature. B-IBI (Weisberg et al., 1997) is dependent on the type of habitat and seasonality.	Benthic index of environmental condition (Engle et al., 1994): Benthic index = $(2,3841 *$ proportion of expected diversity) + $(-0.6728 *$ proportion of total abundance as tubifids) + $(0.6683 *$ proportion of total abundance as bivalves) Coefficient of pollution (Satmasjadis, 1985): Calculation of P is based on several integrated equations. These equations are: $S = s + t/(5 + 0.2s)$ $i_0 = (-0.0187s'2 + 2.63s' - 4)(2.20 - 0.0166h)$ $g' = l/(0.0124i + 1.63)$ $P = g'/[g(i/i_0)^{1/2}]$ $P =$ coefficient of pollution $S' =$ sand equivalent, $s =$ percentage sand, $t =$ percentage silt $i_0 =$ theorical number of individuals, $i =$ number of individuals $h =$ station depth $g =$ theorical number of species, $g =$ number of species

B-IBI (Weisberg et al., 1997):
Eleven metrics are used to calculate the B-IBI (Weisberg et al., 1997):
- Shannon–Wienner species diversity index
- Total species abundance
- Total species biomass
- % abundance of pollution-indicative taxa
- % abundance of pollution-sensitive taxa
- % biomass of pollution-indicative taxa
- % biomass of pollution-sensitive taxa
- % abundance of carnivore and omnivores
- % abundance of deep-deposit feeders
- Tolerance Score
- Tanypodinae to Chironomidae % abundance ratio

The scoring of metrics to calculate the B–IBI is done by comparing the value of a metric from the sample of unknown sediment quality to thresholds established from reference data distributions

Exergy Index (Jørgensen and Mejer, 1979; 1981; Marques et al., 1997):

$Ex = T \times \sum_i \beta_i \times C_i$

Where T is the absolute temperature, C_i is the concentration in the ecosystem of component i (e.g., biomass of a given taxonomic group or functional group), β_i is a factor able to express roughly the quantity of information embedded in the genome of the organisms. Detritus was chosen as reference level, i.e., $\beta_i = 1$ and exergy in biomass of different types of organisms is expressed in detritus energy equivalents

Specific exergy: (Jørgensen and Mejer, 1979; 1981):

$SpEx = Ex_{tot}/Biom_{tot}$

Ascendancy (Ulanowicz, 1986):

$$A = \sum_i \sum_j T_{ij} \log \left[\frac{T_{ij} T_{..}}{T_{.j} T_{i.}} \right]$$

$T_{ij} =$ Trophic exchange from taxon i to taxon j

Thermodynamically oriented and network analysis based indicators

Exergy and specific exergy: Quantitative samples. Data on taxa (higher taxonomic groups) biomasses. Useful not sufficiently tested developmental phase.
Ascendancy: Quantitative benthic samples; Taxa identification; Species density (number of individuals and/or biomass). Knowledge on the food-web structure and system energy through flow.
Objective, powerful, most often impossible to apply due to lack of data.

3.2.2   Biodiversity as Reflected in Diversity Measures

Biodiversity is a widely accepted concept usually defined as biological variety in nature. This variety can be perceived intuitively, which lead to the assumption that it can be quantified and adequately expressed in any appropriated manner (Marques, 2001), although expressing biodiversity as diversity measures had proved to be a difficult challenge. Nevertheless, diversity measures have been possibly the most commonly used approach, which assumes that the relationship between diversity and disturbances can be seen as a decrease in diversity as stress increases.

Looking to a certain systematization, Magurran (1988) classifies diversity measurements into three main categories:

1. Indices that measure the enrichment of the species, such as the Margalef's one, which are, in essence, a measurement in the number of species in a defined sampling unit.
2. Models of the abundance of species, as the K-dominance curves (Lambshead et al., 1983) or the lognormal model (Gray, 1979), which describe the distribution of their abundance, from situations in which there is a high uniformity, to those in which the abundance is very uneven. However, the lognormal model deviation was long time ago rejected by several authors due to the impossibility of finding any benthic marine sample that clearly responded to the lognormal distribution model (Shaw et al., 1983; Hughes, 1984; Lambshead and Platt, 1985).
3. Indices based on the proportional abundance of species aiming to account for species richness and regularity of species distribution in a single expression. Second, these indices can be subdivided into those based on information theory, and the ones accounting for species dominance. Indices derived from the information theory (e.g., Shannon–Wienner) assume that diversity, or information, in a natural system can be measured in a similar way as information contained in a code or message. On the other hand, dominance indices (e.g., Simpson or Berger–Parker) are referred as measurements that account for the abundance of the most common species.

Recently, a measure called "taxonomic distinctness" has been used in some studies (Warwick and Clarke, 1995, 1998; Clarke and Warwick, 1999) to assess biodiversity in marine environments, taking into account taxonomic, numerical, ecological, genetic, and philogenetic aspects of diversity. Nevertheless, it is most often very complicated to meet certain requirements to apply taxonomic distinctness, as it requires a complete list of the species present in the area under study in pristine situations. Moreover, some research has shown that taxonomic distinctness is not more sensitive than other diversity indices that can applied when detecting disturbances (Sommerfield and Clarke, 1997), and consequently this measure has not been widely used on marine environment quality assessment and management studies.

3.2.3   Indicators Based on Ecological Strategies

Some indices try to assess environmental stress effects accounting for the ecological strategies followed by different organisms. That is the case of trophic indices such as the infaunal index proposed by Word (1979), or the polychaetes feeding guilds (Fauchald, 1979), which are based on the different feeding strategies of the organisms. Another example is the nematodes/copepods index (Rafaelli and Mason, 1981), or the copepods/nematodes one (Parker, 1980), which account for the different behavior of two taxonomic groups under environmental stress situations. These ones have been abandoned due to their dependence of parameters such as depth and sediment particle size, as well as because of their unpredictable pattern of variation depending on the type of pollution (Gee et al., 1985; Lambshead, 1986). More recently, other proposals appeared such as the polychaetes/amphipods ratio index (Gómez Gesteira and Dauvin, 2000), or the index of r/K strategies proposed by De Boer et al. (2001), which considers all benthic taxa, although it does emphasize the difficulty of scoring each species precisely through the biological trait analysis.

3.2.4   Indicators Based on Species Biomass and Abundance

Other approaches account for the variation of organism's biomass as a measure of environmental disturbances. Along these lines, we have methods such as Species Abundance and Biomass (SAB) (Pearson and Rosenberg, 1978), which consists of a comparison between the curves resulting from ranking the species as a function of their representativeness in terms of their abundance and biomass. The use of this method is not advisable because it is purely graphical, which leads to a high degree of subjectivity that impedes relating it quantitatively to different environmental factors. The Abundance and Biomass Curves (ABC) method (Warwick, 1986) also involves the comparison between the cumulative curves of species biomass and abundance, from which Warwick and Clarke (1994) derived the W statistic index.

3.2.5   Indicators Accounting for the Whole Environmental Information

From a more holistic point of view, some authors proposed indices capable of integrating the whole environmental information. An approach for application in coastal areas was first developed by Satmasjadis (1982), relating sediment particles size to benthic organism's diversity. Other indices such as the index of biotic integrity (IBI) for coastal systems (Nelson, 1990), the benthic index of environmental condition (Engle et al., 1994), or the Chesapeake Bay B–BI (Benthic-Biotic Integrity) Index (Weisberg et al., 1997) included physicochemical factors, diversity measures, specific richness, taxonomical composition, and the trophic structure of the system. Nevertheless, these indicators are rarely used in a generalized way because they have usually been developed to be applied in a particular system or area, which turns them dependent on the type of habitat and seasonality. On the other hand, their application is problematic because it requires a large amount of data of different nature.

3.2.6 Thermodynamically Oriented and Network Analysis-Based Indicators

In the last two decades, several functions have been proposed as holistic ecological indicators, intending firstly to express emergent properties of ecosystems arising from self-organization processes in the run of their development, and secondly to act as orientors (goal functions) in model development. Such proposals resulted from a wider application of theoretical concepts, following the assumption that it is possible to develop a theoretical framework able to explain ecological observations, rules, and correlations on the basis of an accepted pattern of ecosystem theories (Jørgensen and Marques, 2001). This is the case with ascendancy (Ulanowicz, 1986; Ulanowicz and Norden, 1990) and emergy (Odum, 1983; 1996). Both originated in the field of network analysis, which appear to constitute suitable system-oriented characteristics for natural tendencies of ecosystems development (Marques et al., 1998). Also, Exergy (Jørgensen and Mejer, 1979, 1981), a concept derived from thermodynamics and can be seen as energy with a built -in measure of quality, has been tested in several studies (e.g., Nielsen, 1990; Jørgensen, 1994, Fuliu, 1997, Marques et al., 1997; 2003).

3.3 HOW TO CHOOSE THE MOST ADEQUATE INDICATOR?

The application of a given ecological indicator is always a function of data requirements and data availability. Therefore, in practical terms, the choice of ecological indicators to use in a particular case is a sensible process. Table 3.1 provides a summary of what we consider to be the essential options that have been applied in coastal and transitional waters ecosystems. Table 3.2 exemplifies the process of selecting the most adequate ecological indicators as a function of data requirements and data availability.

In the process of selecting an ecological indicator, data requirements and data availability must be accounted for. Moreover, the complementary use of different indices or methods based on different ecological principles is highly recommended in determining the environmental quality status of an ecosystem.

3.4 CASE STUDIES: SUBTIDAL BENTHIC COMMUNITIES IN THE MONDEGO ESTUARY (ATLANTIC COAST OF PORTUGAL) AND MAR MENOR (MEDITERRANEAN COAST OF SPAIN)

3.4.1 Study Areas and Type of Data Utilized

Different ecological indicators were used in the Mondego estuary, located on the western coast of Portugal, and Mar Menor, a $135\,km^2$

Table 3.2 Application of indices as a function of data requirements and data availability

Data availability		Indicators
Qualitative data	Metadata	
	Rough data	Shannon–Wienner
		Margalef
		Average taxonomic distinctness (Δ^*)
Quantitative data	Populations numeric density data	AMBI
		BENTIX
		Bellan
		Bellan–Santini
		Shannon–Wienner
		Margalef
		Simpson
		Berger–Parker
		K-dominance curves
		Average taxonomic diversity index (Δ)
		Average taxonomic distinctness (Δ^+)
		Benthic index of environmental condition
		Coefficient of pollution
	Numeric density data and biomass data	Individuals identification up to specific level
		AMBI
		BENTIX
		Bellan
		Bellan–Santini
		Shannon–Wienner
		Margalef
		Simpson
		Berger–Parker
		K-dominance curves
		Average taxonomic diversity index (Δ)
		Average taxonomic distinctness (Δ^+)
		Benthic index of environmental condition
		Coefficient of pollution
		Method ABC
		Exergy
		Specific exergy
		Ascendancy
		Individuals identification up to family or higher taxonomic levels
		Shannon–Wienner
		Margalef
		Simpson
		Berger–Parker
		K-dominance curves
		Benthic index of environmental condition
		B-IBI
		Method ABC
		Exergy index
		Specific exergy
		Ascendancy

Mediterranean coastal lagoon located on the southeast coast of Spain. The lagoon is connected to the Mediterranean at some points by channels through which the water exchange takes place with the open sea.

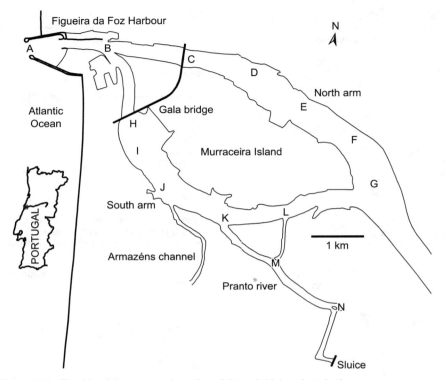

Figure 3.1   The Mondego estuary. Location of the subtidal stations in the estuary.

The Mondego estuary, located on the western coast of Portugal, is a typical, temperate, small intertidal estuary. As for many other regions, this estuary shows symptoms of eutrophication, which have resulted in an impoverishment of its quality. More detailed description of the system is reported elsewhere (e.g., Marques et al., 1993a, 1993b, 1997, 2003; Flindt et al., 1997; Lopes et al., 2000; Pardal et al., 2000; Martins et al., 2001; Cardoso et al., 2002). Regarding the Mondego estuary case study, two different data sets were selected to estimate different ecological indicators.

The first one was provided by a study on the subtidal soft bottom communities, which characterized the whole system with regard to species composition and abundance, taking into account its spatial distribution in relation to the physicochemical factors of water and sediments. The infaunal benthic macrofauna was sampled twice during spring in 1998 and 2000 at 14 stations covering the whole system (Figure 3.1).

The second one proceeded from a study on the intertidal benthic communities carried out from February 1993 to February 1994 in the south arm of the estuary (Figure 3.2). Samples of macrophytes, macroalgae, and associated macrofauna, as well as samples of water and sediments, were taken fortnightly at different sites, during low water, along a spatial gradient of eutrophication symptoms, from a noneutrophied zone, where a macrophyte

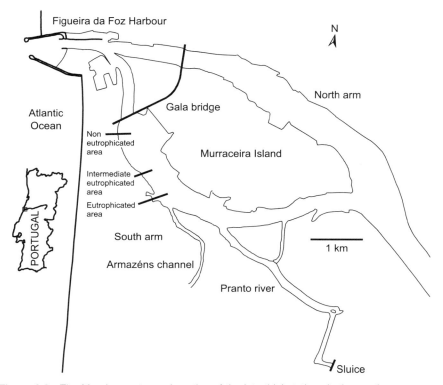

Figure 3.2   The Mondego estuary. Location of the intertidal stations in the south arm.

community (*Zostera noltii*) was present, up to a heavily eutrophied zone, in the inner areas of the estuary, from where the macrophytes disappeared while *Enteromorpha* sp. (green macroalgae) blooms have been observed during the last decade. In this area, as a pattern, *Enteromorpha* sp. biomass normally increases from early winter (February/March) up to July, when an algal crash usually occurs. A second but much less important algal biomass peak may sometimes be observed in September, followed by a decrease up to the winter (Marques et al., 1997).

In both studies, organisms were identified to the species level and their biomass was determined (g/m^2 AFDW). Corresponding to each biological sample the following environmental factors were determined: salinity, temperature, pH, dissolved oxygen, silica, chlorophyll-a, ammonia, nitrates, nitrites, phosphates in water, and organic matter content in sediments. In addition, aiming specifically at estimating ascendancy, data on epiphytes, zooplankton, fish and birds were collected from different sources (e.g., Azeiteiro, 1999; Jorge et al., 2002; Lopes et al., 2000; Martins et al., 2001) taken from April 1995 to January 1998.

Regarding the Mar Menor case study, a single data set was used. In this system, biological communities are adapted to more extreme temperatures and salinities than those found in the open sea. Furthermore, some areas in the

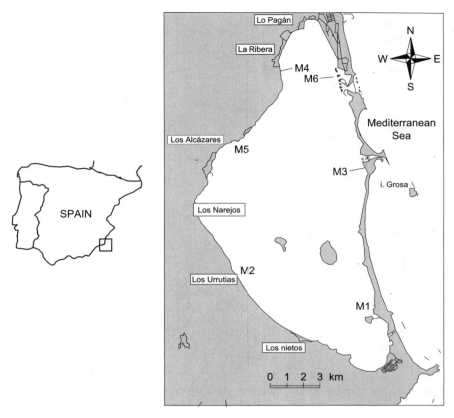

Figure 3.3   Location of the different stations in the Mar Menor.

lagoon present high levels of organic pollution proceeding from direct discharges, while other zones exhibit accumulations of organic materials originated from biological production of macrophyte meadows. Apart from these areas, we can find other communities installed on rocky or sandy substrates that do not present any significant influence of organic matter enrichment.

To estimate different ecological indicators we used data from Pérez-Ruzafa (1989), as they were a complete characterization of the benthic populations in the lagoon with the information needed for a study such as the present one. The subtidal benthic communities were sampled at six stations, located on soft substrates along the lagoon, representative of the different biocoenosis and the main polluted areas (Figure 3.3). In station M3, samples were taken in July (A), February (B), and May (D).

Likewise the Mondego estuary case study, organisms were identified to the species level and their biomass was determined (g/m^2 AFDW). The environmental factors taken into account were salinity, temperature, pH, and dissolved oxygen, as well as sediment particle size, organic matter and heavy metal contents.

3.4.2   Selected Ecological Indicators

In each case we selected ecological indicators representative of each of the groups characterized above and capable of evaluating the system from different perspectives. The discussion with regard to their applicability in each system was based on the potential of each ecological indicator to react positively to different stress situations.

The following ecological indicators were used in both case studies: AMBI, polychaetes/amphipodes ratio, Shannon–Wienner index, Margalef index, ABC method (by means of W statistic), exergy, and specific exergy. (Table 3.1). To estimate exergy and, subsequently, specific exergy from organism biomass we used a set of weighing factors (β), as discussed in chapter 2. For reasons of comparison between different case studies, all of them dated from the previous 10 years, and exergy estimations are still expressed taking into account the old β values. In fact, in terms of environmental quality evaluation, the relative differences between values obtained using the new β values or the old ones are minor, although the absolute differences are significant. Finally, in the case of the Mondego estuary, we estimated ascendancy at three intertidal sampling areas along the eutrophication gradient in the south arm. Possible relations between values of the different indicators utilized and the ecological status of ecosystems is provided in Table 3.3.

3.4.3   Summary of Results

3.4.3.1   Mondego Estuary

We focused in first place on the analysis of the subtidal communities from both arms of the estuary (first data set). As a whole, based on the comparison between results from the 1998 and 2000 sampling campaigns, all the indicators estimated, with the exception of the polychaetes/amphipods index (which could not have been applied to most of the stations anyway), indicated in a few cases some changes in the system, corresponding to a different pattern of species spatial distribution (Table 3.4a and Table 3.4b).

The Margalef index was the only one to be significantly correlated to the others, with the exception of the AMBI and the exergy indices. The Shannon–Wienner index, apart from being well correlated to the Margalef index, showed a pattern of variation similar to the one of the W statistic. The AMBI values appeared as negatively correlated with the specific exergy (Table 3.5). This suggests that most of the information expressed by specific exergy was related to the dominance of taxonomic groups usually absent in environmentally stressed situations. This uneven relationship between different indices can be recognized in the following cases:

1. Following the temporal variation of the communities at the different stations, while the diversity indices and the W statistics show, with regard to station A, that there is a worsening of the system between 1998 and 2000 (Table 3.4a and Table 3.4b), the AMBI, the exergy index and specific

Table 3.3 Possible relations between indicators values and environmental quality status of ecosystems

Indices	Ecological status
AMBI	Unpolluted: 0–1 Slightly polluted: 2 Meanly polluted: 3–4 Heavily polluted: 5–6 Extremely polluted: 7
Polychaetes/amphipodes ratio	≤1: nonpolluted >1: polluted
Shannon–Wienner index	Values most often vary between 0 and 5 bits · individual^{-1}. Resulting from many observations, an example of a possible relation between values of this index and environmental quality status could be: 0–1: bad status 1–2: poor status 2–3: moderate status 3–4: good status >4: very good status This is of course subjective and must be considered with extreme precaution.
Margalef index	High values are usually associated to healthy systems. Resulting from many observations, an example of a possible relation between values of this index and environmental quality status could be: <2.5: bad to poor status 2.5–4: moderate status >4: good status This is of course subjective and must be considered with extreme precaution.
W statistic	The index can take values from +1, indicating a nondisturbed system (high status) to −1, which defines a polluted situation (bad status). Values close to 0 indicate moderate pollution (moderate status).
Exergy index and specific exergy	Higher values are usually associated to healthy systems, but there is not any rating relationship between values and ecosystem status.
Ascendancy	Higher values are usually associated to healthy systems, but there is not any rating relationship between values and ecosystem status.

exergy suggest, on the contrary, an improvement. In fact, in 1998 the AMBI reveals co-dominance among species of the group I (54.2%), group II (10.8%) and group III (35.0%), while in 2000 only group I (51.3%) and group II (48.7%) had been represented. The decrease in environmental quality described by the other indices is basically due to dominance of *Elminius* sp. in station A during 2000. Actually, although these species

Table 3.4a Values of the different indices estimated at the 14 sampling stations in the Mondego estuary, campaigns from 1998

Station	Polychaete/ amphipode ratio	AMBI	Shannon– Wienner	Margalef	W statistics	Exergy index	Specific exergy
A	—	1.21	2.64	2.32	0.27	214.08	99.75
B	—	1.90	2.45	1.08	0.40	31.59	218.84
C	—	3.10	1.36	0.89	0.21	21.39	122.61
D	0.82	2.70	2.77	1.99	0.59	3416.39	230.27
E	—	1.70	2.14	1.26	0.30	59.59	59.13
F	—	1.60	2.61	1.55	−0.05	6.33	202.32
G	—	3.00	0.87	0.60	0.18	3.55	222.38
H	—	7.00	0.00	0.00	−1.00	5.76	450
I	—	2.00	1.43	0.94	−0.15	6.53	159.35
J	—	3.13	2.03	1.07	−0.06	33.29	165.58
K	—	2.02	1.91	1.25	0.22	15.31	10.98
L	—	3.00	1.66	0.81	−0.04	310.90	119.26
M	—	2.94	1.32	0.98	−0.20	72.35	179.68
N	—	3.00	0.63	0.72	−0.18	3.131	146.37

Table 3.4b Values of the different indices estimated at the 14 sampling stations in the Mondego estuary, campaigns from 2000

Station	Polychaete/ amphipode ratio	AMBI	Shannon– Wienner	Margalef	W statistics	Exergy index	Specific exergy
A	—	0.73	0.90	1.44	−0.19	3528.27	276.30
B	2.38	3.5	3.44	4.01	0.20	3424.53	217.43
C	—	3.9	2.40	1.52	0.23	4.52	50.90
D	—	2.3	1.84	0.89	0.39	1.95	220.86
E	—	2.4	0.65	0.27	−0.50	2.48	321.82
F	—	0.75	1.37	0.66	0.20	2.30	145.61
G	—	3	2.03	1.23	0.19	16.16	175.20
H	—	2.3	2.55	1.73	0.45	15.04	65.44
I	1.60	2.6	2.92	1.99	0.50	31.09	348.10
J	—	3	2.51	1.34	0.24	427.15	215.13
K	—	2.9	1.46	1.02	0.06	307.04	200.52
L	—	3	2.39	1.43	0.11	85.18	82.85
M	—	2.8	1.68	1.14	−0.09	7.22	69.52
N	—	3	1.38	0.79	0.24	1.67	1.82

does not indicate any kind of pollution, its abundance caused a decrease in diversity values, as the Shannon–Wienner index depends on species richness and evenness. Also, the W statistics were influenced by the dominance of *Elminius* sp. because, by coincidence, these species are very small in size. The increase in the values of the exergy index and specific exergy was fundamentally due to the increase in the biomass of species from groups such as molluscs and equinoderms, which have higher β factors.

2. Additionally, according to the diversity indices and W statistics, in stations B and C the environmental quality of the system should be improving (Table 3.5), while AMBI shows a worsening. In the case of

Table 3.5 Pearson correlations between the values of the different indicators estimated in 1998 and 2000 at the 14 sampling stations located in the two arms of the Mondego estuary

	AMBI	Shannon–Wienner	Margalef	W statistic	Exergy index
Shannon–Wienner	+0.36				
Margalef	+0.20	+0.83**			
W statistic	−0.18	+0.75**	+0.72*		
Exergy	+0.22	+0.46	+0.68**	+0.27	
Specific exergy	−0.76**	−0.23	−0.46	−0.60**	+0.15

*$P \leq 0.05$; **$P \leq 0.01$.

station B, the decline occurs drastically (from 1.98 in 1998 to 3.5 in 2000), changing from what could be considered an unbalanced community, in which species belonging to ecological group I prevailed (42.9%), to a transitional pollution state, revealed by the dominance of species of ecological groups III (43.8%) and IV (41.6%). Station C also changed to a transitional pollution or even meanly polluted situation (AMBI: 3.9) as a function of the dominance of ecological groups III (48.8%), IV (41.5%) and V (9.7%). With regard to the exergy index and specific exergy, results point to an improvement in station B, this coincides with the information provided by diversity measures and the W statistic indices, while they revealed a worsening in station C, similarly to AMBI.

By applying a one way ANOVA to 1998 results (Table 3.6), we can verify that diversity indices and the W statistic were efficient in distinguishing between stations from the north and south arms of the estuary, although values

Table 3.6 Values obtained after the application of a one-way Anova test considering the sampling stations located in the two arms of the Mondego estuary in 1998

	n	Mean	F	P
Shannon–Wienner				
North arm	6	2.32	10.47	0.007
South arm	8	1.23		
Margalef				
North arm	6	1.51	8.40	0.013
South arm	8	0.79		
W statistic				
North arm	6	0.28	6.53	0.025
South arm	8	−0.15		
Exergy index				
North arm	6	536.13	4.74	0.34
South arm	8	63.89		
Specific exergy				
North arm	6	165.04	4.74	0.89
South arm	8	175.89		
AMBI				
North arm	6	2.03	2.65	0.13
South arm	8	3.38		

Table 3.7 Values obtained after the application of an one-way Anova test considering the sampling stations located in the two arms of the Mondego estuary during 2000

	n	Mean	F	P
Shannon–Wienner				
North arm	6	1.76	0.65	0.43
South arm	8	2.15		
Margalef				
North arm	6	1.46	0.07	0.79
South arm	8	1.33		
W statistic				
North arm	6	0.05	1.23	0.28
South arm	8	0.21		
Exergy index				
North arm	6	997.17	1.84	0.20
South arm	8	124.91		
Specific exergy				
North arm	6	201.16	1.16	0.30
South arm	8	140.48		
AMBI				
North arm	6	2.26	1.39	0.26
South arm	8	2.82		

estimated for the south arm consistently indicated a higher disturbance, which is contradictory to our knowledge regarding the system reality. With regard to AMBI, exergy index and specific exergy, differences between both arms of the estuary were not statistically significant. On the other hand, regarding the 2000 results, none of ecological indicators was able to capture the differences between stations of both arms (Table 3.7).

With regard to the relationship between physicochemical factors and the variation of ecological indicators, we may observe that salinity and temperature were significantly correlated with the values of the Shannon–Wienner index ($r = 0.81$; $P < 0.01$, with salinity), Margalef index ($r = 0.78$; $P < 0.05$, with salinity), and AMBI ($r = +0.9$; $P < 0.01$, with salinity, and $r = -0.93$; $P < 0.01$ with temperature).

Let us consider now the intertidal communities along the gradient of eutrophication symptoms in the south arm of the estuary (Figure 3.4). In this case, despite different patterns of variation, with the exception of the AMBI and the polychaetes/amphipods ratio, the indicators used were able to differentiate between the three sampling areas along the south arm, as showed by a one-way-ANOVA (Table 3.8). The Margalef index, as well as the exergy index and specific exergy behaved as expected, exhibiting higher values at the Z. noltii beds and lower values in the inner areas of the south arm. However, contrary to expectations, the Shannon–Wienner and the W statistics, showed higher values in the most heavily eutrophied zone ($x = 1.69$; $x = 0.48$; $x = 0.04$, respectively) than in the Z. noltii beds ($x = 0.78$; $x = 0.79$; $x = -0.01$, respectively).

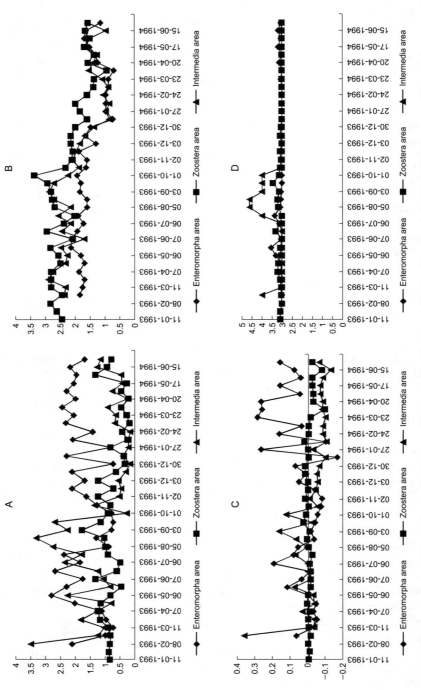

Figure 3.4 Temporal and spatial variation of the Shannon–Wienner index (A), Margalef index (B), W statistic (C), AMBI (D), exergy (E), specific exergy (F), and polychaetes/amphipods ratio (G) in the south arm of Mondego estuary.

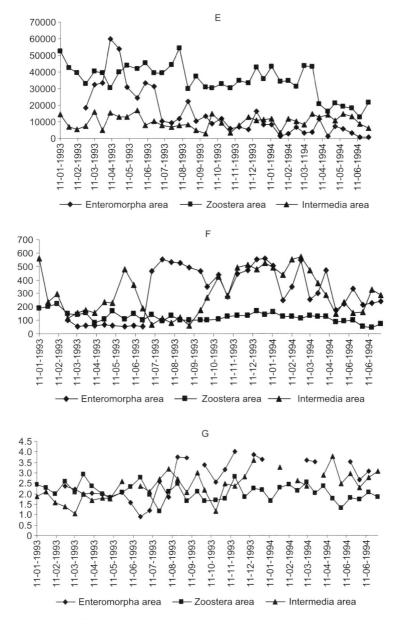

Figure 3.4   Continued.

Regarding ascendancy, we could recognize a similar pattern of spatial variation along the gradient of eutrophication in the south arm of the estuary, exhibiting a higher value in the noneutrophied area (16549 g AFDW/m/y bits; 42.3% of the total development capacity), followed by the heavily eutrophied zone (3976 g AFDW/m/y bits; 36.7% of the total development capacity).

Table 3.8   Values obtained after the application of one-way Anova test considering the three sampling areas located along the spatial gradient of eutrophication symptoms in the south arm of the Mondego estuary, in 1993 to 1994

	n	Mean	F	P
Shannon–Wienner				
Noneutrophicated area	35	0.78	17.12	0.00003
Eutrophicated area	35	1.69		
Intermedia area	35	1.14		
Margalef				
Noneutrophicated area	35	2.17	13.78	0.00004
Eutrophicated area	35	1.52		
Intermedia area	35	1.86		
Polychaete/amphipode ratio				
Noneutrophicated area	35	2.08	6.46	0.0002
Eutrophicated area	35	2.67		
Intermedia area	35	2.39		
W statistic				
Noneutrophicated area	35	−0.01	6.27	0.002
Eutrophicated area	35	0.04		
Intermedia area	35	−0.02		
Exergy index				
Noneutrophicated area	35	14893.58	6.23	0.0006
Eutrophicated area	35	35048.9		
Intermedia area	35	10143.89		
Specific exergy				
Noneutrophicated area	35	120.96	20.28	0.00008
Eutrophicated area	35	308.54		
Intermedia area	35	296.99		
AMBI				
Noneutrophicated area	35	3.07	3.36	0.06
Eutrophicated area	35	3.07		
Intermedia area	35	3.23		

The lowest values were found in the intermediate eutrophied area (1731 g AFDW/m/y bits; 30.4% of the total development capacity).

As mentioned previously, the AMBI index was unable to distinguish these three areas since estimated values of the AMBI were close to 3, which indicates slightly polluted scenarios, where species of the ecological group III are expected to be dominant (Borja et al., 2000). Rarely, AMBI values between 4 and 5 were estimated from 22 July to 1 October at station B (Figure 3.4D), located in the intermediate eutrophied zone, which indicates a meanly polluted situation. Moreover, the AMBI showed an opposite pattern of variation in relation to the other indicators used, as demonstrated by Pearson correlations estimated (Table 3.9). This contrasted exactly with the subtidal communities, when AMBI showed a similar response to the Shannon–Wienner index and specific exergy.

As for the polychaete/amphipods ratio, it expressed the existence of an eutrophication gradient, exhibiting lower values in the *Z. noltii* beds and higher

Table 3.9 Pearson correlations between the values of the different indices estimated in 1993/ 1994 at the three sampling areas along the spatial gradient of eutrophication symptoms in the south arm of the Mondego estuary

	AMBI	Shannon–Wienner	Margalef	W statistics	Exergy index	Specific exergy
Shannon–Wienner	+0.35*					
Margalef	−0.40**	+0.31*				
W statistic	+0.22*	0.82**	+0.21*			
Exergy	−0.16	−0.36**	+0.36**	−0.21*		
Specific exergy	+0.21	+0.28**	−0.14	+0.14	−0.59**	
Polychaete/amphipode ratio	−0.02	+0.04	+0.17	+0.06	+0.16	−0.13

Table 3.10 Values obtained after the application of the Pearson correlations between the different indexes and indexes and the different environmental variables in south arm of Mondego estuary

	Temperature	NH_4^+	NO_2^-	NH_3^-	PO_4^-
Polychaetes/amphipods ratio	+0.13	−0.06	−0.15	−0.002	−0.06
AMBI	+0.03	−0.10	−0.20*	−0.15	−0.15
Shannon–Wienner	+0.11	−0.27**	−0.26**	−0.34**	+0.11
Margalef	−0.03	−0.19*	−0.33**	−0.04	−0.34**
W statistics	−0.06	−0.31**	−0.15	−0.20*	−0.01
Exergy index	−0.11	−0.16	−0.01	+0.10	−0.25*
Specific exergy	0.05	+0.14	+0.10	−0.15	+0.22*

$*P \leq 0.05$; $**P \leq 0.01$.

values in the intermediate and most eutrophied areas, but has not been sensitive enough to distinguish between these ones ($P \leq 0.05$, see Table 3.8). Finally, with regard to relationships with physicochemical factors, the Shannon–Wienner index and W statistic showed significant correlation with ammonium, nitrite and nitrate concentrations in the water column, while the Margalef index, the exergy index and specific exergy were significantly correlated with phosphate concentration levels (Table 3.10).

3.4.3.2 Mar Menor

The values of the different environmental parameters analyzed showed that the areas mostly affected by organic enrichment correspond to stations M1 and M3, where organic matter content in sediments reaches values higher than 5%. They also have in common dominance of the polychaetes taxonomic group, with *Heteromastus filiformis* being the most abundant species. We should then expect there to be an occurrence of lower values for the exergy index and specific exergy, diversity measures, and W statistics, as well as higher ones for AMBI and the polychaetes/amphipods.

This was confirmed for all indices in station M3, but it was not in station M1, where only W statistics and Margalef indices gave lower values (Table 3.11). As well as this, the latter index is the only capable of detecting

Table 3.11 Values of the different indices estimated at the sampling stations in the Mar Menor

Station	Polychaete/ amphipode ratio	AMBI	Shannon–Wienner (bits/indvs)	Marga-lef	W statistic	Exergy index (g*m^{-2}det. energy equiv)	Specific exergy (ex/ unit. Biomass)
M1	—	2.10	2.75	5.34	−0.3	15762446	603402
M2	1.86	3.42	2.06	7.79	0.25	211020	1592
M3A	—	3.5	0	1.28	0.27	285182	14990
M3C	—	3.33	1.44	4.61	−0.15	94659	92702
M3D	2.20	3.32	1.9	8.49	−0.01	1555244	109065
M4	1.32	2.08	3.54	10.05	0.31	70381987	67160
M5	—	2	2.75	8.32	−0.11	2523455	14250
M6	—	3.45	3.75	12.16	0.24	28713101	78518

Table 3.12 Values obtained after the application of one-way Anova test considering the nonpolluted and polluted stations in the Mar Menor

	n	Mean	F	P
Shannon–Wienner				
Nonpolluted area	4	1.52	4.69	0.07
Polluted area	4	3.02		
Margalef				
Nonpolluted area	4	4.93	6.98	0.04
Polluted area	4	9.58		
W statistic				
Nonpolluted area	4	0.17	6.01	0.06
Polluted area	4	−0.15		
Exergy index				
Nonpolluted area	4	4424382.75	1.57	0.25
Polluted area	4	25457390.8		
Specific exergy				
Nonpolluted area	4	205039.75	1.47	0.27
Polluted area	4	40380		
AMBI				
Nonpolluted area	4	2.73	0.39	0.55
Polluted area	4	3.06		

significant differences between polluted and nonpolluted areas (Table 3.12). AMBI values are similar in all the stations, a slight disturbance in stations M1, M4, and M5 and moderate for the rest of the stations. This similarity in results for all the stations has made it unable to distinguish between different situations of disturbance in that area. In spite of this, it shows a positive response to the organic matter content in the sediment, but in a not significant manner ($r = 0.41$; $P > 0.05$). Meanwhile, the polychaetes/amphipods ratio could only be applied in stations M3 (during spring), M4 and M2. In the other stations, the absence of amphipods originated values tended toward infinity, therefore had no meaning.

Table 3.13 Pearson correlations between the values of the different indices estimated in the Mar Menor

	AMBI	Shannon–Wienner	Margalef	W statistics	Exergy index
Shannon–Wienner	−0.50				
Margalef	−0.15	+0.88**			
W statistics	+0.33	+0.01	+0.25		
Exergy	−0.49	+0.66	+0.52	+0.40	
Specific exergy	−0.44	+0.20	−0.16	−0.66	+0.05

*$P \leq 0.05$; **$P \leq 0.01$.

The W statistic gives rather confusing results as station M1, which contained organic matter, presents lower values than M3, which is seen as the most polluted one ($W = -0.3$).

In general terms, we could only recognize a similar pattern of variation between diversity measures and the exergy index, which showed positive and significant correlations (Table 3.13). These indicators were also negative and significantly correlated ($P < 0.05$) with the organic matter content in the sediment, as well as with other structuring factors of the system such as salinity or, in the case of the Margalef and Shannon–Wienner indices, sediment particle size (Table 3.14a and Table 3.14b). Specific exergy showed a clear positive

Table 3.14a Values obtained after the application of the Pearson correlations between the different indexes and the content of organic matter and heavy metals in stations of the Mar Menor

	Organic matter	Cd	Pb	Cu	Mn	Zn
Exergy index	−0.49*	0.33	−0.36	−0.31	−0.17	−0.42
Specific exergy	0.30	+0.35	+0.81**	+0.60	+0.50	+0.71**
Shannon	−0.67*	0.14	−0.22	−0.09	0.06	−0.34
Margalef	−0.68*	0.11	−0.44	−0.17	−0.13	−0.48
AMBI	+0.35	+0.34	+0.07	+0.39	−0.29	+0.28
W statistics	−0.48	+0.17	−0.52	−0.25	−0.27	−0.46

*$P \leq 0.05$; **$P \leq 0.01$.

Table 3.14b Values obtained after the application of the Pearson correlations between the different indexes and indexes and the different environmental variables in stations of the Mar Menor

	Salinity	Tempe-rature	Hydro-dynamism	Dissolved oxygen	Suspension material	Gravel (%)	Sand (%)	Silt (%)	Clay (%)
Exergy index	−0.60*	+0.02	+0.10	−0.06	−0.22	+0.02	+0.12	−0.39	−0.59
Specific exergy	+0.27	+0.50	−0.47	−0.20	−0.38	+0.92*	+0.23	−0.32	−0.14
Shannon	−0.61*	+0.40	+0.34	−0.09	+0.26	+0.27	+0.60	−0.70*	−0.76**
Margalef	−0.60*	+0.28	+0.51	−0.22	+0.50	+0.006	+0.57	−0.77*	−0.71**
AMBI	−0.47	+0.16	+0.21	+0.22	+0.29	−0.32	−0.36	+0.29	+0.43
W statistics	−0.31	+0.16	+0.14	+0.09	+0.16	−0.49	+0.12	−0.38	−0.42

*$P \leq 0.05$; **$P \leq 0.01$.

correlation with the presence of certain heavy metals as lead ($r = +0.89$; $P \le 0.05$) and zinc ($r = 0.71$; $P \le 0.05$), which does not correspond to what we should expect. For instance, stations M2D, which presented the highest concentration of these two heavy metals, exhibited also the higher value of specific exergy.

Regarding the exergy index values, the influence of biomass losses or gains, which are related to numerical changes in the dominant populations in environmentally stressed situations, is much more important than fluctuations in the β factors, which are related to the quality of the biomass in the system. In the case of specific exergy, the influence of such biomass fluctuations is very much diluted, as the β factors related to the quality of the biomass take precedence. In this sense, the loss of the taxonomic groups affected by toxic substances, as a function of different degrees of tolerance, will deeply affect specific exergy values. Now, mollusks are known by their ability to bio-accumulate heavy metals, which is not the case for, for example, polychaetes, crustaces and equinodermes. Since the β factor for molluscs is higher than the other groups (see chapter 2), it becomes immediately easy to understand why specific exergy was found to be higher in areas affected by heavy metals pollution.

3.5 WAS THE USE OF THE SELECTED INDICATORS SATISFACTORY IN THE TWO CASE STUDIES?

In order to compare the efficiencies of the ecological indicators utilized in each of the case studies, we considered suitable to evaluate a basic property: their ability to reflect different stress situations.

In the light of our results, we may reasonably consider that all the indicators, with the exception of the polychaetes/amphipods ratio, worked satisfactorily when considered separately, although in several cases the information provided was contradictory.

3.5.1 Application of Indicators Based on the Presence vs. Absence of Species: AMBI

As a whole, the AMBI worked reasonably well, although it was inefficient in discriminating among areas with clearly different eutrophication symptoms along the spatial gradient in the south arm of the Mondego estuary (e.g., dominance of Z. noltii vs. Enteromorpha sp. as main primary producers), or between stations affected by organic enrichment (covered by Cymodocea–Caulerpa meadows) from those presenting sediments with low organic matter content in the Mar Menor. In the case of the Mondego estuary this fact may perhaps be explained if we consider that eutrophication effects at the level of primary producers, which are clearly visible, are still not strong enough at other trophic levels to be detected by AMBI. In fact, although a number of shifts in species composition are already recognizable, in qualitative terms, the benthic community structure in the three zones along the spatial gradient still exhibits, in a certain extent, a reasonably alike arrangement regarding the macrofaunal species (Marques et al., 2003). Nevertheless, regarding other impact sources, such as outfalls, oil platforms, etc., it has been demonstrated that AMBI clearly shows a stress gradient (Borja et al., 2003).

The AMBI values estimated in the Mondego estuary were similar at the three sampling areas due to the common dominance of *Hydrobia ulvae*, which belongs to ecological group III. Besides, all the other indicators were strongly affected by large abundances of *H. ulvae* and *Cerastoderma edule*, the dominant species, although such dominance does not have anything to do with pollution, being rather related to the availability of higher resources (Pardal et al., 2000). In the case of Mar Menor, a sampling station such as M1, with a remarkable amount of organic matter in the sediment, was evaluated as slightly polluted because of co-dominance between species belonging to the genus *Bittium*, belonging to ecological group I, and polychaete species, belonging to group IV. The reason for this is that *Bittium* species have a herbivorous trophic strategy due to the available food resources provided by the presence of *Caulerpa–Cymodocea*, while polychaetes tend usually to be favored by sediment organic enrichment. Apart from that, in both study sites AMBI did not vary with time, therefore was less influenced by seasonal changes in abundance than the other indicators.

3.5.2 Indices Based on Ecologic Strategies: Polychaetes/Amphipods Ratio

The polychaetes/amphipods ratio was able to reflect correctly the existence of an eutrophication gradient in the case of the intertidal communities in the south arm of the Mondego estuary, but in the other two studies it was impossible to apply it simply due to the absence of amphipods in the samples. The ratio in such case would reflect an extremely polluted scenario, which we knew for certain was not the case. This indicator has been successfully used to detect the effects of organic and oil pollution on subtidal communities at the Bay of Morlaix (Mediterranean Sea) and at the Ría de Area and Betanzos (Atlantic Ocean), but in the present case studies the best results were provided when applied to intertidal data. As, for example, in the nematodes/copepods ratio used in relation to meiobenthic communities, this indicator is probably influenced by a large spectrum of ecological factors in which some types of pollution are included, meaning that this simplistic ratio is inadequate and difficult to relate to environmental quality.

3.5.3 Biodiversity as Reflected in Diversity Measures: Margalef and Shannon–Wienner Indices

Regarding diversity measures, results showed that the old Margalef index performed better, despite its simplicity, as it is more better at distinguishing between different eutrophication levels, as in the case of the Mondego estuary, and in detecting organic enrichment situations, as in the case of the Mar Menor marine lagoon. In fact, the more complex Shannon–Wienner index has been influenced too much by the dominance of certain species (e.g., *H. ulvae* in the Mondengo estuary, or *Bittium reticulatum* in the Mar Menor lagoon), whose presence has no relation with any type of disturbance or pollution phenomenon, being rather favored by abundant food resources.

3.5.4 Indicators Based on Species Biomass and Abundance: *W* statistic

Finally, the *W* statistic was capable, to a certain extent, of distinguishing between nondisturbed, slightly disturbed and strongly disturbed situations. Moreover, it does not depend on previously known reference values. Nevertheless, the dominance of certain species with small-size individuals is characteristic of nonpolluted environments, which is not unusual (the Mondego estuarine benthic community is a good example), but will lead to erroneous evaluations regarding the environmental quality status. This problem was in fact perceived in several case studies (Ibanez and Dauvin, 1988; Beukema, 1988; Weston, 1990; Craeymeersch, 1991). The reason why this indicator was not very successful in detecting organic pollution in the Mar Menor lagoon may rely on the fact that it was exclusively developed to evaluate the impact of organic pollution. In Mar Menor, although sediment organic enrichment is a concern, there are other types of pollution (e.g., metallic pollution) and different types of environmental stress.

3.5.5 Thermodynamically Oriented and Network Analysis-Based Indicators: Exergy Index, Specific Exergy and Ascendancy

3.5.5.1 Exergy and Specific Exergy

As a whole, our results suggest that the exergy index is able to capture useful information about the state of the community. In fact, more than a simple description of the environmental state of a system, the spatial and temporal variations of the exergy index may provide us a much better understanding of the system development in the scope of a broader theoretical framework. However, at the present stage, in simple snapshots the exergy index can barely provide an explicit evaluation regarding disturbed (i.e., polluted) vs. nondisturbed scenarios. For instance, in the case of the Mar Menor marine lagoon, despite responding to sediment organic enrichment, both the exergy index and specific exergy were unable to distinguish between areas affected by organic pollution and areas that are not. Nevertheless, regarding the intertidal communities of the south arm of the Mondego estuary, both exergy-based indicators were able to distinguish between areas with different eutrophication symptoms. Differences in efficiency in both case studies might have been due to the fact that in the Mar Menor lagoon the effects of organic pollution are to a certain extent diluted among other system-structuring factors, while in the south arm of the Mondego estuary eutrophication is undoubtedly the major driving force behind the ongoing changes.

Finally, it is interesting to note that the specific exergy appeared positively correlated to heavy-metal contamination (such as lead and zinc), while the exergy index was not. This was due to their different response to biomass fluctuations in the community. In fact, the influence of such fluctuations on specific exergy values is far less important because the weighting factors (β) expressing the quality of biomass take precedence.

3.5.5.2 Ascendancy

Ascendancy was tested only with regard to the Mondego estuarine intertidal communities, but the results nevertheless suggest that it is able to capture useful information about the system, namely distinguishing between areas along the eutrophication gradient. It was possible to observe that the *Z. noltii*-dominated community clearly showed the highest value, which was in accordance with theoretical expectations. Nevertheless, no test of statistical significance can (until now) be applied to the differences between the values pertaining to different areas, due to the complexity of comparing information–theoretic combinations. Although this approach is a powerful theoretical tool, this inconvenient limits *a posteriori* demonstration that there are statistically significant differences to interpret.

3.5.6 Brief Conclusions

When selecting an ecological indicator, we must in first take into account its dependence on external factors that escape our control, such as the need for reference values that often do not exist, or particular characteristics regarding the habitat type. As a result, we may reasonably conclude that no indicator will be valid in all situations. Therefore different ranges of values will have distinct significance in different scenarios, meaning that diverse classification schemes should be applied to different habitat types.

When evaluating the health status of an ecosystem, our task can be greatly facilitated if we select indicators that do not depend on any reference conditions. The polychaetes/amphipods ratio, as well as diversity measures, exergy index or specific exergy, and ascendancy — all tested by the present authors — do not fulfil this requirement, but the AMBI and W statistic do. Together with the fact that they are also independent from the type of habitat make them at first glance very suitable indicators.

The inconvenient of AMBI is that the classification of species as indicators of different grades of pollution, which constitutes its base, often contains very subjective elements, and interpreting the meaning of the presence of a given species may be ambiguous. For instance, *Chaetozone setosa*, depending on the authors, is considered an indicator of moderate pollution (Bellan, 1967; Solís–Weiss, 1982) or of intense pollution (Glemarec and Hily, 1981; Glemarec et al., 1982; Majeed, 1987). Also, *Spiochaetopterus costarum* is considered by Bellan (1967) as an indicator of slightly polluted environments, and by López Jamar (1985) as characteristic of highly polluted areas. Similarly, *Nereis caudata* is considered as an indicator of intense pollution by Bellan (1967), Zabala et al. (1983) and Lardicci et al. (1993), and simply tolerant by Glemarec and Hily (1981), Glemarec et al. (1982) and Majeed (1987). On the other hand, the *W* statistic has the inconvenience of being strongly affected by the dominance of certain small-sized species characteristic of nonpolluted environments, which leads to inevitable bias.

Our application of diversity measures, as expression biodiversity, was only partially successful, with the simpler Margalef index performing better than the more complex Shannon–Wienner. We tested only two indicators, but anyway many other tries proposing new ways to estimate diversity couldn't provide any tangible conceptual progress (see Magurran, 1988), and there is probably no conceivable "diversity index" capable to express the dynamics of mixed populations, exhibiting stabilized values through space and time. The difficulties may be summarized as follows (Marques et al., 2003):

1. The increase of diversity through time is inevitably gradual, more often than not associated with the emergence and transformation of an organized system, but the decrease of diversity is most frequently abrupt.
2. Looking to the spatial characteristics of ecosystems, we are forced to conclude that it is impossible to have stabilized variance, which may lead us to favor any kind of spectral expression taking into account the way diversity may shift as a function of the space considered. The problem in this case is that each spatial enlargement provides a different spectrum as a function of the characteristics of new sites added to the sample.
3. Since the biosphere is continuous, it is not adequate to set apart "local" diversity (called α diversity) from diversity estimated by pooling discontinuous patches (β diversity) or measured at larger spatial scales (γ diversity), although to a certain extent such descriptions may be useful in assessing the state of an ecosystem.

The exergy index and specific exergy theoretically constitute more complex ecological indicators, aiming at integrating empirical biological data and ecological observations in terms of a comprehensive thermodynamic hypothesis (Jørgensen and Marques, 2001), instead of interpreting results according to a number of nonuniversal generalizations. This point is important because despite the little respect accorded to it by those in other fields of science, ecology deals with some of the most complex phenomena encountered in modern science. Ecosystem analyses must encompass several disciplines in a coordinated fashion to answer specific questions concerning how large, multidimensional systems work (Livingston et al., 2000; Jørgensen and Marques, 2001). Besides, both indicators have been applied to structurally dynamic models of shallow lakes, appearing to represent a promising approach. Theoretically, exergy storage is assumed to become optimized during ecosystem development, and ecosystems are supposed to self organize toward a state of an optimal configuration of this property (Marques et al., 1997). Since the exergy index and specific exergy may respond differently to environmental stress, as well as to a system's seasonal dynamics, they provide different spatial and temporal pictures, and it is advisable to use both. There is nevertheless an obvious need for the determination of more accurate (discrete) weighing factors to estimate the exergy index from the biomass of organisms, which presently constitutes a very powerful constraint to the application of these indicators.

We tested ascendancy in the only case where the available data was enough to estimate it — the Mondego estuarine intertidal communities, which was obviously a very circumscribed application. Data difficulties notwithstanding, network analysis appeared to provide a systematic approach to apprehending what is happening at the whole-system level, which is obviously powerful from the theoretical point of view. Moreover, the current study on the Mondego estuarine ecosystem seems to have provided an example of how the measures coming out of network analysis can lead to an improved understanding of the eutrophication process itself (Patrício et al., in press). Nevertheless, there is a major inconvenience regarding its use, namely the considerable time and labor needed to collect all the data necessary to perform network analysis, which severely limits its application.

Summarizing, we can say that a single approach does not seem appropriate due to the complexity inherent in assessing the environmental quality of a system. Rather, this should be evaluated by combining a suite of ecological indicators, which may provide complementary information. This very same message has intermittently been conveyed to the scientific community working on environmental quality assessment (e.g., Dauer et al., 1993), together with an increasing concern regarding the need for a deeper understanding of ecological processes and for the development of a theoretical network able to explain observations, rules, and correlations on the basis of an accepted pattern of ecosystem theories (Jørgensen and Marques, 2001; Marques and Jørgensen, 2002). In other words, nature is too complex to be successfully described by simple ecological indicators (Marques, 2001). Will the ecological scientific community be able to overcome the challenge? Time is ripe to try it.

REFERENCES

Azeiteiro, U.M.M. Ecologia Pelágica do Braço Sul do Estuário do Rio Mondego. Ph.D Thesis, University of Coimbra, 1999.

Bellan, G. Pollution et peuplements benthiques sur substrat meuble dans la region de Marseille. 1.- Le secteur de Cortiou. *Rev. Intern. Oceanogr. Med.* 6, 53–87, 1967.

Bellan, G. Annélides polychétes des substrats solids de troits mileux pollués sur les côrtes de Provence (France): Cortiou, Golfe de Fos, Vieux Port de Marseille. *Téthys* 9, 260–278, 1980.

Bellan, G. and Santini, D. Relationship between populations of amphipods and pollution. *Mar. Pollut. Bull.* 11, 224–227, 1980.

Beukema, J.J. An evaluation of the ABC method abundance/biomass comparison as applied to macrozoobenthic communities living on tidal flats in the Dutch Wadden Sea. *Mar. Biol.* 99, 425–433, 1988.

Borja, A., Franco, J., and Pérez, V. A marine biotic index to establish the ecological quality of soft-bottom benthos within European estuarine and coastal environments. *Mar. Pollut. Bul.* 40, 1100–1114, 2000.

Borja, A., Muxika, I., and Franco, J. The application of a Marine Biotic Index to different impact sources affecting soft-bottom benthic communities along European coasts. *Mar. Pollut. Bul.* 46, 835–845, 2003.

Cardoso, P.G., Lillebø, A.I., Pardal, M.A., Ferreira, S.M., and Marques J.C. The effect of different primary producers on *Hydrobia ulvae* population dynamics: a case study in a temperate intertidal estuary. *J. Exp. Mar. Biol. Ecol.* 277, 173–195, 2002.

Craeymeersch, J.A. Applicability of the abundance/biomass comparison method to detect pollution effects on intertidal macrobenthic communities. *Hydrobiol. Bull.* 24, 133–140, 1991.

Clarke, K.R. and Warwick, R.M. The taxonomic distinctness measure of biodiversity: weighting of step lengths between hierarchical levels. *Mar. Ecol. Prog. Ser.* 184, 21–29, 1999.

Dauer, D.M., Luckenbach, M.W., and Rodi, A.J. Abundance-biomass comparison ABC method: effects of an estuarine gradient, anoxic/hypoxic events and contaminated sediments. *Mar. Biol.* 116, 507–518, 1993.

De Boer, W.F., Daniels, P., and Essink, K. Towards ecological quality objectives for North Sea benthic communities. National Institute for Coastal and Marine Management (RIKZ), Report no. 2001-11, Haren, The Netherlands. Contract RKZ 808, 2001, 64 p.

Engle, V., Summers, J.K., and Gaston, G.R. A benthic index of environmental condition of Gulf of Mexico. *Estuaries* 172, 372–384, 1994.

Fauchald, K. The polychaete worms. Definitions and keys to the orders, families and genera. *Sci. Ser.* 28, 21–30, 1979.

Flindt, M.R., Kamp-Nielsen, L., Marques, J.C., Pardal, M.A., Bocci, M., Bendoricho, G., Nielsen, S.N., and Jørgensen., S.E. Description of the three shallow water estuaries: Mondego River (Portugal), Roskilde Fjord (Denmark) and the Lagoon of Venice (Italy). *Ecol. Model.* 102, 17–31, 1997.

Gee, J.M., Warwick, R.M., Schaanning, M., Berge, J.A., and Ambrose, W.G., Jr. Effects of organic enrichment on meiofaunal abundance and community structure in sublittoral soft sediments. *J. Exp. Mar. Biol. Ecol.* 91, 247–262, 1985.

Gómez Gesteira, J.L. and Dauvin, J.C. Amphipods are good bioindicators of the impact of oil spills on soft bottom macrobenthic communities. *Mar. Poll. Bul.* 40, 1017–1027, 2000.

Glemarec, M. and Hily, C. Perturbations apportées a la macrofaune benthique de la Baie de Concarneau par les effluents urbains et portuaires. *Acta Oecol. Appl.* 2, 139–150, 1981.

Glemarec, M., Hussenot, E., and Le Moal, Y. Utilization of biological indicators in hypertrophic sedimentary areas to describe dynamic process after the Amoco Cádiz oil-spill. In: Ning Labbish Chao and William Kirby-Smith, Eds. Proceedings of the International Symposium on Utilization of Coastal

Ecosystems: Planning, Pollution and Productivity, 21-27 November 1982. Rio Grande, Brazil, 1982, pp. 1–18.

Gray, J.S. Pollution-induced changes in populations. *Phil. Trans. R. Soc. London* 286, 545–561, 1979.

Hughes, R.G. A model of the structure and dynamics of benthic marine invertebrate communities. *Mar. Ecol. Prog. Ser.* 15, 1–11, 1984.

Ibanez, F. and Dauvin, J.C. Long-term changes 1977–1987 in a muddy fine sand Abra alba-Melina palmata community from the Western Channel: multivariate time series analysis. *Mar. Ecol. Prog. Ser.* 19, 65–81, 1988.

Jorge, I., Monteiro, C.C., and Lasserre, G. "Fish community of the Mondego estuary: space-temporam organization," in *Aquatic Ecology of the Mondego River Basin. Global Importance of Local Experience*, Pardal, M.A., Marques, J.C., and Graça, M.A.S., Eds. Imprensa da Universidade de Coimbra, Coimbra, 2002, pp. 199–217.

Jørgensen, S.E. Development of models able to account for changes in species composition. *Ecol. Model.* 62, 195–208, 1992.

Jørgensen, S.E. Review and comparison of goal functions in system ecology. *Vie Mileu* 44, 11–20, 1994.

Jørgensen, S.E. and Mejer, H. A holistic approach to ecological modelling. *Ecol. Model* 7, 169–189, 1979.

Jørgensen, S.E. and Mejer, H. "Exergy as a key function in ecological models," in *Energy and Ecological Modelling. Developments in Environmental Modeling*, Mitsch, W., Bosserman, R.W. and Klopatek, J.M., Eds. Elsevier. Amsterdam, 1981, pp. 587–590.

Jørgensen, S.E. and Marques, J.C. Thermodynamics and ecosystem theory, case studies from hydrobiology. *Hydrobiologia* 445, 1–10, 2001.

Jørgensen, S.E., Marques, J.C., and Nielsen, S.N. Structural changes in an estuary, described by models and using exergy as orientor. *Ecol. Model.* 158, 233–240, 2002.

Lambshead, P.J.D. Sub-catastrophic sewage and industrial waste contamination as revealed by marine nematode faunal analysis. *Mar. Ecol. Prog. Ser.* 29, 247–260, 1986.

Lambshead, P.J.D., Platt, H.M., and Shaw, K.M. The detection of differences among assemblages of marine benthic species based on an assessment of dominance and diversity. *J. Nat. Hist.* 17, 859–847, 1983.

Lambshead, P.J.D. and Platt, H.M. Structural patterns of marine benthic assemblages and their relationship with empirical statistical models. *Proceedings of the Nineteenth European Marine Biology Symposium*, Plymouth, 1985, Gibbs, P.E. Ed. pp.16–21.

Lardicci, C., Abbiati, M., Crema, R., Morri, C., Bianchi, C.N., and Castelli, A. The distribution of polychaetes along environmental gradients: an example from the Ortobello Lagoon, Italy. *Mar. Ecol.* 14, 35–52, 1993.

Livingston, R.J., Lewis, F.G., III, Woodsum, G.C., Niu, X., Galperin, B., Huang, W., Christensen, J.D., Monaco, M.E., Battista, T.A., Klein, C.J., Howell, R.L., IV, and Ray, G.L. Use of coupled physical and biological models: response of oyster population dynamics to freshwater input. *Estuar. Coast. Shelf. Sci*, 50, 655–672, 2000.

Lopes, R.J., Pardal, M.A., and Marques, J.C. Impact of macroalgal blooms and water predation on intertidal macroinvertebrates: experimental evidence from the Mondego estuary Portugal. *J. Exp. Mar. Biol. Ecol.* 249, 165–179, 2000.

López-Jamar, E. Distribución espacial del poliqueto *Spiochaeterus costarum* en las Rías Bajas de Galicia y su posible utilización como indicador de contaminación orgánica en el sedimento. *Bol. Inst. Esp. Oceanogr.* 2, 68–76, 1985.

Magurran, A.E. Ecological *Diversity and its Measurement*. Croom Helm, London, 1988, 179 p.

Majeed, S.A. Organic matter and biotic indices on the beaches of North Brittany. *Mar. Pollut. Bull.* 18, 490–495, 1987.

Marques, J.C., Rodrigues, L.B., and Nogueira, A.J.A. Intertidal macrobenthic communities structure in the Mondego estuary (Western Portugal): reference situation. *Vie Mileu.* 43, 177–187, 1993a.

Marques, J.C., Maranhão, P., and Pardal, M.A. Human impact assessment on the subtidal macrobenthic community structure in the Mondego estuary (Western Portugal). *Estuar. Coast. Shelf. Sci.* 37, 403–419, 1993b.

Marques, J.C, Pardal, M.A, Nielsen, S.N., and Jørgensen, S.E. Analysis of the properties of exergy and biodiversity along an estuarine gradient of eutrophication. *Ecol. Model.* 102, 155–167, 1997.

Marques, J.C., Pardal, M.A., Nielsen, S.N., and Jørgensen, S.E. "Thermodynamic orientors: exergy as a holistic ecosystem indicator: a case study," in *Ecotargets, Goal Functions and Orientors. Theorical Concepts and Interdisciplinary Fundamentals for an Integrated, System Based Environmental Management*, Müller, F. and Leupelt, M. Eds. Springer-Verlag, Berlin, 1998, chapter 2.5, pp. 87–101.

Marques, J.C. Diversity, biodiversity, conservation, and sustainability. *The Scientific World*, 1, 534–543, 2001.

Marques, J.C. and Jørgensen, S.E. Three selected observations interpreted in terms of a thermodynamic hypothesis. Contribution to a general theoretical framwork. *Ecol. Model.* 158, 213–222, 2002.

Marques, J.C., Nielsen, S.N., Pardal M.A., and Jørgensen, S.E. Impact of eutrophication and river management within a framework of ecosystem theories. *Ecol. Model.* 166, 147–168, 2003.

Martins, I., Pardal, M.A., Lillebø, A.I., Flindt, A.R., and Marques, J.C. Hydrodynamics as a major factor controlling the occurrence of green macroalgal blooms in a eutrophic estuary: a case study on the influence of precipitation and river management. *Estuar. Coast. Shelf. Sci.* 52, 165–177, 2001.

McLusky, D.S. The *Estuarine Ecosystem*. 2nd ed. Blackie, New York, 1989.

Nelson, W.G. Prospects for development of an index of biotic integrity for evaluating habitat degradation in coastal systems. *Chem. Ecol.* 4, 197–210, 1990.

Nielsen, S.N. Application of exergy in structural-dynamical modelling. *Vehr. Int. Ver. Limnol.* 24, 641–645, 1990.

Nielsen, S.N. Strategies for structural dynamics modelling. *Ecol. Model.* 63, 91–101, 1992.

Nielsen, S.N. Modelling structural dynamic changes in a Danish Shallow Lake. *Ecol. Model.* 73, 13–30, 1994.

Nielsen, S.N. Optimisation of exergy in a structural dynamic model. *Ecol. Model.* 77, 111–112, 1995.

Odum, H.T. Systems *Ecology: An Introduction*. Wiley, Toronto, 1983.

Odum, H.T. *Environmental Accounting. Emergy and Environmental Decision Making*. Wiley, New York, 1996.

Pardal, M.A., Marques, J.C., Metelo, I., Lillebø, A.I., and Flindt, A.R. Impact of eutrophication on the life cycle population dynamics and production of

Amphitoe valida (Amphipoda) along an estuarine spatial gradient (Mondego estuary, Portugal). *Mar. Ecol. Progr. Ser.* 196, 207–219, 2000.

Parker, J.G. Effects of pollution upon the benthos of Belfast Lough. *Mar. Pollut. Bulle.* 11, 80–83, 1980.

Patrício, J., Ulanowicz, R.E., Pardal, M.S.A., and Marques, J.C. Ascendancy as ecological indicator: a case study on estuarine pulse eutrophication. *Estuar. Coast. Shelf. Sci.* 60, 23–35, 2004.

Pearson, T.H. and Rosenberg, R. Macrobenthic succession in relation to organic enrichment and pollution of the marine environment. *Oceanogr. Mar. Biol. Ann. Rev.* 16, 229–331, 1978.

Pérez-Ruzafa, A. Estudio ecológico y bionómico de los poblamientos bentónicos del Mar Menor (Murcia, SE de España). Ph.D Thesis, University of Murcia, 1989.

Perillo, G., Piccolo, M.C., and Freije, R.H. The Bahía Blanca Estuary. *Ecol. Stud.* 144, 205–217, 2001.

Rafaelli, D.G. and Mason, C.F. Pollution monitoring with meiofauna: using the ratio nematodes/copepods. *Mar. Pollut. Bul.* 12, 158–163, 1981.

Roberts, R.D., Gregory, M.G., and Foster, B.A. Developing an efficient macrofauna monitoring index from an impact study — a dredge spoil example. *Mar. Pollut. Bul.* 36, 231–235, 1998.

Satsmadjis, J. Analysis of benthic fauna and measurement of pollution. *Revue Internationale Oceanographie Medicale*, 66–67, 103–107, 1982.

Shaw, K.M., Lambshead, P.J.D., and Platt, H.M. Detection of pollution-induced disturbance in marine benthic assemblages with special reference to nematodes. *Mar. Ecol. Progr. Ser.* 11, 195–202, 1983.

Simboura, N. and Zenetos A. Benthic indicators to use ecological quality classification of Mediterranean soft bottom marine ecosystems, including a new biotic index. *Med. Mar. Sci.* 3, 77–110, 2002.

Solís-Weiss, V. Aspectos ecológicos de la contaminación orgánica sobre el macrobentos de las cuencas de sedimentación en la bahía de Marsella Francia. *An. Inst. Cienc. Mar. Limnol.* 9, 19–44, 1982.

Sommerfield, P.J. and Clarke, K.R. A comparison of some methods commonly used for the collection of sublittoral sediments and their associated fauna. *Mar. Environ. Res.* 43, 145–156, 1997.

Straškraba, M. Cybernetic formulation of control in ecosystems. *Ecol. Model.* 18, 85–98, 1983.

Ulanowickz, R.E. *Growth and Development Ecosystems Phenomenology*. Springer-Verlag, New York, 1986.

Ulanowickz, R.E. and Norden, J.S. Symmetrical overhead in flow and networks. *Intern. J. Syst. Sci.* 21, 429–437, 1990.

Warwick, R.M. A new method for detecting pollution effects on marine macrobenthic communities. *Mar. Biol.* 92, 557–562, 1986.

Warwick, R.M. Environmental impact studies on marine communities: pragmatical considerations. *Aust. J. Ecol.* 18, 63–80, 1993.

Warwick, R.M. and Clarke, K.R. Relearning the ABC: taxonomic changes and abundance/biomass relationships in disturbed benthic communities. *Mar. Biol.* 118, 739–744, 1994.

Warwick, R.M. and Clarke K.R. New "biodiversity" measures reveal a decrease in taxonomic distinctness with increasing stress. *Mar. Ecol. Prog. Ser.* 129, 301–305, 1995.

Warwick, R.M. and Clarke, K.R. Taxonomic distinctness and environmental assessment. *J. Appl. Ecol.* 35, 532–543, 1998.

Weisberg, S.B., Ranasinghe, J.A., Dauer, D.M., Schaffner, L.C., Díaz, R.J., and Frithsen, J.B. An estuarine benthic index of biotic integrity (B-IBI) for the Chesapeake Bay. *Estuaries*, 20, 149–158, 1997.

Weston, D.P. Quantitative examination of macrobenthic community changes along an organic enrichment gradient. *Mar. Ecol. Progr. Ser.* 61, 233–244, 1990.

Word, J.Q. The infaunal trophic index. *Calif. Coast. Wat. Res. Proj. Annu. Rep.* 19–39, 1979.

Xu, Fu-Liu. Exergy and structural exergy as ecological indicators for the state of the Lake Chalou ecosystem. *Ecol. Model.* 99, 41–49, 1997.

Zabala, K., Romero, A., and Ibáñez, M. La contaminación marina en Guipuzcoa: I. Estudio de los indicadores biológicos de contaminación en los sedimentos de la Ría de Pasajes. *Lurralde* 3, 177–189, 1983.

CHAPTER 4

Development and Application of Ecosystem Health Indicators in the North American Great Lakes Basin

H. Shear, P. Bertram, C. Forst, and P. Horvatin

Assessing the health of the North American Great Lakes Basin ecosystem is a significant challenge. The lakes themselves contain one-fifth of the world's fresh surface water with over 17,000 km of shoreline. The basin consists of over 520,000 km² of land with about 33.5 million people living there. The basin (including the St. Lawrence River) is governed by two nations, eight states, two provinces, and hundreds of municipal and local governments. A set of Great Lakes Basin ecosystem health indicators will enable the Great Lakes community to work together within a consistent framework to assess and monitor changes in the state of the ecosystem. Data collected through various government and nongovernment programs can be analyzed, interpreted, and ecosystem health information characterized within a series of such indicators. A consensus agreement by environmental management agencies and other interested stakeholders about what information is necessary and sufficient to characterize the state of the Great Lakes ecosystem's health, and to measure progress toward ecosystem goals, will facilitate more efficient monitoring and

1-56670-665-3/05/$0.00 + $1.50
© 2005 by CRC Press

reporting programs. This chapter will present the process for indicator selection or development, with some examples of indicator reporting.

4.1 INTRODUCTION

4.1.1 Background on the Great Lakes Basin

The purpose of this chapter is to present information on some of the indicators representing various ecosystem components of the Great Lakes Basin.

Assessing the health of the Great Lakes Basin ecosystem is a significant challenge. The Great Lakes St. Lawrence River Basin consists of over 520,000 km^2 of land, and the lakes themselves contain one-fifth of the world's fresh surface water surrounded by over 17,000 km of shoreline. About 33.5 million people reside within the basin. Natural resources within the basin supply tens of millions of people with drinking water, support a multibillion dollar recreation and tourism industry, provide habitat for thousands of plant and animal species, offer transportation and manufacturing opportunities, and support an extensive agricultural industry.

Governance in the basin is complex. Political jurisdictions include two nations, eight states, two provinces, dozens of tribes and First Nations, and hundreds of municipal and local governments. Within each of the governance structures are multiple agencies, offices, and other organizations that exercise jurisdiction over one or more Great Lakes Basin ecosystem components.

The U.S. and Canada have collectively spent billions of dollars and uncounted hours attempting to reverse the effects of cultural eutrophication, toxic chemical pollution, overfishing, habitat destruction, introduced species, and other human-induced pressures on the Great Lakes. To assess if past programs have been successful and if future or continuing programs will result in environmental improvement commensurate with the resources expended, a consensus agreement is desired by environmental management agencies and other interested stakeholders about what information is necessary and sufficient to characterize the state of Great Lakes Basin ecosystem. Having agreed on information requirements, the relative strengths of the various agencies and organizations can then be utilized to improve the timeliness and quality of the data collection and the availability of the information to multiple users.

The ecosystem approach is the defining mechanism by which management agencies are meant to carry out their research and deliver their regulatory programs (United States and Canada, 1987; Vallentyne and Beeton, 1988; Hartig and Vallentyne, 1989; Hartig and Zarull, 1992; International Joint Commission, 1995). In real terms, the ecosystem approach has come to mean a comprehensive approach to environmental issues, considering the interacting living (including humans) and nonliving components of the Great Lakes Basin.

The Great Lakes Water Quality Agreement (GLWQA) between the U.S. and Canada was signed in 1972 by Prime Minister Pierre Trudeau and President Richard Nixon in order "to restore and maintain the chemical, physical, and biological integrity of the waters of the Great Lakes Basin Ecosystem" (United States and Canada, 1987). Its primary intent was to decrease external loadings of phosphorus to the lakes. The agreement was revised in 1978, to establish (among other things) annual loading targets of phosphorus for each of the Great Lakes. In 1987 the agreement was amended, placing much more emphasis on control and elimination of toxic contaminants and on a broad ecosystem approach to solving the Great Lakes ecosystem problems.

The 1987 revisions to the agreement also introduced some far-reaching concepts that have had significant impact on management planning and implementation for the Great Lakes. For instance, ecosystem objectives are to be developed for each of the Great Lakes, and indicators are to be identified to measure progress towards those objectives. The governments of Canada and the U.S. also agreed to report on progress toward the agreement goals and objectives every two years.

The State of the Lakes Ecosystem Conference (SOLEC) was established by the governments of Canada and the United States in 1992 in response to reporting requirements of the GLWQA. It is held every two years, and is designed to report on the condition of the Great Lakes Basin ecosystem regarding progress toward the goals and objectives of the Great Lakes Water Quality Agreement. The conferences are science-based rather than programmatic, and they are a result of consultation and collaboration between U.S. and Canada, and between federal, state, provincial and local government agencies, environmental groups, industry and the public.

No one organization has the resources or the mandate to examine the state of all the Great Lakes ecosystem components, but dozens of organizations and thousands of individuals routinely collect data, analyze them, and report on parts of the ecosystem. Through the SOLEC process, however, data and other information are collected from multiple sources and are reported as a series of indicators on Great Lakes ecosystem components.

4.1.2 Indicator Selection

An indicator is a parameter or value that reflects the condition of an environmental (or human health) component, usually with a significance that extends beyond the measurement or value itself (Canada and the United States, 1999). Used alone or in combination, indicators provide the means to assess progress towards one or more objectives: are conditions improving so that the objective is closer to being met, or are conditions deteriorating? The achievement of these objectives then leads towards the achievement of higher order goals and a vision for the ecosystem.

Application of the indicators requires two pieces of information: the measurement and some sort of reference value. The measurement describes the observed state of the ecosystem component. The reference value, or end point, reflects the desired state of the ecosystem component. Qualitative assessments are derived by comparing the observed measurements to the desired states.

With respect to the health of the Great Lakes, scientists and nonscientists have engaged in the development of appropriate ecosystem indicators (Ryder and Edwards, 1985; Edwards and Ryder, 1990; Bertram and Reynoldson, 1992; Bertram et al., 2002). There is a continuum of proposed indicators from the ones that are easily understood by the nonscientific public, to those that are more technical and better understood by the scientific community.

The selection of the Great Lakes indicators through the SOLEC process was based on a few key principles (Bertram et al., 2002), outlined below:

- *Build upon the work of others.* Over 800 existing or proposed indicators of the condition of Great Lakes ecosystem components were identified. From this large pool of indicators, a subset was selected and then further refined, combined or modified to best represent the ecosystem component under consideration.
- *Focus on broad spatial scales.* SOLEC assessments are oriented toward the whole Great Lakes watershed and at the individual lake basins. Other venues are available for reporting on more local conditions, such as specific harbors or wetlands.
- *Select a framework for subdividing the Great Lakes Basin ecosystem.* A combination of geographic areas and nongeographic issues provided an organizing framework. They included offshore and nearshore waters, coastal wetlands, nearshore terrestrial areas, human health, land use, and societal indicators.
- *Select a system for types of indicators.* There are several classification schemes or models for indicators, one of which is the state–pressure– response (human activities) model (OECD, 1993; Bertram and Stadler-Salt, 1999). Indicators were selected to assess the state of the ecosystem components and the pressures or "stressors" that affect the state. Some human activities that affect or result from the stressors were included, but the initial focus of the indicators was on the state and stressor indicators.
- *Identify criteria for indicator selection.* The primary criteria were for each indicator to be necessary — that is, contributes a unique perspective of an ecosystem component; and feasible — that is, practical and able to be implemented. Also, the entire suite of indicators should be sufficient to provide a comprehensive assessment of Great Lakes ecosystem health.

The process for identifying the current Great Lakes indicators required about two years and involved at least 150 people, and the process is continuing. A core working group and an expert panel were created for each of the geographic areas and nongeographic issues under consideration. Both U.S. and Canadian expertise was involved. Each of the working groups mined indicators and indicator ideas from existing sources. The groups then screened their long list and revised, combined or created new indicators as needed.

The work of all the groups was combined into the proposed Great Lakes indicator list and presented at SOLEC 1998. Within the framework categories, indicators for nearshore and open waters, coastal wetlands and nearshore terrestrial environments were all well represented. The human health category has fewer indicators, in part because of the difficulty of identifying human health effects due to the Great Lakes influence. More reliance was therefore placed on indicators of potential human health exposure.

The organization of the list of Great Lakes indicators is flexible and can be easily regrouped. For example, the indicators can be grouped according to environmental compartment (e.g., air, water, land, sediments, biota, humans) or by Great Lakes issues (e.g., contaminants and pathogens, nutrients, non-native species, habitat, climate change, stewardship).

4.1.3 Definition of the Selected Indicators

The working groups applied the three criteria of "necessary," "sufficient," and "feasible," to over 800 potential indicators. Through scientific review and the application of secondary criteria, this number was reduced to 80 in 1998. This list of indicators continues to evolve. Some indicators may be revised, some indicators may get added over time, and some indicators may be removed from the list. This will help the Great Lakes community to ensure that it can respond to new or emerging issues. For example, indicators for groundwater, forests, societal response, agriculture, and others were not included in the original set, but they were proposed and some were developed in conjunction with SOLEC 2002.

For SOLEC 2002, reports were prepared on 43 of the current 82 (reconfigured from the 80 presented in 1998) Great Lakes indicators plus 10 example reports for additional proposed indicators. The 43 indicators were selected and reported on because data were readily available. At the conference, the indicators were discussed in workshop sessions, and comments, criticisms, and suggestions for improvements were noted.

Data for a few of the 43 reported indicators were not available basin-wide. In other cases only some of the data for the indicators were presented. The remaining indicators have yet to be reported because the required data have not been collected. Changes to existing monitoring programs or the initiation of new monitoring programs may be necessary. Additionally, some indicators are still in the development stage. Over time, monitoring and reporting on the full suite of indicators is expected to be implemented.

After the necessary refinements were made to some of the indicator reports, they were organized according to the pressure–state–human response model, summarized, and compiled into a formal report, "State of the Great Lakes 2003" (Canada and the U.S., 2003a). The complete indicator reports as prepared by the expert authors, including literature citations and references to data sources, were similarly compiled into a companion report, "Implementing Indicators 2003 — A Technical Report" (Canada and the U.S., 2003b).

4.2 GENERAL CONSIDERATIONS

4.2.1 Ecological Description of the Great Lakes Basin

The Laurentian Great Lakes are large and complex. The basin's ecology and human land use are driven by two main factors: the geography of the basin, and its geology. The lakes differ in their ecology largely due to the vast geographic area that they occupy. From north to south, the Great Lakes Basin spans almost 11 degrees of latitude. This accounts for a significant climatic variation. In the north, the climate is characterized by short, warm summers and long, cold winters. In the south, the summers are longer and warmer, the winters somewhat shorter and milder. The geology of the basin has also influenced its ecology, and has determined human land uses. The south is dominated by agriculture because of its deep fertile soils, as well as the more favorable climate. The north is generally forested with minimal agriculture because of the thin soils overlaying granite rock, and also because of the cooler climate. Large urban areas are generally confined to the southern part of the basin, with a couple of exceptions in the Lake Superior Basin.

As a result of this land use and human settlement pattern, one can characterize the issues in the basin as follows.

4.2.1.1 Toxic Contaminants

Contaminants in the Great Lakes have shown a significant decrease over the last 20 to 25 years as a result of actions taken by Canada and the U.S. to ban and control contaminants such as mercury, DDT and PCBs from entering the Great Lakes. Contaminants are still an issue in localized areas such as harbors and embayments, and fish consumption advisories exist in many areas of the Great Lakes.

4.2.1.2 Land Use

Major population centers in the Great Lakes Basin include the northwestern part of the Canadian shoreline of Lake Ontario, the south shore of Lake Erie, most of the southern Lake Michigan Basin, and two centers in Lake Superior (Duluth-Superior and Thunder Bay). The St. Clair–Detroit River ecosystem is one of the most highly industrialized areas in the Great Lakes Basin. The major cities of Port Huron and Detroit, Michigan, and Sarnia and Windsor, Ontario, are major petrochemical and manufacturing centers. Lake Huron has the lowest human population density in the entire basin, and as such, has shown fewer environmental problems.

The Lake Erie Basin includes a Carolinian Zone that has been described as Canada's most endangered major ecosystem. The Carolinian Zone sustains at least 18 globally rare vegetation community types; 36 globally rare species; and 108 vulnerable, threatened, and endangered species. The watershed also has habitats that sustain 143 fish species, many of which contribute to a thriving sport and commercial fishery.

Lake Michigan is the second largest of the Great Lakes by volume, has the world's largest area of freshwater sand dunes, and contains 40% of the U.S. Great Lakes coastal wetlands. Recreational and industrial activities have had strong impact on both the natural dynamics of the dunes and on dune and wetland habitats. Wetland loss in the Lake Michigan basin is disproportionately greater than the U.S. average.

4.2.1.3   Invasive Species

There are over 150 species of invasive non-native species in the Great Lakes Basin. Many of these species are aquatic. The impact of non-native species on Great Lakes ecology and economy has been documented for several species, including the sea lamprey and zebra mussel. Today, for example, as a result of zebra mussel infestation in Lake Erie, the entire ecology of the Lake has changed, and phenomena such as type E botulism deaths in fish and wildlife have appeared, and may be linked to zebra mussels. Millions of dollars per year are spent keeping zebra mussels from clogging water intake pipes. Invasive non-native species continue to be a concern for Lake Michigan. In 2002, the non-native fish species, the ruffe, was found in Lake Michigan for the first time. Other non-native species, including zebra mussels and round goby, are continuing to impact Lake Michigan's aquatic ecosystems.

4.2.1.4   Habitat Status Including Wetlands

Much of the Lake Ontario watershed, tributaries, and nearshore lands remain degraded, particularly in the western basin, and new concerns continue to emerge to further complicate recovery efforts.

Wetland areas exist in pockets throughout the Lake St. Clair–St. Clair River region. The largest is in the Walpole Island First Nation Territory at the mouth of the St. Clair River. Walpole Island also has remnant tall grass prairie and oak savannah habitats. A smaller wetland survives in Michigan at the north end of Lake St. Clair.

Lake Huron has over 30,000 islands, contributing to its distinction of having the longest shoreline of any lake in the world. The islands and nearshore areas still support a high diversity of aquatic and riparian species.

Aquatic habitats in the main basin of Lake Huron are in relatively good health. Many of the tributaries in the system, however, are still severely stressed by both development and point and nonpoint source pollution. These stressors are resulting in changes to tributary fish community composition.

4.2.1.5   Lake Ecology

Populations of fish-eating waterbirds in Lake Ontario have recovered and are reproducing normally. Recent data have shown that several other key indicator species such as the bald eagle (within the Lake Ontario basin), otter, and mink are also making a comeback.

In the western basin, increased populations of mayflies (a bottom-dwelling species) are providing forage for many fish species. Trout-perch, another bottom-dwelling species that was in decline in the 1950s, seems to be making a comeback. These changes suggest that the bottom community may be starting to recover.

The health of fish communities is of particular concern in the Great Lakes Basin because of their economic, recreational, and ecological importance. Current stressors to fish communities include continued habitat degradation, loss of food sources due to non-native species, and contamination. The fish community in Lake Superior is considered the healthiest, with the largest self-sustaining lake trout population in the entire system. In Lakes Michigan, Huron and Ontario, the top predator fish community is artificially maintained through hatchery rearing and stocking programs. In the last few years, however, natural reproduction of native lake trout has once again been documented at several locations in Lakes Huron and Ontario.

4.2.1.6 Nutrients

Although significant reductions in nutrient loadings have been achieved, phosphorus concentrations in Lake Erie appear to be increasing again and may be linked to a zone of oxygen depletion in the Central basin. Nutrients may be becoming an issue again in nearshore waters of Lake Ontario, leading to nuisance growths of *Cladophora*.

4.2.2 Data Collection Methods

No single organization has the resources or the mandate to examine the state of all the ecosystem components. Dozens of organizations collect data, analyze them, and report on parts of the ecosystem. A consensus by environmental management agencies and other interested stakeholders about what information is necessary and sufficient to characterize the state of Great Lakes ecosystem health will facilitate more efficient monitoring and reporting.

The SOLEC approach is to use the relative strengths of the agencies to improve the timeliness and quality of the data collection and the availability of the information to multiple users. Subject matter experts are invited to contribute their information about a particular indicator and to make assessments of the relative state of the ecosystem component or pressure being addressed. Quality assurance and quality control programs are implemented by each organization that collects the data. For SOLEC reporting, "Implementing Indicators 2003 — A Technical Report" (Canada and the U.S., 2003b) offers references to data sources to facilitate the tracing of data and exploration of methodologies.

4.3 RESULTS

For SOLEC 2002 and for the State of the Great Lakes 2003 reports (Canada and the U.S., 2003a, 2003b), the indicators were grouped into the categories of state, pressure, and response. Within the state and pressure categories, the indicators were further divided into those with data that were basin-wide and consistent over time (at least within a lake basin) and those whose data represented smaller geographic areas or which were not consistent between areas or over time. There were no response indicators that were consistent over the entire Great Lakes Basin. The following example indicator findings are grouped accordingly.

4.3.1 State Indicators — Complete

4.3.1.1 Hexagenia

The distribution, abundance, biomass, and annual production of the burrowing mayfly *Hexagenia* in mesotrophic Great Lakes habitats are measured directly. These metrics are used as the indicator of ecosystem health because *Hexagenia* is intolerant of pollution, and is therefore a good reflection of water and lakebed sediment quality in mesotrophic Great Lakes habitats. *Hexagenia* was historically the dominant, large, benthic invertebrate in these habitats, and was an important item in the diets of many valuable fishes. Figure 4.1 shows the populations of *Hexagenia* in Western Lake Erie over the past 11 years, with a clear indication of a population recovery (Ciborowski et al., unpublished observations).

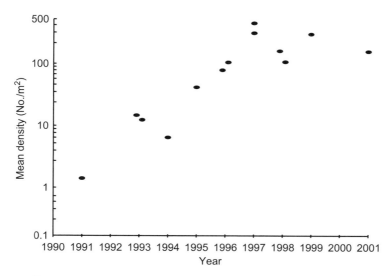

Figure 4.1 Density of *Hexagenia* in western Lake Erie (1991 to 2001). From Ciborowski, J., Krieger, K., Schloesser, D and Corkum, L, unpublished.

4.3.1.2   Wetland Dependent Bird Diversity and Abundance

Assessments of wetland-dependent bird diversity and abundance in the Great Lakes Basin are used to evaluate health and function of coastal and inland wetlands. Breeding birds are valuable components of Great Lakes wetlands and rely on physical, chemical and biological health of their habitats. Because these relationships are particularly strong during the breeding season, presence and abundance of breeding individuals can provide a source of information about wetland status and trends. When long-term monitoring data are combined with an analysis of habitat characteristics, trends in species abundance and diversity can contribute to an assessment of how well Great Lakes coastal wetlands are able to support birds and other wetland-dependent wildlife. Populations of several wetland-dependent birds are believed to be at risk due to continuing loss and degradation of their habitats. Figure 4.2 (Weeber and Vallianatos, 2000) shows results for four bird species.

While five years of data are not enough to draw any definitive conclusions, clearly the trends in these populations are declining or at best remaining static.

4.3.1.3   Area, Quality and Protection of Alvar Communities

This indicator assesses the status of alvars. Alvar communities are naturally open habitats occurring on flat limestone bedrock. Over 67% of known alvar occurrences within the Great Lakes Basin are close to the shoreline.

More than 90% of the original extent of alvar habitats has been destroyed or substantially degraded. Emphasis is focused on protecting the remaining 10%. Approximately 64% of the remaining alvar areas exist within Ontario,

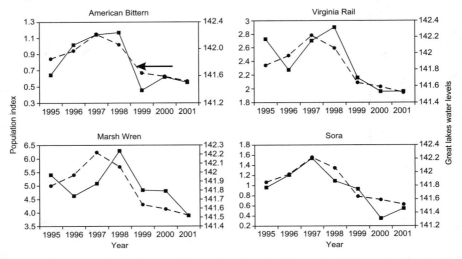

Figure 4.2   Population index trends for selected wetland dependent birds and water levels. From *Implementing Indicators 2003 — A Technical Report* (Canada and the U.S., 2003b).

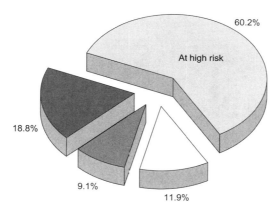

Figure 4.3 Protection status for Great Lakes alvars. From *Implementing Indicators 2003 — A Technical Report* (Canada and the U.S., 2003b).

16% in New York State, and 15% in Michigan and smaller areas in Ohio, Wisconsin and Quebec.

Less than 20% of the nearshore alvar acreage is currently fully protected, while over 60% is at high risk. Michigan has 66% of its nearshore alvar acreage in the "fully protected" category, while Ontario has only 7%. In part, this is a reflection of the much larger total shoreline acreage in Ontario (see Figure 4.3).

4.3.2 State Indicators — Incomplete

4.3.2.1 Native Freshwater Mussels

The purpose of this indicator is to report on the location and status of freshwater mussel (unionid) populations and their habitats throughout the Great Lakes system, with emphasis on endangered and threatened species. The long-term goal for the management of native mussels is for populations to be stable and self-sustaining wherever possible throughout their historical range in the Great Lakes, including the connecting channels and tributaries.

The introduction of the zebra mussel to the Great Lakes in the late 1980s has destroyed unionid communities throughout the system. Unionids were virtually extirpated from the offshore waters of western Lake Erie by 1990 and Lake St. Clair by 1994, with similar declines in the connecting channels and many nearshore habitats. There were on average 18 unionid species found in these areas before the zebra mussel invasion. After the invasion, 60% of surveyed sites had three or fewer native species left, 40% of sites had no native species left, and the abundance of native mussels had declined by 90 to 95% (see Figure 4.4).

Significant communities were, however, recently discovered in several nearshore areas where zebra mussel infestation rates are low. All of the refuge sites discovered to date have two things in common: they are very shallow (less

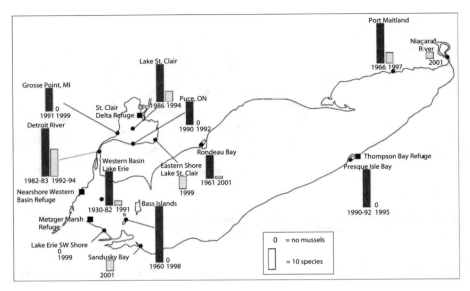

Figure 4.4   Abundance of native mussels in Lake Erie. From *Implementing Indicators 2003 — A Technical Report* (Canada and the U.S., 2003b).

than 1 to 2 m deep), and they have a high degree of connectivity to the lake that ensures access to host fishes. These features appear to combine with other factors to discourage the settlement and survival of zebra mussels.

4.3.3   Pressure Indicators — Complete

4.3.3.1   *Phosphorus Concentrations and Loadings*

This indicator assesses total phosphorus levels in the Great Lakes, and is used to support the evaluation of trophic status and food web dynamics in the Great Lakes. Efforts begun in the 1970s to reduce phosphorus loadings have been successful in maintaining or reducing nutrient concentrations in the lakes, although high concentrations still occur locally in some embayments and harbors. Phosphorus loads have decreased in part due to changes in agricultural practices (e.g., conservation tillage and integrated crop management), promotion of phosphorus-free detergents, and improvements made to sewage treatment plants and sewer systems. Figure 4.5 shows that the average concentrations in the open waters of Lakes Superior, Michigan, Huron, and Ontario are at or below expected levels (Environment Canada and USEPA, unpublished data). Concentrations in the three basins of Lake Erie fluctuate from year to year, and frequently exceed target concentrations. In Lakes Ontario and Huron, although most offshore waters meet the desired guideline, some offshore and nearshore areas and embayments experience elevated levels which could promote nuisance algae growths such as the attached green alga, *Cladophora*.

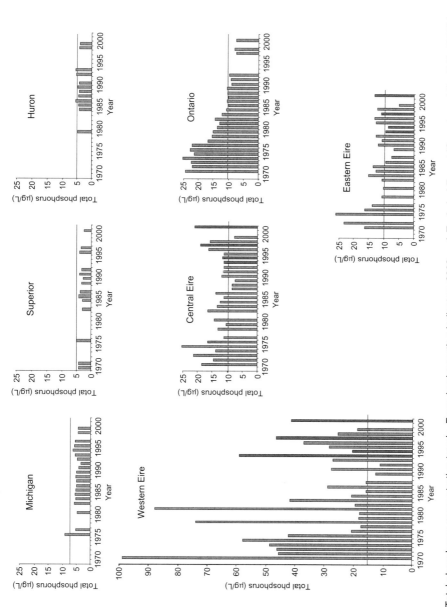

Figure 4.5 Total phosphorus concentration trends. From *Implementing Indicators 2003 — A Technical Report* (Canada and the U.S., 2003b).

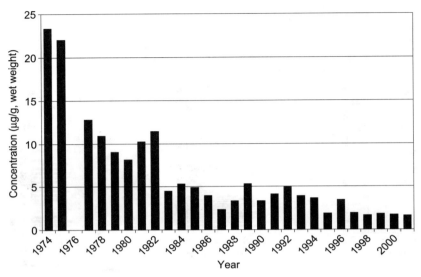

Figure 4.6 Concentration trends for DDE in herring gulls. From *Implementing Indicators 2003 — A Technical Report* (Canada and the U.S., 2003b).

4.3.3.2 *Contaminants in Colonial Nesting Waterbirds*

This indicator assesses current chemical concentration levels and trends as well as ecological and physiological endpoints in representative colonial waterbirds (gulls, terns, cormorants and herons). These features will be used to infer and measure the impact of contaminants on the health of colonial nesting waterbirds. An example of the kind of information provided by this indicator is shown in Figure 4.6 (Environment Canada, Canadian Wildlife Service, unpublished data). Levels of DDE in herring gull eggs are shown for a time series from 1974 to 2001.

4.3.3.3 *Contaminants in Edible Fish Tissue*

This indicator assesses the historical trends of the edibility of fish in the Great Lakes using fish contaminant data and a standardized fish advisory protocol. The approach is illustrated in Figure 4.7 where two of the Great Lakes are shown (USEPA, unpublished data). The various action levels for human consumption of fish are shown as horizontal lines with the corresponding action level noted. Unfortunately data gaps and data variability do not allow one to discern statistically significant trends at this time. Nevertheless, since the 1970s there have been declines in many persistent bioaccumulative toxic (PBT) chemicals in the Great Lakes Basin. However, these chemicals continue to be a significant concern regarding fish consumption.

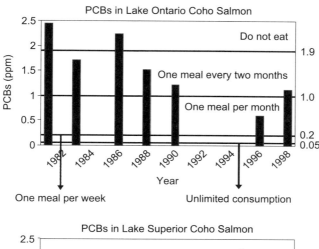

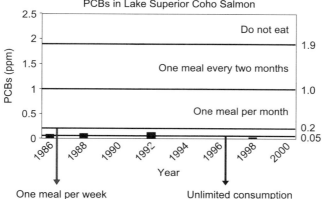

Figure 4.7 PCB Concentrations in edible portions of coho salmon in Lakes Ontario and Superior. From *Implementing Indicators 2003 — A Technical Report* (Canada and the U.S., 2003b).

4.3.4 Pressure Indicators — Incomplete

4.3.4.1 Mass Transportation

The purpose of the indicator is to assess the percentage of commuters using public transportation, and to infer the stress caused by the use of private motor vehicles and their resulting high resource utilization and pollution creation to the Great Lakes ecosystem.

Public transit ridership numbers in U.S. cities and surrounding suburbs remained relatively constant from 1996 to 2000. The majority of transit agencies have not seen more than a 2% change in ridership numbers and less than 10% of the service area population use public transportation. The four agencies that showed the four highest transit use percentages are located in the four largest cities. Of these four, the Chicago Transit Authority, which serves the city of Chicago and surrounding suburbs, had the largest percent of transit

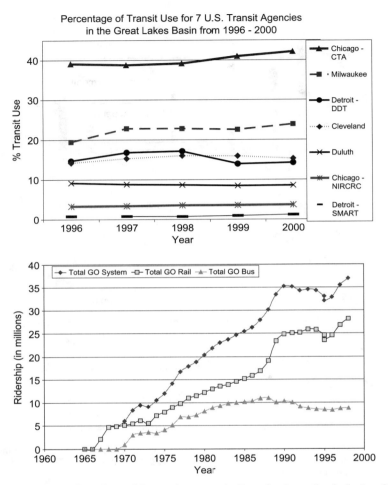

Figure 4.8 U.S. and Canadian public transit use trends. From *Implementing Indicators 2003 — A Technical Report* (Canada and the U.S., 2003b).

use. Percentage of transit use is high where the concentration of people is also the highest.

On the Canadian side, transit ridership is shown for the Greater Toronto Area (GTA) from 1965 to 2000. During that period there was a seven-fold increase in ridership, reflecting increased usage of the system, but also a major increase in population in the GTA (Figure 4.8).

4.3.4.2 Escherichia Coli *and Fecal Coliform Levels in Nearshore Recreational Waters*

This indicator assesses *E. coli* and fecal coliform levels in nearshore recreational waters. These levels act as a surrogate indicator for other

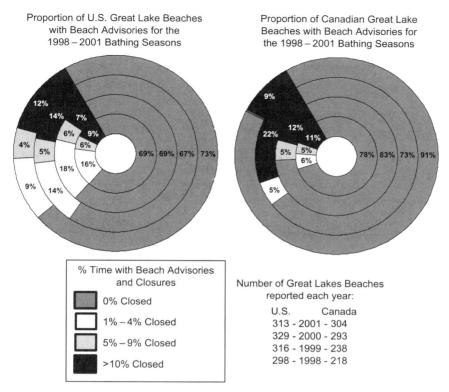

Proportion of U.S. Great Lake Beaches with Beach Advisories for the 1998–2001 Bathing Seasons

Proportion of Canadian Great Lake Beaches with Beach Advisories for the 1998–2001 Bathing Seasons

% Time with Beach Advisories and Closures

0% Closed
1%–4% Closed
5%–9% Closed
>10% Closed

Number of Great Lakes Beaches reported each year:

U.S.   Canada
313 - 2001 - 304
329 - 2000 - 293
316 - 1999 - 238
298 - 1998 - 218

Figure 4.9 U.S. and Canadian beach closures 1998 to 2001. From *Implementing Indicators 2003 — A Technical Report* (Canada and the U.S., 2003b).

pathogen types, in order to infer potential harm to human health through body contact with nearshore recreational waters.

For both the U.S. and Canada, as the frequency of monitoring and reporting increases, more advisories (beach postings) and closings are also observed. Both countries experienced a doubling of beaches that had advisories or closings for more than 10% of the season in 2000. Further analysis of the data may show seasonal and local trends in recreational waters. If episodes of poor recreational water quality can be associated with specific events, then forecasting for episodes of poor water quality may become more accurate. In the Great Lakes Basin, bacteria levels tend to be predictable after storm events (Figure 4.9).

4.3.5   Response Indicators — Incomplete

4.3.5.1   Citizen/Community Place-Based Stewardship Activities

This indicator assesses the number, vitality and effectiveness of citizen and community stewardship activities. Community activities that focus on local

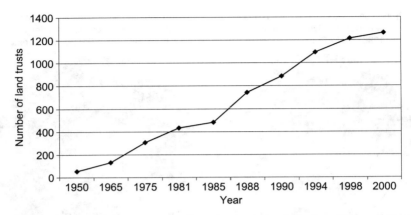

Figure 4.10 Trends in U.S. land trusts. From *Implementing Indicators 2003 — A Technical Report* (Canada and the U.S., 2003b).

landscapes or ecosystems provide a fertile context for the growth of the stewardship ethic and the establishment of a "sense of place."

Land trusts and conservancies are a particularly relevant subset of all community-based groups that engage in activities to promote sustainability within the Great Lakes Basin because of their direct focus on land and habitat protection. Data from the Land Trust Alliance's (LTA's) national land trust census show that the number of land trusts operating at least partly within the Great Lakes Basin increased from 3 in 1930 to 116 in 2000, with half of the increase occurring since 1990. The total area protected by land trusts in the Basin more than doubled between 1990 and 2000, rising from 177,077 to 397,784 acres. Nationally, protected land increased from 1,908,547 acres to 6,479,672 acres, according to LTA. The Nature Conservancy alone had protected an additional 111,725 acres in the Great Lakes Basin (Figure 4.10).

4.4 DISCUSSION

Examination of the most recent findings of the entire suite of Great Lakes indicators led to the identification of particular themes and management issues. Five general themes emerged: land use, habitat degradation, climate change, toxic contamination and indicator development.

4.4.1 Land Use

Current land use decisions throughout the basin are affecting the chemical, physical, and biological aspects of the ecosystem. Each lake and river assessment presented at SOLEC 2002 cited the need for improved land-use decisions to counter the detrimental effects of urban sprawl and increased population growth. One approach to analyzing land use, the "ecological footprint," has been applied to the Great Lakes Basin by the originators of the approach, Mathis Wackernagel and William Rees (1996). They estimated that an area equivalent to 50% of the land mass of the U.S. is needed to support the current lifestyle of Great Lakes Basin citizens. Managers are keenly aware of the importance of using the most current information when making land-use decisions that may contribute to either the sustenance or degradation of the ecosystem.

4.4.2 Habitat Degradation

Many factors, including the spread of non-native species, degrade plant and animal habitats. For example, mussel species are facing extinction due to pressures from non-native zebra and quagga mussels, hydrological alterations are impacting the functioning of wetland habitats, and poorly planned development is degrading or destroying essential habitats. Managers need current data, research to determine appropriate ecological protection and restoration tools and technologies, monitoring programs to understand species trends, and educational programs that provide the public with a broad spectrum of actions.

4.4.3 Climate Change

Climate change has the potential to impact Great Lakes water levels, habitats for biological diversity, and human land uses such as agriculture. In Ohio, for example, a string of mild winters has contributed to an infestation of slugs in corn and soybean crops. Farmers may be faced with a return to tillage plowing or the use of molluscicides to control the infestation. Either choice would reverse some of the most encouraging progress toward controlling nonpoint source pollution. A management challenge is to further research and understand the potential impacts of climate change on the Great Lakes Basin and to adapt to those changes as required.

4.4.4 Toxic Contamination

Although the Great Lakes community has been remediating toxic contamination in water, fish, sediments, air, and people for more than 30 years, problems persist. Loadings of contaminants to the Great Lakes have been greatly reduced from their peak in the 1970s, but pathogens in the water at swimming beaches, for example, are an increasing concern. Controls on industrial emissions of contaminants have been legislated and enforced, resulting in reductions in levels of contaminants in the environment. Nonpoint source runoff reductions are significant, but optimal reductions are not yet being achieved. The approach to dealing with agricultural practices to reduce runoff of pesticides and fertilizers may require a mix of approaches including voluntary measures and incentives. A management challenge is to economically and practically continue to remove toxic contamination and excess nutrients from the ecosystem.

4.4.5 Indicator Development

Given the large number of current and potential indicators, it is difficult to sort and interpret findings in a way that is expedient and productive for managers. Managers and others prefer a few scientifically sound indices based on the suite of indicators so that they can make appropriate management decisions or can better interpret the information presented in the State of the Great Lakes reports. A management challenge is to find a method for aggregating indicators in a way that leads to more informed management decision-making.

4.5 CONCLUSIONS

SOLEC is a framework for organizing ecosystem objectives, for conducting monitoring programs and information management, and for assessing and reporting on the integrity of the Great Lakes. The list of Great Lakes indicators that is reported on through the SOLEC process is dynamic, and indicators may be added or dropped as required by Great Lakes managers.

The entire suite of indicators addresses most of the Great Lakes multiple ecosystem components, yet the overall assessment of the condition of the Great Lakes ecosystem remains incomplete. Some of the reported indicators make use of data that are available, consistent over years and geographic coverage, and reflect basin-wide conditions. For other reported indicators, data are not readily available, are not available as a time series, or are not consistent either over time or across geographic areas. Additionally, within the proposed suite of Great Lakes indicators, there are many that have yet to be reported. They may require further development, refinement of the specific metric being measured, or testing in more local areas before being applied to the Great Lakes Basin. In some cases, existing monitoring programs may require adaptation, or new monitoring efforts need to be initiated.

Several challenges remain to fully implement reporting based on indicators. Networking and collaboration are necessary to decide on a set of indicators, and then various levels of government must ultimately accept the list of indicators. Also, appropriate monitoring and reporting activities must be built into existing Great Lakes programs. Finally, indicators must be reported on in a format that will meet the needs of multiple users. For instance, the managers from Great Lakes government and nongovernmental entities may need specific, scientific details of a particular indicator or group of indicators, whereas upper level administrators, policy makers and the public need a summarized or simplified version of the indicators. One approach to meeting this challenge is to report the same information in different formats and levels of detail. The current reporting mechanism associated with SOLEC includes the State of the Great Lakes report, an accompanying technical report with unabridged indicator reports and complete citations, and fact sheets, which address specific public concerns in a simplified format.

REFERENCES

Bertram, P. and Reynoldson, T.B. Developing ecosystem objectives for the Great Lakes: policy, progress and public participation. *Journal of Aquatic Ecosystem Health* 1, 89–95, 1992.

Bertram, P. and Stadler-Salt, N. *Selection of Indicators for Great Lakes Basin Ecosystem Health.* Version 4. Prepared for the State of the Lakes Ecosystem Conference. U.S. Environmental Protection Agency, Chicago, Illinois U.S.A. and Environment Canada, Burlington, Ontario, 1999. See < http://www. on.ec.gc.ca/solec > or < http://www.epa.gov/glnpo/solec/98/ >.

Bertram, P., Stadler-Salt, N., Horvatin, P., and Shear, H. Bi-national assessment of the Great Lakes: SOLEC Partnerships. *Environmental Monitoring and Assessment* 81, 27–33, 2002.

Bowerman, W., Roe, S.A., Gilberton, M., Best, D.A., Sikarskie, J.G., Mitchell, R.S., and Summer, C.L. Using bald eagles to indicate the health of the Great Lakes environment. *Lakes and Reservoirs: Research and Management, Special Edition.* Shiga–Michigan Joint Symposium 2001, 2002.

Canada and the United States. *State of the Great Lakes 1999,* Toronto and Chicago, 89 p., 1999.

Canada and the United States. *State of the Great Lakes 2001,* Toronto and Chicago, 82 p., 2001. Also online at < http://www.binational.net >.

Canada and the United States. *State of the Great Lakes 2003.* Environment Canada and United States Environmental Protection Agency, Toronto and Chicago, 2003a, 103 p.

Canada and the United States. *Implementing Indicators 2003 — A Technical Report.* Environment Canada and United States Environmental Protection Agency, Toronto and Chicago, 2003b, 161 p.

Ciborowski, J., Krieger, K., Schloesser, D., and Corkum, L. Unpublished observations.

Edwards, C.J. and Ryder, R.A. Biological surrogates of mesotrophic ecosystem health in the Laurentian Great Lakes. *Report to the Science Advisory Board of the International Joint Commission,* Windsor, Ontario, 1990.

Environment Canada and United States Environmental Protection Agency, unpublished data.

Environment Canada, Canadian Wildlife Service, unpublished data.

Great Lakes Fishery Commission. Unpublished sea lamprey data; provided to the State of the Lakes Ecosystem Conference, 2002.

Green Mountain Institute for Environmental Democracy (GMIED). *The Resource Guide to Indicators,* 2nd ed. Montpelier, Vermont, 1998. See < http://www/ gmied.org >.

Hartig, J.H. and Vallentyne, J.R. Use of an ecosystem approach to restore degraded areas of the Great Lakes. *Ambio* 18, 423–428, 1989.

Hartig, J.H. and Zarull, M.A. Towards defining aquatic ecosystem health for the Great Lakes. *Journal of Aquatic Ecosystem Health* 1, 97–107, 1992.

International Joint Commission (IJC). *A Proposed Framework for Developing Indicators of Ecosystem Health for the Great Lakes.* Windsor, Ontario, 1991.

International Joint Commission. *Practical Steps to Implementing the Ecosystem Approach in Great Lakes Management.* Windsor, Ontario, 1995.

International Joint Commission (IJC). *Indicators to Evaluate Progress under the Great Lakes Water Quality Agreement.* Prepared by the Indicators for Evaluation Task Force, Windsor, Ontario, 1996.

Messer, Jay J. "Indicators in regional ecological monitoring and risk assessment," in *Ecological Indicators,* Vol. 1, Makenzie, D.H., Hyatt, D.E., and McDonald, V.J., Eds. Elsevier, New York, 1992, pp. 135–146.

Organization for Economic Cooperation and Development (OECD). Core set of indicators for environmental performance reviews. *Environmental Monographs* 83, 1993, 35 p.

Regier, H.A. "Indicators of ecosystem integrity," in *Ecological Indicators,* Vol. 1, Makenzie, D.H., Hyatt, D.E., and McDonald, V.J., Eds. Elsevier, New York, 1992, pp. 183–200.

Ryder, R.A. and Edwards, C.J. A conceptual approach for the application of biological indicators of ecosystem quality in the Great Lakes Basin. Report to the Science Advisory Board of the International Joint Commission. Windsor, Ontario, 1985.

United States and Canada. *Great Lakes Water Quality Agreement, as amended by Protocol.* Ottawa and Washington, 84 p., 1987.

United States Environmental Protection Agency, unpublished data.

Vallentyne, J.R. and Beeton, A.M. The ecosystem approach to managing human uses and abuses of natural resources in the Great Lakes Basin. *Environmental Conservation* 15, 58–62, 1988.

Wackernagel and Rees. *Our Ecological Footprint.* New Society Publishers, Gabriola Island, British Columbia, Canada, 1996, 160 p.

Weeber, R.C. and Vallianatos, M. The Marsh Monitoring Program 1995–1999: Monitoring the Great Lakes Wetlands and their Amphibian and Bird Inhabitants. Bird Studies Canada, Environment Canada, and the United States Environmental Protection Agency, 47 p., 2000.

CHAPTER 5

Application of Ecological and Thermodynamic Indicators for the Assessment of Lake Ecosystem Health

Fu-Liu Xu

A tentative theoretical frame, a set of ecological and thermodynamic indicators, and three methods have been proposed for the assessment of lake ecosystem health in this chapter. The tentative theoretical frame includes five necessary steps: (1) the identification of anthropogenic stresses; (2) the analysis of ecosystem responses to the stresses; (3) the development of indicators; (4) the determination of assessment methods; and (5) the qualitative and quantitative assessment of lake ecosystem health. A set of ecological and thermodynamic indicators covering lake structural, functional, and system-level aspects were developed, according to the structural, functional, and system-level responses of 59 actual and 20 experimental lake ecosystems to the 5 kinds of anthropogenic stresses: eutrophication, acidification, heavy metals, pesticides, and oil pollution. Three methods are proposed for lake ecosystem health assessment: (1) the direct measurement method (DMM); (2) the ecological modeling method (EMM); and (3) the ecosystem health index method (EHIM). These indicators and methods were successfully applied to the assessment and comparison of ecosystem health for a Chinese lake and 30 Italian lakes.

1-56670-665-3/05/$0.00 + $1.50
© 2005 by CRC Press

5.1 INTRODUCTION

5.1.1 Ecosystem Type and Problem

Lakes are extremely important storage areas for the earth's surface freshwater, with important ecosystem service functions that can keep the development of society and economy sustainable.[1,2] However, eutrophication and acidification, as well as heavy metal, oil and pesticide pollution caused by human activities have deteriorated continuously the healthy status of lake ecosystems. The water in over half of the lakes around the world has been seriously polluted. If this trend continues, it will not only affect human health and socio-economic development, but may also cause the breakup of lake ecosystems altogether.[3,4] Studies on lake ecosystem health therefore have important and practical significance for the restoration of ecosystem health and the maintenance of their ecological service functions.

Since the mid-1980s, studies on lake ecosystem health have begun to attract the attention of environmentalists and ecologists, with increasingly frequent use in academic and government publications as well as the popular media.[5] More and more environmental managers consider the protection of ecosystem health as a new goal of environmental management.[6-11] In the past few years, many national and international environmental programs have been established. One of these leading programs is "Assessing the State of Ecosystem Health in the Great Lakes," supported by the Canadian and U.S. governments.[12] In the U.S., important ongoing programs related to lake ecosystem health include mainly "Assessing Health State of Main Ecosystems,"[11] and "Stresses on Ecosystem Health — Chemical Pollution."[13] In Canada, an ongoing key program related to lake ecosystem health is the "Aquatic Ecosystem Health Assessment Project."[14] In China, special attention has also been paid to lake ecosystem health. Two projects have been carried out, namely "The Effects of Typical Chemical Pollution on Aquatic Ecosystem Health,"[15] and "The Indicators and Methods for Lake Ecosystem Assessment,"[16,17] Ongoing programs supported by the Natural Science Foundation of China (NSFC) include "The Limiting Factors and Dynamic Mechanism for Lake Ecosystem Health"; "Regional Differentia and its Mechanisms for the Ecosystem Health of Large Shallow Lakes"; and "Assessment and Management of Watershed Ecosystem Health."[18]

So far, a number of indicators have been proposed for lake ecosystem health assessment; for example, gross ecosystem product (GEP),[19] ecosystem stress indicators,[20] the index of biotic integrity (IBI),[21] thermodynamic indicators including exergy and structural exergy,[22,23] and a set of comprehensive ecological indicators covering structural, functional and system-level aspects.[15,16] Some methods or procedures have also been proposed for assessing lake ecosystem health; for example, a tentative procedure by Jørgensen,[23] and the direct measurement method (DMM) and ecological model method (EMM) by Xu et al.[16,17] However, owing to the lack of criteria, it causes two major problems using present methods to assess lake ecosystem health. First, we can only assess the relative healthy status — it is extremely difficult to assess the actual health status. Second, it is impossible to make the comparisons of ecosystem health status for different lakes. In order to solve these problems, a new method, the ecosystem health index method (EHIM), is developed in this chapter.

5.1.2 The Chapter's Focus

This chapter focuses on indicators and methods for assessing lake ecosystem health, followed by an examination of two case studies. Also, a tentative theoretical frame or procedure for assessing lake ecosystem health is proposed. The discussions on indicators, methods, and the results of case studies are then presented.

5.2 METHODOLOGIES

5.2.1 A Theoretical Frame

A tentative theoretical frame or procedure for assessing lake ecosystem health is shown in Figure 5.1. It shows that there are five necessary steps in which the development of indicators and the determination of assessment methods are two key steps. However, in order to develop sensitive indicators, the anthropogenic stresses have to be identified, and the responses of lake ecosystems to the stresses have to be analyzed, since the stresses caused by human activities are mainly responsible to the degradation of lake ecosystem health.

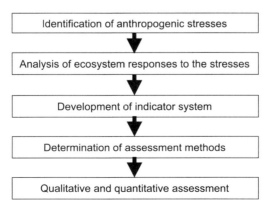

Figure 5.1   A tentative procedure for assessing lake ecosystem health.

5.2.2   Development of Indicators

5.2.2.1   The Procedure for Developing Indicators

The flow chart for developing indicators is shown in Figure 5.2. It can be seen that the anthropogenic stresses identified to the lake ecosystems include eutrophication and acidification, as well as heavy metal, pesticide, and oil pollution. The lake ecosystems studied should include actual and experimental anthropogenic stresses. The response of lake ecosystems to the stresses should be composed of structural, functional, and system-level aspects.

5.2.2.2.   Lake Data for Developing Indicators

The actual lake ecosystems (including 29 Chinese lakes (Figure 5.3) and 30 Italian lakes (Table 5.1)) were applied for eutrophication, while the 20 experimental lake ecosystems were chosen because of their eutrophicated conditions, as well as heavy metals, pesticides and oil pollution (Table 5.2).

It can be seen from Figure 5.3 that 29 Chinese lakes distribute in different regions in China. Their surface areas range from the $3.7\,km^2$ Lake Xuanwu-Hu to $4200\,km^2$ Lake Qinghai-Hu. Their trophic status are from oligotrophic (e.g., Lake Qinghai-Hu) to extremely hypertrophic (e.g., Lake Liuhua-Hu, Lake Dongshan-Hu and Lake Dong-Hu).

Thirty Italian lakes are located on Sicily. About 70% of the lakes are used for irrigation; while 30% lakes are used for drinking. Their mean depths are between 1.5 and 19 m. Their surface area ranges from 1 to $577\,km^2$ with average volume varying from 0.1 to 154 billion m^3.

Experimental ecosystems, including microcosms, mesocosms, and experimental ponds, have been increasingly used in the research on the toxicity and impacts of chemicals on aquatic ecosystems during the last two decades. Experimental ecosystem perturbations allow us to separate the effects of

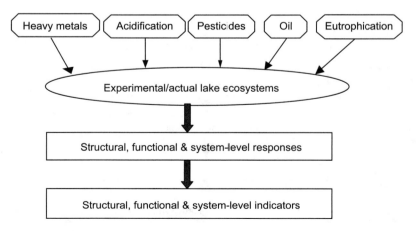

Figure 5.2   A flow chart for developing indicators for lake ecosystem health assessment.

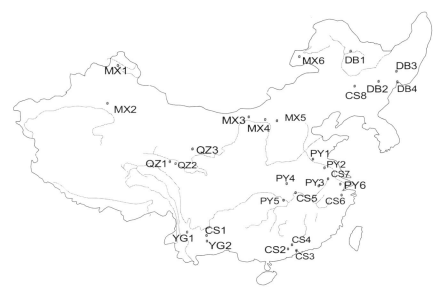

Figure 5.3 Geographic locations of 29 Chinese lakes used for developing indicators. MX1: Lake Wulungu-Hu; MX2: Lake Beshiteng-Hu; MX3: Lake Wuliangshu-Hai; MX4: Lake Huashu-Hai; MX5: Lake Dai-Hai; MX6: Hulun-Hu; DB1: Lake Wudalianchi; DB2: Lake Jingbe-Hu; DB3: Lake Xiaoxingkai-Hu; DB4: Lake Daxingkai-Hu; QZ1: Lake Zhaling-Hu; QZ2: Lake Eling-Hu; QZ3: Lake Qinghai-Hu; YG1: Lake Erhai; YG2: Lake Fuxian-Hu; PY1: Lake Nanshi-Hu; PY2: Lake Hongzhe-Hu; PY3: Lake Chao-Hu; PY4: Lake Baoan-Hu; PY5: Lake Hong-Hu; PY6: Lake Tai-Hu; CS1: Lake Dian-Chi; CS2: Lake Liuhua-Hu; CS3: Lake Dongshan-Hu; CS4: Lake Lu-Hu; CS3: Lake Dong-Hu; CS6: Lake Xi-Hu; CS7: Lake Xuanwu-Hu; CS8: Lake Nan-Hu.

various pollutants, to assess early effects of perturbations in systems with known background properties, and to assess quantitatively the result of known perturbations to whole ecosystems.[25,26] The experimental ecosystems for developing indicators include 2 microcosms, 14 mesocosms, and 4 experimental ponds; and the experimental perturbations include acidification, oil, copper, and organic chemical contamination (Table 5.2).

5.2.2.3 Responses of Lake Ecosystems to Chemical Stresses

Xu et al. examined the structural, functional, and ecosystem-level symptoms resulting from chemical stress, acidification, and copper, oil, and pesticide contamination in lake ecosystems, based on the above-mentioned data on experimental ecosystems.[15] They concluded that the structural responses of freshwater ecosystems to chemical stresses were noticeable in terms of an increase in phytoplankton cell size and phytoplankton and microzooplankton biomass, and a decrease in zooplankton body size, zooplankton and macrozooplankton biomass and species diversity, and in the zooplankton/phytoplankton and macrozooplankton/microzooplankton ratios. The functional responses included decreases in alga C assimilation,

Table 5.1   Basic limnological characteristics for 30 Italian lakes

Lake name	Cond. (mS/cm)	TP (μg/l)	N-NH$_4$ (μg/l)	N-NO$_3$ (μg/l)	SiO$_2$ (mg/l)
Ancipa	0.18	30.66	12	77	2.0
Arancio	0.72	166.65	667	676	4.8
Biviere di Cesro	0.08	46.02	31	76	0.6
Biviere di Gela	2.72	45.15	22	78	2.3
Castello	0.96	109.83	775	263	2.9
Cimia	2.15	49.57	199	803	4.0
Comunelli	2.51	45.33	331	129	3.4
Dirillo	0.53	60.54	60	514	4.1
Disueri	1.21	1093.43	684	2226	3.6
Fanaco	0.53	54.34	199	1143	3.3
Gammauta	0.49	183.07	154	446	2.7
Garcia	0.77	51.36	22	1165	3.6
Gorgo	4.51	80.87	33	65	6.1
Guadalani	0.42	38.89	111	459	0.3
Nicoletti	1.42	35.13	46	66	1.5
Ogliastro	2.72	40.87	173	1710	2.9
Olivo	0.91	38.00	71	69	1.6
Pergusa	33.65	87.97	788	157	1.6
Piana degli Albanesi	0.37	46.77	349	412	0.4
Piana del Leone	0.41	46.85	160	546	2.4
Poma	0.74	51.11	73	994	1.4
Pozzillo	1.13	49.38	91	355	1.6
Prizzi	0.46	52.99	86	503	2.5
Rubino	1.05	28.94	18	711	1.0
San Giovanni	1.49	80.56	658	283	2.7
Santa Rosalia	0.42	55.81	125	279	3.4
Scanzano	0.50	61.65	300	1283	2.3
Soprano	1.85	2962.96	7671	57	12.7
Trinita	1.86	83.24	26	417	3.8
Vasca Ogliastro	0.32	106.69	28	177	3.4
Villarsosa	2.27	64.06	524	276	1.0

resource use efficiency, the P/B (Gross production/Standing crop biomass) and B/E (Biomass supported/unit energy flow) ratios, an increase in community production, and a departure from 1 for the P/R (Gross production/community respiration) ratio (see Equation 5.3 to Equation 5.5 below for definitions). System-level responses included decreases in exergy, structural exergy, and ecological buffering capacities.[15,16]

Xu investigated the structural responses of the Lake Chao to eutrophication.[27] He found that with an increasing eutrophication gradient, algal cell number and biomass were increased, while algal biodiversity, zooplankton biomass and the ratio of zooplankton biomass to algal biomass were decreased.

Xu[28] and Lu[29] studied the structural, functional, and system-level responses of 29 Chinese lakes and 30 Italian lakes to eutrophication, respectively. The results are summarized in Table 5.3 and are very similar to the results from the experimental lake ecosystems stressed by acidification, and heavy metal, oil and pesticide pollution, with the exemption of zooplankton biomass and exergy for lakes with the trophic states from oligo-eutrophication to eutrophication.

Table 5.2 The studies on the responses of lake ecosystems to experimental perturbations

Stressors	Study type*	Location	Duration (days)	Reference**
Acidification	Meso.	West Virginia	75	[54]
Acidification	Meso.	California	35	[55]
Acidification	Meso.	Ohio	10	[56]
Acidification	Meso	Ohio	35	[57]
Copper	Meso.	Ohio	4	[58]
Copper	Meso.	Ohio	14	[59]
Oil	EP	Tennessee	420	[60]
Dursban	EP	California	90	[61]
2,4D-DMA	EP	Missouri	56	[62]
TCP	Meso.	Neuherberg	24	[63]
PCP	Meso.	Neuherberg	24	[63]
Trichloroethylene	Meso.	Southern Germany	44	[64]
TCB	Meso.	Southern Germany	22	[65]
Benzene	Meso.	Western Germany	26	[66]
Atrazine	Micro.	New Mexico	365	[67]
HCBP	Micro	New Mexico	365	[67]
Permethrin	Meso.	Tsukuba, Japan	30	[68]
Hexazinone	Meso.	Ontario	77	[69], [70]
Bifenthrin	EP	New Jersey	8	[71]
Carbaryl	Meso.	Ohio	4	[58]

*Micro. = Microcosms; Meso. = Mesocosms; EP = Experimental Ponds.
**For acidification see [72]–[75]; For oil pollution see [76]–[79]; for copper pollution see [80]–[84]; for pesticide pollution see [85]–[90].
Modified from Xu, et al. *Ecol. Model.* 116, 80, 1999. With permission.

Table 5.3 The structural, functional, and system-level responses of actual lake ecosystems to eutrophication*

		Dynamics in lake trophic states	
	Responses indicators	Oligo-eutrophication — Eutrophication	Eutrophication — Hyper-eutrophication
Structural responses	Phytoplankton cell number[a,t]	Increase	Increase
	Phytoplankton biomass (BA)[a,b]	Increase	Increase
	Phytoplankton cell size[a,b]	Increase	Increase
	Phytoplankton diversity[a]	Decrease	Decrease
	Zooplankton biomass (BZ)[a,b]	Increase	Decrease
	Zooplankton body size[a,b]	Decrease	Decrease
	Zooplankton diversity[a]	Decrease	Decrease
	BZ/BA ratio[a,b]	Decrease	Decrease
	BZmacro./BZmicro. Ratio[a,b]	Decrease	Decrease
Functional responses	Phytoplankton primary production[a]	Increase	Increase
	P/B ratio[a]	≈ 1	<0.5
	P/R ratio[a]	≈ 1	<1.0
System-level responses	Exergy[a,b]	Increase	Decrease
	Structural exergy[a,b]	Decrease	Decrease

*:please see References 28 and 29 for details.
[a]for 29 Chinese lakes; [b]for 30 Italian lakes.

5.2.2.4 Indicators for Lake Ecosystem Health Assessment

Ecological indicators for lake ecosystem health assessment resulting from chemical stress are important for both the early warning signs of ecosystem malfunction and confirmation of the presence of a significant ecosystem pathology.[9,20] Ecological indicators as valid and reliable tools should include structural, functional, and system-level aspects. According to the above-mentioned structural, functional, and system-level responses of actual and experimental lake ecosystems to chemical stress, a set of comprehensive ecological indicators, including structural, functional, and ecosystem-level aspects, for assessing lake ecosystem health can be derived (Table 5.4). Table 5.4 indicates that a healthy ecosystem can be characterized by:

- Small cell size in phytoplankton
- Large body size in zooplankton
- High zooplankton and macrozooplankton biomass levels
- Low phytoplankton and microzooplankton biomass levels
- A high zooplankton/phytoplankton ratio
- A high macrozooplankton/microzooplankton ratio
- High degrees of species diversity.

Table 5.4 The ecological indicators for lake ecosystem health assessment

	Ecological indicators	Relative healthy state		Methods for indicator values
		Good	**Bad**	
Structural indicators	1. Phytoplankton cell size	Small	Large	Measure
	2. Zooplankton body size	Large	Small	Measure
	3. Phytoplankton biomass (BA)	Low	High	Measure
	4. Zooplankton biomass (BZ)	High	Low	Measure
	5. Macrozooplankton biomass (Bmacroz.)	High	Low	Measure
	6. Microzooplankton biomass (Bmicroz.)	Low	High	Measure
	7. BZ/BA ratio	High	Low	Calculate
	8. Bmacroz./Bmicroz. ratio	High	Low	Calculate
	9. Species diversity (DI)	High	Low	Measure and calculate
Functional indicators	10. Algal C assimilation ratio	High	Low	Measure
	11. Resource use efficiency (RUE)	High	Low	Measure and calculate
	12. Community production (P)	Low	High	Measure
	13. P/R ratio	≈ 1	$>$ or < 1	Measure and calculate
	14. P/B ratio	High	Low	Measure and calculate
	15. B/E ratio	High	Low	Measure and calculate
System-level indicators	16. Buffer capacities (β)	High	Low	Calculate
	17. Exergy (Ex)	High	Low	Calculate
	18. Structural exergy (Ex$_{st}$)	High	Low	Calculate

Modified from Xu, et al. Lake ecosystem health assessment: indicators and methods. *Wat. Res.* 35(1), 3159, 2001. With permission.

- High levels of algal C assimilation
- High resource use efficiencies
- Low community production
- High P/B and B/E ratios
- A P/R ratio approaching 1
- High exergy, structural exergy, and buffer capacities.

5.2.3   Calculations for Some Indicators

5.2.3.1   Calculations of Exergy and Structural Exergy

The definitions and calculations of exergy and structural exergy (or specific exergy) are discussed in chapter 2 and in References 22, 23, and 32 to 35.

5.2.3.2   Calculation of Buffer Capacity

The buffer capacity is defined as follows:[32,34,36]

$$\beta = \frac{1}{\delta(\text{state variable})/\delta(\text{forcing function})} \tag{5.1}$$

Forcing functions are the external variables that are driving the system, such as discharge of waste, precipitation, wind, solar radiation, and so on. While state variables are the internal variables that determine the system (e.g., in a lake the concentration of soluble phosphorus, the concentration of zooplankton etc.). The concept should be considered multidimensionally, as all combinations of state variables and forcing functions may be considered. It implies that even for one type of change there are many buffer capacities corresponding to each of the state variables.

5.2.3.3   Calculation of Biodiversity

The definitions and calculations of diversity index (DI) for an ecosystem are discussed in chapter 2 and in References 37 and 38.

5.2.3.4   Calculations of Other Indicators

$$RUE = (\text{zooplankton C assimilation rate})/(\text{algal C assimilation rate}) \times 100\% \tag{5.2}$$

$$P/R = \text{Gross production (P)/Community respiration (R)} \tag{5.3}$$

$$P/B = \text{Gross production (P)/Standing crop biomass (B)} \tag{5.4}$$

$$B/E = \text{Standing crop biomass (B)/unit energy flow (E)} \tag{5.5}$$

5.2.4   Methods for Lake Ecosystem Health Assessment

Three methods have been applied to assess lake ecosystem health: (1) direct measurement method (DMM); (2) ecological model method (EMM); and (3) ecosystem health index method (EHIM). The methods are reviewed in chapter 2, where the general methodology is mentioned. The indicators can be selected from Table 5.4 and Table 5.3.

5.3 CASE STUDIES

5.3.1   Case 1: Ecosystem Health Assessment for Italian Lakes Using EHIM

5.3.1.1   Selecting Assessment Indicators

Assessment indicators are composed of basic and additional indicators. Basic indicators are crucial for lake ecosystem health assessment. Basic indicators have the consanguineous relationships to ecosystem health status, while additional indicators have a less important relationship to ecosystem health status. A lake ecosystem health status can be evaluated mainly on the base of basic indicators; however, the assessment by additional indicators can be considered as the remedies of results from basic indicators.

In most lake ecosystems, the indicators that give the consanguineous relationships to ecosystem health status are phytoplankton biomass (BA) and chlorophyll-a (Chl-a) concentration. The higher BA or Chl-a concentrations in a lake, the worse the lake ecosystem health status. Therefore, BA and Chl-a can service as two basic indicators. According to data availability for Italian lakes, BA are selected as a basic indicator; while zooplankton biomass (BZ), BZ/BA, exergy (Ex) and structural exergy (Exst) are applied as additional indicators.

5.3.1.2   Calculating Sub-EHIs

There are two main steps to calculate sub-EHIs for all selected indicators. The first step is to calculate EHI(BA) for the basic indicator, BA. The second step is to calculate EHI(BZ), EHI(BZ/BA), EHI(Ex) and EHI(Exst) for the additional indicators, BZ, BZ/BA, Ex and Exst, respectively. After the EHI(BA) for the basic indicator being obtained, the sub-EHIs including EHI(BZ), EHI(BZ/BA), EHI(Ex) and EHI(Exst) for the additional indicators can be deduced according to the relationships between the basic indicator (BA) and the additional indicators (BZ, BZ/BA, Ex and Exst).

5.3.1.2.1   EHI(BA) Calculation

For the EHI(BA) calculation, it is assumed that, EHI(BA) = 100 if BA is lowest, which means the best healthy state, and that EHI(BA) = 0 if BA is highest, which means the worst healthy state. Referring Carlson's studies on trophic state index (TSI),[39] the relationship between ecosystem health status and phytoplankton biomass in a lake ecosystem can be described as a logarithmic normal distribution. Therefore EHI(BA) can be calculated from the following equation:

$$EHI(BA) = 100 \times \frac{\ln C_x - \ln C_{min}}{\ln C_{max} - \ln C_{min}} \tag{5.6}$$

where EHI(BA) is sub-EHI for basic indicator BA; C_x is the measured BA value; C_{min} is the measured lowest BA value; C_{max} is the measured highest BA value.

Equation 5.6 can be predigested as the following format:

$$EHI(BA) = 10(a + b \ln C_x) \tag{5.7}$$

where a and b are constants, and can be computed by the following equation:

$$\begin{cases} a = -10 \times \dfrac{\ln C_{min}}{\ln C_{max} - \ln C_{min}} \\ b = 10 \times \dfrac{1}{\ln C_{max} - \ln C_{min}} \end{cases} \tag{5.8}$$

According to the measured data for 30 Italian lakes, $C_{min} = 0.004(mg/l)$, $C_{max} = 150(mg/l)$. Then, $a = 5.2425$, $b = -0.94948$. Thus, the expression for calculating EHI(BA) for Italian lakes can be obtained as follows:

$$EHI(BA) = 10 \times (5.2425 - 0.94948 \times \ln(BA)) \tag{5.9}$$

It can be seen that the equation for calculating EHI(BA) can be deduced from the BA measured data by logarithmic expression for differences between extreme values.

5.3.1.2.2 *EHI(BZ), EHI(BZ/BA), EHI(Ex) and EHI(Exst) Calculations*

The sub-EHIs for additional indicators, EHI(BZ), EHI(BZ/BA), EHI(Ex) and EHI(Exst), can be calculated according to the relationships between the basic indicator (BA) and the additional indicators (BZ, BZ/BA, Ex and Exst). From Lu,[29] there are very simple relationships between BA and BZ/BA and Exst; while there are more complicated relationships between BA and BZ and Ex. Thus, the different ways should be adopted to calculate EHI(BZ/BA), EHI(Exst) and EHI(BZ), EHI(Ex).

For 30 Italian lakes, there are strongly negative relationship between BA and BZ/BA and Exst. The following two expressions can be obtained by means of regression analysis:

$$\ln(BA) = 0.3878 - 0.7742 \times \ln(BZ/BA) \tag{5.10}$$

$$\ln(BA) = 5.1119 - 0.0688 \times (Exst) \tag{5.11}$$

Thus, the equations for calculating EHI(BZ/BA) and EHI (Exst) can be deduced from Equation 5.14 to Equation 5.16:

$$EHI(BZ/BA) = 10 \times (5.2425 - 0.94948 \times (0.3878 - 0.7742 \times \ln(BZ/BA))) \tag{5.12}$$

$$EHI(Exst) = 10 \times (5.2425 - 0.94948 \times (5.1119 - 0.0688 \times (Exst))) \tag{5.13}$$

According to Lu,[29] there are three kinds of relationships between BA, BZ and Ex in 30 Italian lakes, owing to the dynamics in phytoplankton community structure, the toxic effects of phytoplankton species, and the food sources of zooplankton. The first type of relationship between BA, BZ, and Ex is that BZ and Ex apparently increase with the BA increase. The second type of relationship is that BZ and Ex decrease with the BA increase. The third type of relationship is that BZ and Ex slowly increase with the BA increase. The first and the third type of relationship between BA, BZ, and Ex are more obvious than the second type of relationship. However, this second type is less obvious since there are many lakes and BA is different in each lake when BZ and Ex start to decrease. This second type can be considered as the transition from the first to the third type of relationship.

In order to better describe these relationships, two linear expressions are used to simulate the first and the third relationships, respectively. By means of fuzzy mathematics, each data point in the second type of relationship and in the first and the third kind of relationships can be determined to belong to the first or to the third type of relationship, through the comparison of its attributability to the first type of relationship with its attributability to the third type of relationship.

For the first and the third relationships between BA and BZ, two linear expressions can be obtained using regression analysis:

$$f_1: \ln(BA) = 0.1036 + 0.7997 \times \ln(BZ), (N = 95, r = 0.702, p < 0.01) \tag{5.14}$$

$$f_2: \ln(BA) = 2.7359 + 0.6766 \times \ln(BZ), (N = 19, r = 0.563, p < 0.01) \tag{5.15}$$

For the first and the third relationships between BA and Ex, two linear expressions are as follows:

$$f_3: \ln(BA) = -4.0256 + 0.8236 \times \ln(Ex), (N = 95, r = 0.717, p < 0.01) \qquad (5.16)$$

$$f_4: \ln(BA) = -2.5380 + 0.9899 \times \ln(Ex), (N = 19, r = 0.829, p < 0.01) \qquad (5.17)$$

Thus, four expressions for calculating EHI(BZ) and EHI(Ex) can be deduced from Equation 5.14 to Equation 5.17, and Equation 5.9, respectively:

$$EHI(BZ)_1 = 10(5.2425 - 0.94948 \times (0.1036 + 0.7997 \times \ln(BZ))) \qquad (5.18)$$

$$EHI(BZ)_2 = 10(5.2425 - 0.94948 \times (2.7359 + 0.6766 \times \ln(BZ))) \qquad (5.19)$$

$$EHI(Ex)_1 = 10(5.2425 - 0.94948 \times (-4.0256 + 0.8236 \times \ln(Ex))) \qquad (5.20)$$

$$EHI(Ex)_2 = 10(5.2425 - 0.94948 \times (-2.538 + 0.9899 \times \ln(Ex))) \qquad (5.21)$$

Equation 5.18 to Equation 5.21 can be synthesized as the following two comprehensive expressions:

$$EHI(BZ) = \begin{cases} EHI(BZ)_1 \ (BA, BZ) \in \alpha \\ EHI(BZ)_2 \ (BA, BZ) \in \beta \end{cases} \qquad (5.22)$$

$$EHI(Ex) = \begin{cases} EHI(Ex)_1 \ (BA, Ex) \in \gamma \\ EHI(Ex)_2 \ (BA, Ex) \in \delta \end{cases} \qquad (5.23)$$

In Equation 5.22, α and β are the attributability of measured data (BA, BZ) to the two linear expressions, Equation 5.14 and Equation 5.15, which can be calculated from the following attributable functions:

$$\alpha(BA, BZ) = \begin{cases} 1 & 0 < BA \leq 2.5 \\ 1 - \dfrac{(\ln(BA) - f_1(BZ))^2}{(\ln(BA) + f_1(BZ))^2} & 2.5 < BA \leq 50 \\ & 50 < BA \leq 150 \\ 0 \end{cases} \qquad (5.24)$$

$$\beta(BA, BZ) = \begin{cases} 0 & 0 < BA \leq 2.5 \\ 1 - \dfrac{(\ln(BA) - f_2(BZ))^2}{(\ln(BA) + f_2(BZ))^2} & 2.5 < BA \leq 50 \\ & 50 < BA \leq 150 \\ 1 \end{cases} \qquad (5.25)$$

where BA is the measured values; $f_1(BZ)$ and $f_2(BZ)$ are the calculated BA values from Equation 5.14 and Equation 5.15, respectively; 2.5 is the minimum

BA value in the β set which expresses the third kind of relationship; 50 is the maximum BA value in the α set which expresses the first kind of relationship.

It can be seen from Equation 5.24 and Equation 5.25 that for the measured data point (BA, BZ), if $\alpha(\text{BA}, \text{BZ}) \geq \beta(\text{BA}, \text{BZ})$, then $(\text{BA}, \text{BZ}) \in \alpha$, its EHI(BZ) can be calculated from Equation 5.18; if $\alpha(\text{BA}, \text{BZ}) < \beta(\text{BA}, \text{BZ})$, then $(\text{BA}, \text{BZ}) \in \beta$, its EHI(BZ) can be calculated from Equation 5.19.

In Equation 5.23, γ and δ are the attributability of actual data (BA, Ex) to the two linear expressions 5.16 and 5.17, which can be calculated from the following attributable functions:

$$\gamma(\text{BA}, \text{Ex}) = \begin{cases} 1 & 0 < \text{BA} \leq 2.5 \\ 1 - \dfrac{(\ln(\text{BA}) - f_3(\text{Ex}))^2}{(\ln(\text{BA}) + f_3(\text{Ex}))^2} & 2.5 < \text{BA} \leq 50 \\ 0 & 50 < \text{BA} \leq 150 \end{cases} \tag{5.26}$$

$$\delta(\text{BA}, \text{Ex}) = \begin{cases} 0 & 0 < \text{BA} \leq 2.5 \\ 1 - \dfrac{(\ln(\text{BA}) - f_4(\text{Ex}))^2}{(\ln(\text{BA}) + f_4(\text{Ex}))^2} & 2.5 < \text{BA} \leq 50 \\ 1 & 50 < \text{BA} \leq 150 \end{cases} \tag{5.27}$$

where BA is the measured values; $f_3(\text{Ex})$ and $f_4(\text{Ex})$ are the calculated BA values by Equation 5.16 and Equation 5.17, respectively; 2.5 is the minimum BA value in the δ set which expresses the third kind of relationship; 50 is the maximum BA value in the γ set which expresses the first kind of relationship.

It can be see from Equations 5.26 and Equation 5.27 that for the sample point (BA, Ex), if $\gamma(\text{BA}, \text{Ex}) \geq \delta(\text{BA}, \text{Ex})$, then $(\text{BA}, \text{Ex})\gamma$, its EHI(Ex) can be calculated from Equation 5.20; if $\gamma(\text{BA}, \text{Ex}) < \delta(\text{BA}, \text{Ex})$, then $(\text{BA}, \text{Ex})\delta$, its EHI(Ex) can be calculated from Equation 5.21.

5.3.1.3 Determining Weighting Factors (ω_i)

There are many factors that affect lake ecosystem health to different extents. It is therefore necessary to determine weighting factors for all indicators. Basic indicators have a consanguineous relationship to ecosystem health status; while additional indicators have a less important relationship to ecosystem health status. A lake ecosystem health status can therefore be evaluated mainly on the base of basic indicators; however, the assessment by additional indicators can be considered as the remedies of results from basic indicators. So, the method of relation–weighting index can be used to determine the weighting factors for all indicators — that is, the relation ratios between BA and other indicators can be used to calculate the weighting factors for all indicators. The equation is as follows:

$$\omega_i = \frac{r_{i1}^2}{\sum_{i=1}^{m} r_{i1}^2} \tag{5.28}$$

Table 5.5 Statistic correlative ratios between BA and other indicators

Relative indicators	ln(BA) — ln(BA)	ln(BA) — ln(BZ)*	ln(BA) — ln(BZ)†	ln(B — ln(BZ/BA)	ln(BA) — ln(Ex)*	ln(BA) — ln(Ex)†	ln(BA) — (Exst)
Sample number	114	95	19	114	95	19	114
r_{ij}	1	0.702	0.563	−0.731	0.717	0.829	−0.699
r_{ij}^2	1	0.4928	0.3170	0.5344	0.5141	0.6872	0.4886

*expresses the first kind of relationship between BA and BZ or Ex; †expresses the third kind of relationship between BA and BZ or Ex.

where ω_i is the weighting factor for the ith indicator; r_{i1} is the relation ratio between the ith indicator and the basic indicator (BA); m is the total number of assessment indicators, here $m = 5$.

The statistic correlative ratios between the basic indicator (BA) and other indicators are shown in Table 5.5. Considering two kinds of relationships between BA and additional indicators BZ and Ex, there are two steps to calculate the weighting factors for BZ and Ex. First, the kind of relationship between BA and BZ or Ex has to be determined; and second, the calculations of weighting factors can be done using Equation 5.33 and the corresponding correlative ratios.

5.3.1.4 Assessing Ecosystem Health Status for Italian Lakes

5.3.1.4.1 EHI and Standards for Italian Lakes

According to the sub-EHI calculation equations for all selected indicator, the responding standards for all indicators to the numerical EHI on a scale of 0 to 100 can be obtained (Table 5.6).

Table 5.6. Ecosystem health index (EHI) and its associated parameters as well as their standards for Italian lakes

EHI	Health status	BA (mg/L)	BZ (mg/L)*	BZ (mg/L)†	BZ/BA	Ex (J/L)*	Ex (J/L)†	Exst (J/mg)
0		150		60.7	0.001319		3434.7	1.47
10	Worst	52.3		12.81	0.004576		1185.3	16.78
20		18.3	62.9	2.71	0.01588	8385.6	409.02	32.10
30	Bad	6.37	16.84	0.5713	0.0551	2334.3	141.15	47.42
40		2.22	4.512	0.1206	0.191	649.8	48.71	62.73
50	Middle	0.775	1.209		0.663	180.9		78.05
60		0.271	0.324		2.30	50.36		93.36
70	Good	0.094	0.0868		7.98	14.02		108.68
80		0.033	0.0233		27.7	3.9023		124.00
90	Best	0.011	0.00623		96.1	1.0863		139.31
100		0.004	0.00167		333	0.3024		154.63

*expresses the first kind of relationship between BA and BZ or Ex; †expresses the third kind of relationship between BA and BZ or Ex.

5.3.1.4.2 Ecosystem Health Status

The measured data from summer 1988 for 30 Italian lakes, and the data from four seasons during 1987 to 1988 for Lake Soprano were used for assessing and comparing ecosystem health status. The results for 30 Italian lakes and for Lake Soprano are presented in Table 5.7 and Table 5.8, respectively.

It can be seen from Table 5.7 that the synthetic EHI in summer 1988 for Italian lakes ranges from 60.5 to 12, indicating ecosystem health status from "good" to "worst". Ecosystem health state in Lake Ogliastro was "good" with a maximum EHI of 60.5; while that in Lake Disueri was "worst" with a minimum EHI of 12. Of 30 lakes, 20 had a "middle" health status, 6 lakes had a "bad" health status, 3 lakes had a "worst" health status, and only one lake had a "good" health status.

Table 5.8 shows that, in Lake Soprano, the synthetic EHI ranges from 41.3 to 15.3, expressing ecosystem health status from "middle" to "worst". In winter, the lake ecosystem had a "middle" health status, and by the summer, the lake ecosystem had a "worst" health status.

Table 5.7 Assessment and comparison of ecosystem health status for Italian lakes in the summer, 1988

Lake name	EHI (BA)	EHI (BZ)	EHI (BZ/BA)	EHI (Ex)	EHI (Exst)	EHI	Health state	Order (good-bad)
Ogliastro	63.6	52.3	61.5	52.6	69.4	60.5	Good	1
Fanaco	60.4	49.6	61.7	49.8	69.9	58.6	Middle	2
Ancipa	60.6	56.1	55.0	56.5	52.8	56.9	Middle	3
Prizzi	55.3	43.2	64.1	43.2	74.7	56.0	Middle	4
Vasca	55.3	44.5	62.8	44.5	72.2	55.7	Middle	5
Comunelli	56.5	50.6	57.4	50.8	59.5	55.2	Middle	6
Nicoletti	54.2	50.0	56.0	50.2	55.6	53.4	Middle	7
Garcia	52.8	48.3	56.7	48.4	57.5	52.8	Middle	8
Cesaro	50.6	41.9	61.6	41.9	69.5	52.7	Middle	9
Poma	50.0	45.1	57.7	45.2	60.2	51.4	Middle	10
Pozzillo	51.4	52.4	51.2	52.5	42.0	50.2	Middle	11
Villarosa	45.6	42.5	56.7	42.5	57.4	48.4	Middle	12
Rosalia	48.6	52.9	48.2	53.0	33.8	47.6	Middle	13
Trinita	42.5	37.6	59.2	37.5	64.0	47.3	Middle	14
Dirillo	46.0	51.6	47.4	51.6	31.7	45.8	Middle	15
Gela	48.1	59.7	40.6	59.5	17.4	45.6	Middle	16
Olivo	47.3	57.0	42.7	56.9	21.3	45.5	Middle	17
Libanesi	40.2	43.7	50.8	43.6	41.0	43.3	Middle	18
Castello	34.7	31.2	59.4	30.8	64.6	42.6	Middle	19
Rubino	41.4	51.6	43.5	51.3	22.8	42.1	Middle	20
Guadalami	34.9	36.3	54.2	36.1	50.6	41.3	Middle	21
Cimia	33.0	40.2	48.5	39.9	34.6	38.3	Bad	22
Scanzano	33.8	42.9	46.3	42.6	29.1	38.2	Bad	23
Giovanni	33.9	45.5	43.6	45.1	23.0	37.6	Bad	24
Leone	31.4	41.7	45.5	41.3	27.2	36.6	Bad	25
Gorgo	34.6	26.4	37.9	28.5	13.5	29.4	Bad	26
Gammauta	28.2	21.1	39.2	20.9	15.2	25.7	Bad	27
Arancio	11.8	29.6	14.6	21.9	2.0	15.3	Worst	28
Soprano	11.8	24.4	21.2	19.2	3.0	15.3	Worst	29
Disueri	6.2	23.2	18.0	15.2	2.4	12.0	Worst	30

Table 5.8 Assessment and Comparison of Ecosystem Health Status for Lake Soprano in 1987 to 1988

Season	EHI (BA)	EHI (BZ)	EHI (BZ/BA)	EHI (Ex)	EHI (Exst)	EHI	Health state	Order (good to bad)
Winter	35.6	41.2	49.6	40.9	44.1	41.3	Middle	1
Fall	40.2	52.2	41.9	51.9	19.7	41.1	Middle	2
Spring	27.8	22.1	37.6	22.0	13.1	25.3	Bad	3
Summer	11.8	24.4	21.2	19.2	3.0	15.3	Worst	4

5.3.2. Case 2: Ecosystem Health Assessment for Lake Chao Using DMM and EMM

Lake Chao is located in central Anhui Province of the southeastern China. It is characterized by a mean depth of 3.06 m, a mean surface area of 760 km^2, a mean volume of 1.9 billion m^3, a mean retention time of 136 days, and a total catchment area of 13,350 km^2. It provides a primary water resource for domestic, industrial, agricultural, and fishery use for a number of cities and counties, including Hefei, the capital of Anhui Province. As the fifth largest freshwater lake in China, it was well known for its scenic beauty and richness of its aquatic products before the 1960s. However, over the past decades, following population growth and economic development in the drainage area, nutrient-rich pollutants from wastewater and sewage discharge, agricultural application of fertilizers, and soil erosion, have contributed to an increasing discharge into the lake, and the lake has been seriously polluted by nutrients. The extremely serious eutrophication has already caused severe negative effects on the lake ecosystem health, sustainable utilization, and management. Since 1980, some studies focusing on the investigation and assessment of pollution sources and water quality, eutrophication mechanism, and ecosystem health, as well as on ecological restoration and environmental management, have been carried out.[16,17,40-47]

5.3.2.1 Assessment Using Direct Measurement Method (DMM)

The data measured monthly from April 1987 to March 1988 are used for the Lake Chao ecosystem health assessment. According to data availability, the ecological indicators for the assessment were phytoplankton biomass (BA), zooplankton biomass (BZ), the BZ/BA ratio, algal primary productivity (P), algal species diversity (DI), the P/BA ratio, exergy (Ex), structural exergy (Exst), and phytoplankton buffering capacity ($\beta_{(TP)(Phyto.)}$). The values of these ecological indicators for different periods and the assessment results are presented in Table 5.9. A relative order of health states for the Lake Chao ecosystem proceeding from good to poor was obtained as follows: January to March 1988 > November to December 1987 > June to July 1987 > April to May 1987 > August to October 1987.

Table 5.9 The ecological indicators and their measured values in different period in the Lake Chao (from April 1987 to March 1988)

Ecological indicators*	Measured indicator values in different period**					Relative order of health state in different period (good to poor)
	A	B	C	D	E	
BA	4.5	1.31	21.82	0.60	0.58	E > D > B > A > C
BZ	0.33	0.34	1.76	4.15	13.54	E > D > C > B > A
BZ/BA	0.073	0.26	0.081	6.92	23.24	E > D > B > C > A
P	1.42	1.38	7.03	0.74	0.21	E > D > B > A > C
P/B	0.292	1.053	0.322	1.233	0.363	D > B > E > C > A
DI	1.59	1.62	0.28	1.83	1.97	E > D > B > A > C
Ex	112.0	98.5	606.3	1075.1	3350.9	E > D > C > A > B
Exst	25.33	52.8	48.0	213.6	238.6	E > D > B > C > A
β((TP)(Phyto.))	−0.014	6.45	0.04	0.92	−0.371	B > D > C > A > E
Comprehensive results						E > D > B > A > C

*BA: Phytoplankton biomass (g m^{-3}); BZ: Zooplankton biomass (g m^{-3}); P: Algal primary productivity (gC m^{-2} d^{-1}); DI: Algal diversity index; Ex: Exergy (MJ m^{-3}); Exst: Structural exergy (MJ mg^{-1}); β((TP)(Phyto.)): Phytoplankton buffer capacity to total phosphorus.

**A: Apr.–May 1987; B: Jun.–Jul. 1987; C: Aug.–Oct. 1987; D: Nov.–Dec. 1987; E: Jan.–Mar. 1988.

The numbers are mean values of 31 sampling points' data measured monthly.

5.3.2.2 Assessment Using Ecological Model Method (EMM)

5.3.2.2.1 The Analysis of Lake Ecosystem Structure

In the early 1950s, the lake was covered with macrophytes appearing from the open waters to the shore as floating plants, submerged plants, leaf floating plants, and emergent plants, respectively. More than 190 species of zooplankton were identified. The lake was rich in large benthic animals and in fishery resources dominated by piscivorous fish. Phytoplankton populations were intensely suppressed to low densities by aquatic macrophytes, with diatoms as the dominant form. However, for the past few decades, the lake's ecosystem has been seriously damaged by eutrophication. From the early 1950s to the early 1990s, the coverage of macrophytes decreased significantly from 30% to 2.5% of the lake's total area. Now, as a result of this reduction, more than 90% of the lake's primary productivity is from phytoplankton. At the same time, the fraction of large fish also dramatically decreased from 66.7% to 23.3%. Herbivorous fish also decreased from 38.4% to 3.5%, while carnivorous fish increased significantly from 32.6% to 83%.[45]

5.3.2.2.2 The Establishment of a Lake Ecological Model

5.3.2.2.2.1 Conceptual Diagram. Given the ecosystem structure of Lake Chao, an ecological model describing nutrient cycling within the food web seemed reasonable. The model's conceptual framework is shown in Figure 5.4. The model contains six sub-models relative to nutrients, phytoplankton, zooplankton, fish, detritus, and sediments. The model's state variables include

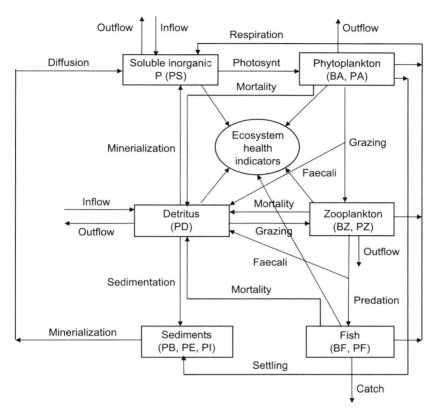

Figure 5.4   The conceptual diagram for the Lake Chao ecological model. (From Xu et al., *Water Res.* 35, 3160, 2001. With permission.)

phytoplankton biomass (BA), zooplankton biomass (BZ), fish biomass (BF), the amount of phosphorus in phytoplankton (PA), the proportion of phosphorus in zooplankton (FPZ), the proportion of phosphorus in fishes (FPF), the amount of phosphorus in detritus (PD), the amount of phosphorus in the biologically active sediment layer (PB), the amount of exchangeable phosphorus in sediments (PE), the amount of phosphorus in interstitial water (PI), and the amount of soluble phosphorus in the lake's water (PS). The model's forcing functions given as a timetable (Table 5.10) include the inflow from tributaries (QTRI), the soluble inorganic P concentration in the inflow (PSTRI), the detritus P concentration in the inflow (PDTRI), precipitation amounts to the lake (QPREC), outflow from the lake (Q), lake volume (V), lake depth (D), lake water temperature (T), and surface light radiation ($I0$).

5.3.2.2.2.2 Model Equations. The equations for the state variables are presented in Table 5.11. See References 17 and 44 for other equations for the process rates and limiting factors.

Table 5.10   The model forcing functions during April 1987 to March 1998*

Month	T (°C)	I0 (Kcal/m²)	D (m)	V (10⁸ m³)	QTRI (10⁶ m³/d)	PDTRI (mg/l)	PSTRI (mg/l)	Q (10⁶ m³/d)	QPREC (10⁶ m³/d)
April 1987	23.80	4063.7	2.27	17.20	29.17	0.028	0.013	23.15	3.50
May	24.03	3794.2	2.28	19.20	13.15	0.022	0.022	24.48	1.82
June	27.40	4200.0	2.80	21.40	56.85	0.040	0.022	10.26	8.55
July	32.25	4500.0	4.30	33.70	39.36	0.067	0.022	17.73	4.43
August	28.90	3491.6	4.30	33.40	2.40	0.026	0.022	39.08	0.34
September	24.00	3506.7	3.37	25.90	13.29	0.024	0.022	31.94	2.99
October	18.08	2074.6	3.07	23.50	8.73	0.021	0.022	26.91	1.52
November	17.90	1788.9	2.28	17.20	0.87	0.018	0.022	24.32	0.00
December	6.21	2051.6	1.99	14.80	0.82	0.019	0.022	1.66	0.47
January 1988	5.90	1480.5	1.99	14.80	7.59	0.024	0.024	0.90	2.32
February	5.20	1541.3	2.29	17.20	9.66	0.021	0.024	16.66	1.89
March	8.40	2244.6	2.11	15.70	3.96	0.021	0.024	8.13	0.82

*The model forcing functions include inflow from tributaries (10⁶m³/d) (QTRI), soluble inorganic P concentration in inflow (mg/l) (PSTRI), detritus P concentration in inflow (mg/l) (PDTRI), precipitation to the lake (10⁶ m³/d) (QPREC), outflow (10⁶ m³/d) (Q), lake volume (10⁸ m³) (V), lake depth (m) (D), temperature of lake water (°C) (T), light radiation on the surface of lake water (kcal/m².d) (I0).

5.3.2.2.2.3 Model Parameters.

The parameters determined from the literature, experiments, and calibrations are listed in Table 5.12.

5.3.2.2.3 The Calibration of the Ecological Model

The comparisons of the simulated and the observed values of important state variables and process rates are presented in Figure 5.5, including phytoplankton rates for growth, respiration, mortality, and settling; internal phosphorus concentration in phytoplankton cells; phytoplankton biomass; and zooplankton and fish growth rates. It can be seen from Figure 5.7 that there were very good agreements between observations and simulations of the growth rates, respiration rates, mortality rates, settling rates, internal phosphorus and biomasses of phytoplankton, as well as zooplankton growth rate, with R^2 being over 0.8. There are also good agreements between the simulated and the observed values for fish growth rates, with R^2 being 0.6316.

The results of model calibration suggested that the model could reproduce the most of important state-variable concentrations and process rates using model equations and coefficients, and would represent pelagic ecosystem structure and function in Lake Chao. It can therefore be applied to the calculation of ecological health indicators.

5.3.2.2.4 The Calculation of Ecosystem Health Indicators

The ecosystem health indicators used in the model include phytoplankton biomass (BA), zooplankton biomass (BZ), zooplankton/phytoplankton ratio

Table 5.11   Differential equations for state variables of the Lake Chao model

(1)
$$\frac{d}{dt} BA = (GA - MA - RA - SA - GZ/Y0 - Q/V) * BA$$

(2)
$$\frac{d}{dt} PA = AUP * BA - (MA + RA + SA + GZ/Y0 + Q/V) * PA$$

(3)
$$\frac{d}{dt} BZ = (MYZ - RZ - MZ - Q/V) * BZ - (PRED1/Y1) * BF$$

(4)
$$\frac{d}{dt} FPZ = MYZ * (FPA - FPZ) = MYZ * ((PA/BA) - FPZ)$$

(5)
$$\frac{d}{dt} BF = (GF - RF - MF - CATCH)$$

(6)
$$\frac{d}{dt} FPF = (PREDY1/Y1) * (FPZ - FPF)$$

(7)
$$\frac{d}{dt} PD = ((1/YO)-1) * GZ * PA - ((1/YO)-1) * PRED1 * PZ + MA * PA$$
$$+ MZ * PZ + MF * PF + QPDIN - (KDP + SD + Q/V) * PD$$

(8)
$$\frac{d}{dt} PB = ((QSED * D)/(DB * DMU)) - QBIO - QDSORP$$

(9)
$$\frac{d}{dt} PE = ((D * KEX * (SA * PS - QSED + SD * PD))/(LUL * DMU)) - KE * PE$$

(10)
$$\frac{d}{dt} PI = (AE/AI) * KE * PE - (QDIFF/AI), AI = LUL * (1 - DMU)/D$$

(11)
$$\frac{d}{dt} PS = RA * PA + RZ * PZ + RF * PF + QPSIN + KDP * PD + QDIFF$$
$$+ ((DB/D) * DMU) * (QBIO + QDSORP) - AUP * BA - (Q/V) * PS$$

---

(1) BA-Phytoplankton biomass (g/m^3), GA-phytoplankton growth rate (1/d), MA-phytoplankton mortality rate (1/d), RA-phytoplankton respiration rate (1/d), SA-phytoplankton mortality rate (1/d), GZ-zooplankton grazing rate (1/d), Y0-assimilation efficiency for zooplankton grazing, Q-outflow(m^3/d), V-lake volume(m^3);

(2) PA-PA in phytoplankton (g/m^3), AUP- phosphorus uptake rate (1/d)

(3) BZ-zooplankton biomass (g/m^3), MYZ-zooplankton growth rate (1/d), RZ-zooplankton respiration rate (1/d), MZ-zooplankton mortality rate (1/d), PRED1-fish predation rate (1/d), Y1-assimilation efficiency for fish predation;

(4) FPZ-P proportion in zooplankton (kg P/kg BZ),

(5) BF-fish biomass (g/m^3), GF-fish growth rate (1/d), RF-fish respiration rate (1/d), MF-fish mortality rate (1/d), CATCH-catch rate of fish (1/d);

(6) FPF-P proportion in fish (kg P/kg BF),

(7) PD-phosphorus in detritus (g/m^3), QPDIN- PD from inflow (mg/L), KDP-PD decomposition rate (1/d), SD-PD settling rate (1/d);

(8) PB-P in biologically active layer (g/m^3), QSED-sediment material from water, D-lake depth (m), DB-depth of biologically active layer in sediment (m), DMU- Dry matter weight of upper layer in sediment (kg/kg), QBIO-demineralization rate of PB (1/d), QDSOPD-sorption and desorption of PB (1/d);

(9) PE-exchangeable P (g/m^3), KEX-ratio of exchangeable P to total P in sediments, LUL-depth of unstable layer in sediments (m), KE-PE mineralization rate (1/d);

(10) PI-P in interstitial water (g/m^3), QDIFF-diffusion coefficient of PE;

(11) PS-Soluble inorganic P (g/m^3), QPSIN-PS from inflow (mg/L).

Table 5.12   Parameters for the Lake Chao ecological mode

Symbol	Description	Unit	Literature range	Value used	Sources
Phytoplankton submodel					
Gamax	Maximum growth rate of phytoplankton	1/d	1–5	4.042	Measurement
MAmax	Maximum mortality rate of phytoplankton	1/d		0.96	Measurement
RAmax	Maximum respiration rate of phytoplankton	1/d	0.005–0.8	0.6	Measurement
AUPmax	Maximum P uptake rate of phytoplankton	1/d	0.0014–0.01	0.003	Calculation
TAopt	Optimal temperature for phytoplankton growth	°C		28	Measurement
TAmin	Minimum temperature for phytoplankton growth	°C		5	Measurement
FPAmax	Maximum kg P per kg phytoplankton biomass	—	0.013–0.03	0.013	[91]
FPAmin	Minimum kg P per kg phytoplankton biomass	—	0.001 - 0.005	0.001	[91]
KI	Michaelis constant for light	kcal/m^2.d	173–518	400	[91]
KPA	Michaelis constant of P uptake for phytoplankton	mg/l	0.0005–0.08	0.06	Measurement
SVS	Settling velocity of phytoplankton	m/d	0.1–0.8	0.19	[91]
α	Extinction coefficient of water	1/m		0.27	[92]
β	Extinction coefficient of phytoplankton	l/m		0.18	[92]
θ	Temperature coefficient for phytoplankton settling	—		1.03	[92]
Zooplankton submodel					
MYZmax	Maximum growth rate of zooplankton	1/d	0.1–0.8	0.35	[91]
MZmax	Maximum basal mortality rate of zooplankton	1/d	0.001–0.125	0.125	[91]
TOXZ	Toxic mortality rate	1/d		0.075	Calibration
Ktoxz	Toxic mortality adjustment coefficient	—		0.5	Calibration
RZmax	Maximum respiration rate of zooplankton	1/d	0.001–0.036	0.02	[91]
PRED1max	Maximum feeding rate of fish on zooplankton	1/d	0.012–0.06	0.04	Calibration
TZopt	Optimal temperature for zooplankton growth	°C		28	Measurement
TZmin	Minimum temperature for zooplankton growth	°C		5	Measurement
KZ	Michaelis constant for fish predation	mg/l		0.75	[93]

(*Continued*)

Table 5.12 Continued

Symbol	Description	Unit	Literature range	Value used	Sources
KSZ	Threshold zooplankton biomass for fish predation	mg/l		0.75	[94]
KA	Michaelis constant for zooplankton grazing	mg/l	0.01–2	0.5	[91]
KSA	Threshold phytoplankton biomass for zooplankton	mg/l	0.01–0.2	0.2	[95]
KZCC	Zooplankton carrying capacity	mg/l		30	Calculation
Y0	Assimilation efficiency for zooplankton grazing	—	0.5–0.8	0.63	[91]
Fish submodel					
GFmax	Maximum growth rate of fish	1/d		0.015	Measurement
MFmax	Maximum basal mortality rate of fish	1/d		0.003	[93]
Ktoxf	Toxic mortality rate	1/d		0.005	Calibration
TOXF	Toxic mortality adjustment coefficient	—		0.015	Calibration
RFmax	Maximum respiration rate of fish	1/d	0.00055–0.0055	0.002	Calculation
TFopt	Optimal temperature for fish growth	°C		22	Measurement
TFmin	Minimum temperature for fish growth	°C		5	Measurement
CATCH	Catch rate of fish	1/d		0.001	Calibration
KFCC	Fish carrying capacity	mg/l		40	Calculation
Y1	Assimilation efficiency for fish predation	—		0.5	Calibration
Detritus, sediments, and soluble inorganic phosphorus submodel					
DB	Depth of biologically active layer in sediment	m		0.005	Measurement
LUL	Depth of unstable layer in sediments	m		0.16	Measurement
DMU	Dry matter weight of upper layer in sediment	kg/kg		0.3	Measurement
KE20	Mineralization rate of PE at 20 C	1/d		0.0673	Measurement
KDIFF	Diffusion coefficient of P in interstitial water	—		1.21	Jørgensen, 1976
KEX	Ratio of exchangeable P to total P in sediments	—		0.18	Measurement
SVD	Settling velocity of detritus	m/d		0.002	Jørgensen, 1976
KDP10	Decomposition rate of detritus P at 10 C	L/d		0.1	Calculation
Φ	Temperature coefficient for detritus degradation			1.072	Jørgensen, 1976
θ	Temperature coefficient for PE decomposition Soluble inorganic P			1.03	Chen and Orlob, 1975

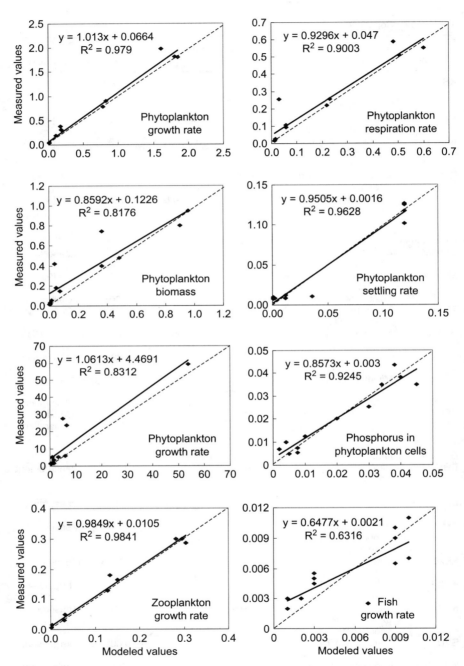

Figure 5.5   The comparisons of the modeled and measured state variables and process rates. (The solid lines are trend lines; the dashed lines are "1:1" lines.)

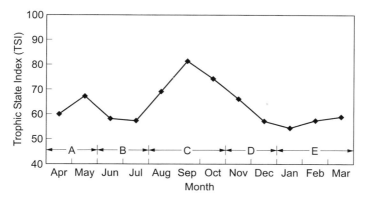

Figure 5.6 Dynamics of trophic state in the Lake Chao during April 1987 to March 1988.

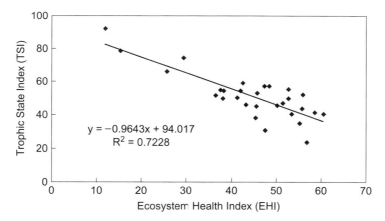

Figure 5.7 Relationships between EHI and TSI in 30 Italy lakes in the summer 1988.

(R_{BZBA}), exergy (Ex), and structural exergy (Ex_{st}). The calculated results of ecosystem health indicators are presented in Table 5.13.

5.3.2.2.5 *The Assessment of Lake Ecosystem Health*

Relative to contaminated ecosystems, a healthy ecosystem will have a higher zooplankton biomass, lower phytoplankton biomass, higher zooplankton/phytoplankton ratio, and higher exergy and structural exergy (see Table 5.4 and Reference 15). According to these principles and the calculated values for the ecological health indicators presented in Table 5.13, the results of the lake ecosystem health assessment of Lake Chao are presented in Table 5.13. The relative health states in terms of timespan have been arranged from "good" to "poor" as: January to March 1988 > November to December 1987 > June to July 1987 > April to May 1987 > August to October 1987. These results are the same as the results using DMM.

Table 5.13 The modeled values of ecological indicators and relative health state in different period in the Lake Chao ecosystem (from April 1987 to March 1988)

Ecological indicators	Time periods*					Relative health state (good to poor)
	(A)	(B)	(C)	(D)	(E)	
BA (mg × m⁻³)	4.21	4.00	35.31	1.34	0.01	E > D > B > A > C
BZ (mg × m⁻³)	0.90	0.91	3.52	12.52	12.77	E > D > C > B > A
BZ/BA	0.21	0.23	0.10	9.34	127.70	E > D > B > A > C
Ex (MJ × m⁻³)	1312.88	1413.52	1951.99	3256.01	3358.11	E > D > C > B > A
Exst (MJ mg⁻¹)	152.52	158.20	57.97	184.01	196.02	E > D > B > A > C
Comprehensive results						E > D > B > A > C

*A: Apr. to May 1987; B: Jun. to Jul. 1987; C: Aug. to Oct. 1987;
D: Nov. to Dec. 1987; E: Jan. to Mar. 1988.
From Xu, et al. *Wat. Res.* 35(1), 3165, 2001. With permission.

5.4 DISCUSSIONS

5.4.1 About Assessment Results

5.4.1.1 Assessment Results for Lake Chao

The results obtained using two assessment methods, the direct measurement method (DMM) and the ecological modeling method (EMM), are very similar. Namely, the relative health states from good to poor in Lake Chao during April 1987 to March 1988 are as follows: January to March 1988 > November to December 1987 > June to July 1987 > April to May 1987 > August to October 1987. This means that the worst healthy state occurred between August and October of 1987, followed by April to May of 1987; while the best healthy state occurred between January and March 1988. These results corresponded well with the observed eutrophic states and the results of eutrophication assessment at the Lake Chao.

In terms of the observations made from April 1987 to March 1988, the most serious algal bloom occurred between August and October of 1987 (summer–autumn bloom). Another algal bloom occurred between April and May 1987 (spring bloom). These two algal blooms are typical symptoms of the eutrophicated conditions at the Lake Chao. Both events severely impaired the lake's ecosystem health.

The results for the lake's trophic state index (TSI) calculations are illustrated in Figure 5.6. The calculations were carried out using the same time period data and six indicators: total phosphorus, total nitrogen, chemical oxygen demand, Secchi disk depth, Chl-a concentration and phytoplankton biomass (see Reference 27 for details).

The average TSI levels were 56.7 for January to March 1988; 61.4 for November to December 1987; 57.7 for June to July 1987; 63.4 for April to May 1987; and 74.4 for August to October 1987. This indicates the most serious eutrophication event occurred between August and October of 1987, followed

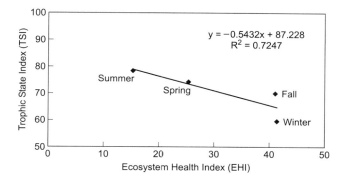

Figure 5.8 Relationships between EHI and TSI in the Lake Soprano in the summer 1987 to 1988.

by the April to May 1987 period. The lowest eutrophication levels happened between January and March of 1988. The assessment results of the lake's ecosystem health obtained using both the DMM and EMM procedures correspond closely with the lake's existing trophic states.

5.4.1.2. Assessment Results for Italian Lakes

The relationships between ecosystem health index (EHI) and trophic state index (TSI) for 30 Italian lakes and for Lake Soprano are demonstrated in Figure 5.7 and Figure 5.8, respectively. The TSI calculations were made using the same time period data and the various indicators (total phosphorus, Secchi disk depth and chlorophyll-a concentration) with the EHI calculations.

It can be seen from both Figure 5.7 and Figure 5.8 that for the 30 Italian lakes and for Lake Soprano, there are strongly negative relationships between EHI and TSI with R^2 being over 0.72. This means that the lakes with a higher TSI have a lower EHI — that is, the lake ecosystem healthy states become worse with increasing trophic states. For instance, in the 30 Italian lakes the TSI values for Lakes Disueri, Soprano and Arancio in the summer 1988 are 92.3, 78.5, and 78.5, respectively, which are the higher values; and correspondingly, the EHI values for these three lakes in the same time period are 12.0, 15.3, and 15.3, respectively, which are the lower values. This means that the more serious the eutrophication, the worst the health status becomes. The same situation also can be found in Lake Soprano in the summer 1988 (see Figure 5.8). It can therefore also be concluded that the assessment results of ecosystem health states for the 30 Italian lakes and for Lake Soprano obtained by EHIM procedures accord closely with the lake's existing trophic states.

5.4.2. About Assessment Indicators

The ecological and thermodynamic indicators presented in this chapter cover structural, functional and system-level aspects of lake ecosystem health.

In order to provide a fully informative assessment of the health condition of a lake ecosystem, it is also necessary to apply the indicators simultaneously. Structural changes represent the first response of a lake ecosystem to external anthropogenic stresses. These are then followed by the functional and system-level changes. Structural and functional changes can be described using structural and functional ecological indicators; while system-level changes can be described using thermodynamic indicators: exergy and structural exergy.

It has been proved that exergy as the ecosystem health indicator consists of:

1. Homeostasis
2. Absence of disease, partly
3. Diversity or complexity
4. Stability or resilience
5. Vigor or scope for growth of Costanza's definition of ecosystem health.

Structural exergy as an ecosystem health indicator consists of biodiversity (point 3) and the balance between system components (point 6). Ecological buffering capacity as an ecosystem health indicator coincides with description point (4) in Costanza's concept definition of ecosystem health. Therefore the combination of exergy, structural exergy, and ecological buffering capacity provides an appropriate system-level description of lake ecological health.[22,23,27,48] It is suggested that more lake biological components such as bacteria, phytoplankton, zooplankton, benthos, macrophyte and fish should be used to calculate exergy and structural exergy if the data are available, so that the more reasonable and practical results can be obtained. However, in the case studies presented in this chapter, only phytoplankton and zooplankton were used to calculate exergy and structural exergy, since the data for other biological components than phytoplankton and zooplankton are very limited.

The indicators are also applicable to the other sites, since the indicators are retrieved from the worldwide experimental and actual research on the responses of lake ecosystems to chemical stresses, including eutrophication and acidification, as well as heavy metal, oil, and pesticide contamination, not just from a single research source.

5.4.3. About Assessment Methods

Three methods — DMM, EMM and EHIM — have been suggested in this chapter for the assessment of lake ecosystem health. The combinations of the three methods can be applied for the relative and absolute diagnosis as well as the prediction of lake ecosystem health status. The results from the case studies show that the DMM can be used for the relative assessment of a single lake only; while the EHIM can be used for the absolute assessment of more than one lake.

The DMM can be used for the rough assessment of a lake ecosystem health in the case of insufficient data. However, owing to the lack of criteria, there are two major problems with using DMM to assess lake ecosystem health. First, it can only assess the relative healthy status (it is extremely difficult to assess the actual health status). Second, it is impossible to make the comparisons of ecosystem health status for different lakes.

The results from the case studies show that the EHIM can solve the above-mentioned two problems. This method offers a numerical scale from 0 to 100, which can easily make the quantitative assessment and comparison of ecosystem health states for single and different lakes. The criteria for different healthy states can also be obtained using EHIM (see Table 5.6). The EHIM is a valuable method with the advantage of uncomplicated principles, simple calculations, and reliable and intuitionistic results. It is expected that the EHIM can be widely used for the quantitative assessment and comparison of ecosystem health states.

It is quite novel to apply the ecological model method (EMM) to ecosystem health assessment, although ecological modeling has gone though a long history since the Lotka–Volterra and Streeter–Phelps models were developed in the 1920s.[49] This may be partly because ecosystem health is a relatively new research field with only about 10 years of history. Compared with the experiment and monitoring method, the ecological model method (EMM) is less time-consuming and laborious. However, only a tentative procedure for the practical assessment of ecosystem health using EMM was suggested by Jørgensen.[22] It is necessary to give more documentation and more case studies for ecosystem health assessment using EMM to promote the development of this new field. The key problems with the EMM approach to ecosystem health assessment are how to develop a reliable lake ecological model and how to integrate it with ecosystem health indicators. Previous efforts to develop and apply lake ecological models have indicated that the most important steps are a pre-examination of the lake ecosystem, a determination of the proposed model's complexity, an estimation of the model's parameters, and a calibration of the model's results.[34,50]

In order to predict the changes of lake ecosystem health following changes in environmental conditions, lake models have to be either validated or designed with a dynamic structure. In the first instance, different data sets will have to be used for the model validations in order for the most suitable parameters to be found.[51–53] However, these validated models can only be applied to the ecological health prediction of lakes whose biological structures have remained unchanged, or only slightly changed, since such models have both a given structure and a set of fixed parameters. In the second case, the models can be called ecological structural dynamic models (ESDM), where goal functions are used to determine how to change the current parameters to express changes in the lake's biological structure following changes in environmental conditions.[51–53] The most appropriate use of such ESDMs would seem to be in predicting changes to lake ecosystem health as a result of changes in environmental conditions. If accomplished, this would represent an important step in ecosystem health assessment through ecological modeling methods (EMM).

5.5 CONCLUSIONS

A tentative theoretical frame, a set of ecological and thermodynamic indicators, and three methods have been proposed for the assessment of lake ecosystem health in this chapter. A tentative theoretical frame is composed of five necessary steps:

1. The identification of anthropogenic stresses
2. The analysis of ecosystem responses to the stresses
3. The development of indicators
4. The determination of assessment methods
5. The qualitative and quantitative assessment of lake ecosystem health.

In order to develop indicators for lake ecosystem health assessment, five kinds of anthropogenic stresses including eutrophication and acidification, as well as heavy metal, pesticide and oil pollution were identified; and 59 actual and 20 experimental lakes were used as the studied lake ecosystems. A set of ecological and thermodynamic indicators covering lake structural, functional, and system-level aspects were developed, according to the structural, functional, and system-level responses of lake ecosystems to the five kinds of anthropogenic stresses. The structural indicators included:

- Phytoplankton cell size and biomass
- Zooplankton body size and biomass
- Species diversity
- Macro- and micro-zooplankton biomass
- The zooplankton to phytoplankton ratio
- The macrozooplankton to microzooplankton ratio.

The functional indicators encompassed the algal C assimilation ratio, resource use efficiency, community production, gross production to respiration (i.e., P/R) ratio, gross production to standing crop biomass (i.e., P/B) ratio, and standing crop biomass to unit energy flow (i.e., B/E) ratio. The ecosystem-level indicators consisted of ecological buffer capacities, exergy, and structural exergy.

Three methods, the direct measurement method (DMM), the ecological modeling method (EMM) and the ecosystem health index method (EHIM), are proposed for lake ecosystem health assessment.

The DMM procedures were designed to:

1. Identify key indicators
2. Measure directly or calculate indirectly the selected indicators
3. Assess ecosystem health on the basis of the indicator values.

The EMM procedures were designed to:

1. Determine the structure and complexity of the ecological model according to the lake's ecosystem structure
2. Establish an ecological model by designing a conceptual diagram, establishing model equations, and estimating model parameters

3. Compare the simulated values of important state variables and process rates with actual observations
4. Calculate ecosystem health indicators using the ecological model
5. Assess lake ecosystem health according to the values of the ecological indicators

The EHIM which based on the four-season measured data from 30 Italian lakes possessed three major steps. First, a numerical EHI on a scale of 0 to 100 was developed. Second, in order to calculate the specific and synthetic EHI, phytoplankton biomass (BA) was selected to serve as a basic indicator, while zooplankton biomass (BZ), the ratio of BZ to BA (BZ/BZ), exergy and structural exergy were used as additional indicators. Third, the specific and synthetic EHI were calculated based on these indicators. The quantitative assessment results for lake ecosystem health could then be obtained according to the synthetic EHI values.

The results from Lake Chao demonstrated that both DMM and EMM provided very similar results. A relative order of health states from poor to good was found: August to October 1987 > April to May 1987 > June to July 1987 > November to December 1987 > January to March 1988. This result reflected well the actual situation of the Lake Chao. Also, the EHI method was successfully applied to the assessment and comparison of ecosystem health for Italian lakes.

REFERENCES

1. Costanza, R., dArge, R., deGroot, R., Farber, S., Grasso, M., Hannon, B., Limburg, K., Naeem, S., Oneill, R.V., Paruelo, J., Raskin, R.G., Sutton, P., vandenBelt, M. The values of the world's ecosystem services and natural capital. *Nature* 387, 253, 1997.
2. Westman, W.E. What are nature's services worth? *Science* 197, 960, 1997.
3. Cairns, J., Jr. Sustainability, ecosystem services, and health. *Int. J. Sustain. Dev. World Exol.* 4, 153, 1997.
4. Daily, G.C. *Nature's Services: Societal Dependence on Natural Ecosystems.* Island Press, Washington, D.C. 1997, 18 p.
5. Xu, F.L. and Tao, S. On the study of ecosystem health: the state of the art. *J. Envir. Sci. (China)* 12, 33, 2000.
6. Schaeffer, D.J., Herricks, E.E., and Kerster, H.W. Ecosystem health: 1. measuring ecosystem health. *Environ. Manage.* 12, 445, 1988.
7. Costanza, R., Norton, B.G., and Haskell, B.D., Eds. *Ecosystem Health: New Goals for Environmental Management.* Island Press, Washington, D.C., 1992.
8. Haskell, B.D., Norton, B.G., and Costanza, R. "What is ecosystem health and why should we worry about it?" in *Ecosystem Health: New Goals for Environmental Management,* Costanza, R., Norton, B.G., and Haskell, B.D., Eds. Island Press, Washington, D.C., 1992, 3 p.
9. Rapport D.J. Ecosystem health: exploring the territory. *Ecosyst. Health* 1, 5, 1995.
10. Rapport D.J., Richard, L., and Paul R.E., Eds. *Ecosystem Health.* Blackwell Science Inc., Malden, MA, 1998.

11. Rapport, D.J., Costanza, R., and McMichael, A.J. Assessing ecosystem health. *Trends Ecol. Evol.* 13, 397, 1998.
12. Shear, H. The development and use of indicators to assess ecosystem health state in the Great Lake. *Ecosyst. Health*, 2, 241, 1996.
13. WRI (Water Research Institute). Stresses on Ecosystem Health — Chemical Pollution, 2002. See < http://www.wri.org/health/ecohealt.html >.
14. NWRI (National Water Research Institute). Aquatic Ecosystem Health Assessment Project, 2002. See < http://www.cciw.ca/nwri/aecb/aquatic-eco-health.html >
15. Xu, F.L, Jørgensen, S.E., and Tao S. Ecological indicators for assessing freshwater ecosystem health. *Ecol. Model.*, 116, 77, 1999.
16. Xu, F.L., Tao, S., Dawson, R.W., Li, B.G., Cao, J. Lake ecosystem health assessment: indicators and methods. *Water Res.*, 35, 3157, 2001.
17. Xu, F.L., Dawson, R.W., Tao, S., Cao, J., Li, B.G. A method for lake ecosystem health assessment: an Ecological Modeling Method and its application. *Hydrobiology*, 443, 159–175, 2001.
18. Xu, F.L., Lam, K.C., Zhao, Z.Y., Zhan, W., Chen, Y.D., Tao, S. Marine coastal ecosystem health assessment: a case study of Tolo Harbour, Hong Kong, China. *Ecol. Model.* 173, 355–370, 2004.
19. Hannon, B., Ecosystem Flow Analysis, *Can. Bul. Fish. & Aqu. Sci.* 213, 97, 1985.
20. Rapport, D.J., Regier, H.A., and Hutchinson, T.C., Ecosystem Behavior Under Stress, *American Naturalist*, 125, 617, 1985.
21. Karr, J.R. et al. *Assessing Biological Integrity in Running Waters: A Method and its Rationale Champaign*. Illinois Natural History Survey, Special Publication 5, 1986.
22. Jørgensen, S.E. Exergy and ecological buffer capacities as measures of ecosystem health. *Ecosyst. Health* 1, 150, 1995.
23. Jørgensen, S.E. The application of ecological indicators to assess the ecological condition of a lake. *Lakes & Reservoirs: Res. Manage.* 1, 177, 1995.
24. Calvo, S., Barone, R., Naselli Flores, L., Frada Orestano, C., Dongarra, G., Lugaro, A., Genchi, G. *Limnological Studies on Lakes and Reservoirs of Sicily*. Naturalisla Sicil., S. IV, Vol. XVII (Supplemento) 1993.
25. Lundgren, A. Model ecosystem as a tool in freshwater and marine research. *Arch. Hydrobiol.* suppl., 70, 157, 1985.
26. Schindler, D.W. Detecting ecosystem responses to anthropogenic stress. *Can. J. Fish. Aquat. Sci.* 44, 6, 1987.
27. Xu, F.L. Exergy and structural exergy as ecological indicators for the development state of the Lake Chao ecosystem. *Ecol. Model.* 99, 41, 1997.
28. Xu, F.L. Lake ecosystem health assessment: indicators and methods. Postdoctoral Report, Peking University, 2000.
29. Lu Xiao-Yan. The dynamics of lake ecological structure and its effects on lake ecosystem health. Masters thesis, Peking University, 2001.
30. Jørgensen, S.E., and Mejer, H. Ecological buffer capacity. *Ecol. Model.* 3, 39, 1977.
31. Jørgensen, S.E., and Mejer, H.F. A holistic approach to ecological modeling. *Ecol. Model.* 7, 169, 1979.
32. Jørgensen, S.E. *Integration of Ecosystem Theories: A Pattern*, 1st ed. Kluwer, Dordrecht, 1992, chap. 6.
33. Jørgensen, S.E. Parameters, ecological constraints and exergy. *Ecol. Model.* 62, 163, 1992.
34. Jørgensen, S.E. Review and comparison of goal functions in system ecology. *Vie Milieu* 44, 11, 1994.

35. Jørgensen, S.E., Nielson, S.N., and Mejer, H.F. Emergy, environ, exergy and ecological modelling. *Ecol. Model.* 77, 99, 1995.
36. Jørgensen, S.E. "Exergy and buffering capacity in ecological system," in *Energetics and Systems*, Mitsch, W. et al., Eds. Ann Arbor Science, Ann Arbor, MI, 1982, 61.
37. Washington, H.G. Diversity, biotic and similarity indices. *Water Res.*, 18, 653, 1984.
38. Margalef, R. Communication of structure in plankton population. *Limn. Oceanogr.* 6, 124, 1961.
39. Carlson, R.E. A trophic state index for lakes. *Limnol. Oceanog.* 22, 361, 1977.
40. Tu, Q.Y., Gu, D.X., Yi, C.Q., Xu, Z.R., Han, G.Z. *The Researches on the Lake Chao Eutrophication.* University of Science and Technology of China, Hefei, 1990.
41. Wang, S.Y., Jin, C.S., Meng, R.X., Xu, F.L. "Environmental research for the Lake Chao in Anhui Province," in *Lakes in China*, Vol. 1, Jin, X.C., Ed. China Ocean Press, Beijing, 189, 1995.
42. Xu, F.L. Scientific decision-making system for environmental management of the Lake Chao watershed. *Environ. Protection* 21, 36, 1994.
43. Xu, F.L. Ecosystem health assessment of Lake Chao, a shallow eutrophic Chinese lake. *Lakes & Reservoirs: Res. Manage.* 2, 101, 1996.
44. Xu, F.L., Jørgensen, S. E., and Tao, S. Modellling the effects of ecological engineering on ecosystem health of a shollow eutrophic Chinese lake (Lake Chao). *Ecol. Model.* 117, 239, 1999.
45. Xu, F.L., Tao, S., and Xu, Z.R. The restoration of riparian wetlands and macrophytes in the Lake Chao, an eutrophic Chinese lake: possibility and effects. *Hydrobiology* 405, 169, 1999.
46. Xu, F.L., Tao, S., Dawson, R.W., Li, B.G. A GIS-based method of lake eutrophication assessment. *Ecol. Model.* 144, 231–244, 2001.
47. Xu, F.L., Tao, S., Dawson, R.W., Xu, Z.R. The distributions and effects of nutrients in the sediments of a shallow eutrophic Chinese lake. *Hydrobiology* 492, 85, 2003.
48. Xu, F.L., Tao, S., and Dawson, R.W. System-level responses of lake ecosystems to chemical stresses using exergy and structural exergy as ecological indicators, *Chemosphere* 46, 173, 2002.
49. Xu, F.L., Tao S., and Dawson, R.W. Lake ecological model: history and development. *J. Environ. Sci. (China)* 14, 255, 2002.
50. Jørgensen, S.E. "Ecological modelling of lakes," in *Mathematical Modelling of Water Quality: Streams, Lakes, and Reservoirs*, Orlob, G.T., Ed. J. Wiley, New York, 1983, 116.
51. Jørgensen, S.E. Structural dynamic model. *Ecol. Model.* 31, 1, 1986.
52. Jørgensen, S.E., Use of models as experimental tools to show that structural changes are accompanied by increased exergy. *Ecol. Model.* 41, 117, 1988.
53. Jørgensen, S.E. Development of models able to account for changes in species composition, *Eco. Model.* 62, 195, 1992.
54. Havens, K.E., and DeCosta, J. Freshwater plankton community succession during experimental acidification. *Arch. Hydrobiol.* 111, 37, 1987.
55. Barmuta, L.A. et al. Responses of zooplankton and zoobenthos to experimental acidification in a high-evaluation lake (Sierra Nevada, California, USA). *Freshwater Biol.* 23, 571, 1990.
56. Havens, K.E. Acidification effects on the algal–zooplankton interface. *Can. J. Fish. Aquat. Sci.* 49, 2507, 1992.

57. Havens, K.E. and Heath, R.T. Acid and aluminium effects on freshwater zooplankton: an in situ mesocosm study. *Environ. Pollut.* 62, 195, 1989.
58. Havens, K.E. An experimental comparison of the effects of two chemical stress on a freshwater zooplankton assemblage. *Environ. Pollut.* 84, 245, 1994.
59. Havens, K.E. Structural and functional responses of a freshwater community to acute copper stress. *Environ. Pollut.* 86, 259, 1994.
60. Giddings, J.M. et al. Effects of chronic exposure to cool-derived oil on freshwater ecosystems: II. Experimental ponds. *Environ. Toxicol. Chem.* 3, 465–488, 1984.
61. Hurlbert, S.H., Mulla, M.S., and Wilson, H.R. Effects of organophosphorous insecticide on the phytoplankton, zooplankton, and insect populations of freshwater ponds. *Eco. Mono.* 42, 269, 1972.
62. Boyle, T.R. Effects of the aquatic herbicide 2,4-DMA on the ecology of experimental ponds. *Environ. Pollut.* 21, 35, 1980.
63. Schauerte, W. et al. Influence of 2,4,6-trichlorophenol and pentachlorophenol on the biota of aquatic system. *Chemosphere* 11, 71, 1982.
64. Lay, J.P., Schauerte, W., and Klein, W. Effects of trichloroethylene on the population dynamics of phyto- and zoo-plankton in compartments of a natural pond. *Environ. Pollut.* 33, 75, 1984.
65. Lay, J.P. et al. Influence of benzene on the phytoplankton and on *Daphinia pulex* in compartments of a experimental pond. *Ecotoxicol. Environ. Safety* 10, 218, 1985.
66. Lay, J.P. et al. Long-term effects of 1,2,4-trichlorobenzene on freshwater plankton in an outdoor-model-ecosystem. *Bull. Environ. Contam. Toxicol.* 34, 761, 1985.
67. Lynch, T.R., Johnson, H.E., and Adams, W.J. Impact of atrazine and hexachlorobiphenil on the structure and function of model stream ecosystem. *Environ. Toxicol. Chem.* 4, 399, 1985.
68. Yasuno, M., Hanazata, T., Iwakuma, T., Takamura, K., Ueno, R., Takamura, N. Effects of permethrin on phytoplankton and zooplankton in an enclosure ecosystem in a pond. *Hydrobiology* 159, 247, 1988.
69. Thompson, D.G., Holmes, S.B., Wainio-Keizer, K., MacDonald, L., Solomon, K.R. Impact of hexazinone and metilsulfuron methyl on the phytoplankton community of a boreal forest lake. *Environ. Toxicol. Chem.* 12, 1695, 1993.
70. Thompson, D.G., Holmes, S.B., Wainio-Keizer, K., MacDonald, L., Solomon, K.R. Impact of hexazinone and metilsulfuron methyl on the zooplankton community of a boreal forest lake. *Environ. Toxicol. Chem.* 12, 1709, 1993.
71. Drenner, et al. Effects of sediment-bound bifenthrin on gizzard shad and plankton in experimental tank mesocosms. *Environ. Toxicol. Chem.* 12, 1297, 1993.
72. Havens, K.E., and Hanazato, T. Zooplankton community responses to chemical stress: a comparison of results from acidification and pesticide contamination research. *Environ. Pollut.* 82, 277, 1993.
73. Schindler, D.W., Mills, K.H., Malley, M.A., Findlay, D.L., Schearer, J.A., Davies, I.J. Long-term ecosystem stress: the effects of years of experimental acidification of a small lake. *Science* 228, 1395, 1985.
74. Schindler, D.W. Experimental perturbations of whole lakes as tests of hypotheses concerning ecosystem structure and function. *Oikos* 57, 25, 1990.
75. Webster, K.E., Frost, T.M., Waters, C.J., Swenson, W.A., Gonzatez, M., Gerrison, P.J. Complex biological responses to the experimental acidification of Little Rock Lake, Wisconsin, USA. *Environ. Pollut.* 78, 73, 1992.
76. Giddings, J.M. Effects of the water-soluble fraction of a coal-derived oil on pond microcosms. *Arch. Environ. Contam. Toxicol.* 11, 735, 1982.

77. Cowser, K.E. Life sciences synthetic fuels semi-annual progress report for the period ending December 31, 1981. ORNL/TM-8229. Oak Ridge National Laboratory, Oak Ridge, TN, 1982.

78. Cowser, K.E. Life sciences synthetic fuels semi-annual progress report for the period ending June 30, 1982. ORNL/TM-8441. Oak Ridge National Laboratory, Oak Ridge, TN, 1982.

79. Franco, P.J. et al. Effects of chronic exposure to cool-derived oil on freshwater ecosystems: I, Microcosms. *Environ. Toxicol. & Chem.* 3, 447, 1984.

80. McNight, D.M. Chemical and biological processes controlling the response of a freshwater ecosystem to copper stress: a field study of the $CuSO_4$ treatment of Mill Pond Reservoir, Burlington, Massachusetts. *Limnol. Oceanogr.* 26, 518, 1981.

81. McNight, D.M., Chisholm, S.W., and Harleman, D.R.F. $CuSO_4$ treatment of nuisance algal blooms in drinking water reservoirs. *Environ. Manage.* 7, 311, 1983.

82. Moore, M.V. and Winner, R.W. Relative sensitivities of *Ceriodaphnia dubia* laboratory tests and pond communities of zooplankton and benthos to chronic copper stress. *Aquatic Toxicol.* 15, 311, 1989.

83. Taub, F.B. et al. Effects of seasonal succession and grazing on copper toxicity in aquatic microcosms. *Verh. Int. Ver. Limnol.* 24, 2205, 1989.

84. Winner, R.W. and Owen, H.A. Seasonal variability in the sensitivity of freshwater phytoplankton communities to a chronic copper stress. *Aquatic. Toxicol.* 19, 73, 1991.

85. Past, M.H. and Boyer, M.G. Effects of two organophosphorous insecticides on chlorophyll a and pheopigment concentrations of standing ponds. *Hydrobiology* 69, 245, 1980.

86. Hughes, D.N. Boyer, M.G., Past, M.H., Fowle, C.D. Persistence of three organophosphorus insecticides in artificial ponds and some biological implication. *Arch. Environ. Contam. Toxicol.*, 9, 269, 1980.

87. Kaushik, N.K., Stephenson, G.L., Solomon, K.R., Day, K.E. Impact of permethrin on zooplankton communities in limnocorrals, *Can. J. Fish. Aquat. Sci.* 42, 77, 1985.

88. Hanazato, T. and Yasuno, M. Effects of a carbonate insecticide, Carbaryl, on summer phyto- and zoo-plankton communities in ponds. *Environ. Pollut.* 48, 145, 1987.

89. Hanazato, T. and Yasuno, M. Influence of time of application of an insecticide on recovery patterns of a zooplankton community in experimental ponds. *Arch. Environ. Contam. Toxicol.* 19, 77, 1990.

90. Hanazato, T. Effects of repeated application of Carbaryl on zooplankton communities in experimental ponds with or without the predator Chaoborus. *Environ. Pollut.* 73, 309, 1991.

91. Jørgensen, S.E. A eutrophication model for a lake. *Ecol. Model.*, 2, 147, 1976.

92. Chen, C.W. and Orlob, G.T. "Ecological simulation of aquatic environments," in *Systems Analysis and Simulation in Ecology*, Vol. 3., Patten B.C., Ed. Academic Press, New York, 1975, 476 p.

93. Jørgensen, S.E., Nielsen, S.N., and Jørgensen, L.A. *Handbook of Ecological Parameters and Ecotoxicology.* Elsevier, Amsterdam, 1991, 380 p.

94. Steele, J.H. *The Structure of the Marine Ecosystems.* Blackwell, Oxford, 1974, 128 p.

95. Biermen, V.J., Verhoff, F.H., Poulson, T.C., Tenney, M.W. "Multinutrient dynamic models of algal growth and species competition in eutrophic lakes," in *Modeling the Eutrophication Process*, Middlebrooks, E., Falkenberg, D.H., and Maloney T.E., Eds. Ann Arbor Science, Ann Arbor, MI, 1974, 89.

CHAPTER 6

Ecosystem Health Assessment and Bioeconomic Analysis in Coastal Lagoons

J.M. Zaldívar, M. Austoni, M. Plus, G.A. De Leo,
G. Giordani, and P. Viaroli

In order to study the management options in a coastal lagoon with intensive shellfish (*Tapes philippinarum*) farming and macroalgal (*Ulva* sp.) blooms, a biogeochemical model has been developed. The model considers the nutrient cycles and oxygen in the water column as well as in the sediments, phytoplankton, zooplankton, and *Ulva* sp. dynamics. Furthermore, a discrete stage-based model for the growth of *Tapes philippinarum* has been coupled with this continuous biogeochemical model. By studying the growth of clams, it considers the nutrient contents in the water column as well as its temperature, including the effects of harvesting and the mortality due to anoxic crisis. The results from 1989 to 1999 show that the model is able to capture the essential dynamics of the lagoon, with values in the same order of magnitude as the measurements from experimental campaigns and with data on clam productivity. The model has therefore been used to assess the effects

1-56670-665-3/05/$0.00 + $1.50

of *Ulva*'s mechanical removal on the lagoon's eutrophication level using the exergy and specific exergy, as well as economic factors in terms of operating vessel costs and averaged prices for clams as optimization parameters. The results show that a combination of ecosystem models and health indicators constitute a sound method for optimizing the management in such complex systems.

6.1 INTRODUCTION

Coastal lagoons are subjected to strong anthropogenic pressures. This is partly due to freshwater inputs rich in organic and mineral nutrients derived from urban, agricultural, or industrial effluents and domestic sewage, but also due to the intensive shellfish farming some of them support. For example, the Thau lagoon in southern France is an important site for the cultivation of oysters (*Crassostrea gigas*) and mussels (*Mytilus galloprovincialis*) (Bacher et al., 1995). The Adriatic lagoons in northern Italy — the namely the Venice, Scardovari and Sacca di Goro lagoons — supported a production of around 58,000 metric tonnes of clams (*Tapes philippinarum*) in 1995 (Solidoro et al., 2000), etc. The combination of all these anthropic pressures call for an integrated management that considers all the different aspects, from lagoon fluid dynamics, ecology, nutrient cycles, river runoff influence, shellfish farming, macro-algal blooms, sediments, as well as the socio-economical implications of different possible management strategies. However, histori-cally, coastal lagoons have been suffering from multiple and uncoordinated modifications undertaken with only limited sectorial objectives in mind — for example, land-use modifications on the watershed affecting the nutrient loads into the lagoon; modifications in lagoon bathymetry by dredging or changing the water circulation in the lagoon, and so on. All these factors are responsible for important disruptions in ecosystem functioning characterized by eutrophic and dystrophic conditions in summer (Viaroli et al., 2001), algal blooms, oxygen depletion and sulfide production (Chapelle et al., 2001).

Obviously, to carry out such an integrated approach, biogeochemical models that take into account the different mechanisms and important variables in the ecosystem are fundamental. These models are able to handle the complex link between human activities and the ecosystem functioning, something that is not possible to capture with more traditional statistical tools. However, in order to analyze the model results, it is necessary to use ecological indicators that will allow a comparison of the health of the ecosystem from several scenario analyses. Historically, the health of an ecosystem has been measured using indices of particular species or components; for example, macrophytes and zooplankton. Such indices are generally inadequate because they are not broad enough to reflect the complexity of ecosystems. It is therefore necessary for the indicators to include structural, functional, and system-level aspects. To cope with these aspects, new indices have been developed (for a recent review see Rapport (1995)). Exergy and related values — that is, structural exergy, specific exergy, etc. — have recently been used to assess ecosystem health in freshwater ecosystems (Xu et al., 1999) as well as marine ecosystems (Jørgensen, 2000).

We have studied, based on previously developed models (Zaldívar et al., 2003a, 2003b) for Sacca di Goro, the effects of *Ulva*'s mechanical removal on the lagoon's eutrophication level using specific exergy (Jørgensen, 1997), and costs and benefits (De Leo et al., 2002; Cellina et al., 2003). The costs are associated with the normal operation of the vessels and with the disposal of the collected *Ulva* biomass whereas the benefits consider the increased productivity of shellfish as well as the decrease in mortality due to anoxic crises. For analyzing the ecosystem health we used specific exergy calculated in terms of biomass of the different model's variables and its information content (Jørgensen, 1997). The comparison between both approaches has allowed us to develop a management strategy that improves the ecosystem health in Sacca di Goro and at the same time reduces the economic losses associated with clam mortality during anoxic crises.

6.2 STUDY AREA: SACCA DI GORO

The Sacca di Goro (see Figure 6.1) is a shallow-water embayment of the Po Delta ($44° 47'$ to $44° 50'$ N and $12° 15'$ to $12° 20'$ E). The surface area is 26 km^2 and the total water volume is approximately $40 \times 10^6 \text{ m}^3$. Numerical models (O'Kane et al., 1992) have demonstrated a clear zonation of the lagoon with the low-energy eastern area separated from two higher-energy zones, including both the western area influenced by freshwater inflow from the Po di Volano and the central area influenced by the Adriatic Sea. The eastern zone (called Valle di Gorino) is very shallow (with a maximum depth of 1 m) and

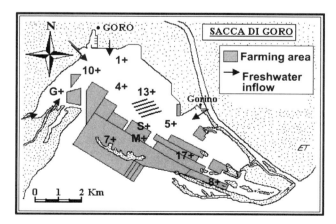

Figure 6.1   General layout of Sacca di Goro with the main farming areas indicated in gray and freshwater inflows by arrows.

accounts for one-half of the total surface area and for one fourth of the water volume of the lagoon.

The bottom is flat and the sediment is composed of typical alluvial mud with a high clay and silt content in the northern and central zones, while sand is more abundant near the southern shoreline, and sandy mud is predominates in the eastern area.

The watershed, Burana-Volano, is a lowland, flat basin located in the Po Delta and covering an area of about $3000 \, \text{km}^2$. On the northern and eastern side it is bordered by a branch of the Po River entering the Adriatic Sea. A large part of the catchment area is below sea level with an average elevation of $0 \, \text{m}$, a maximum elevation of $24 \, \text{m}$ and a minimum of $-4 \, \text{m}$. About 80% of the watershed is dedicated to agriculture. All the land is drained (irrigated) through an integrated channel network and various pumping stations. Point and nonpoint pollution sources discharge a considerable amount of nutrients in the lagoon from small tributaries and drainage channels (Po di Volano and Canal Bianco).

The catchment is heavily exploited for agriculture, while the lagoon is one of the most important aquacultural systems in Italy. About $10 \, \text{km}^2$ of the aquatic surface are exploited for Manila clam (*Tapes philippinarum*) farming, with an annual production of about 8000 metric tons (Figure 6.2). Fish and

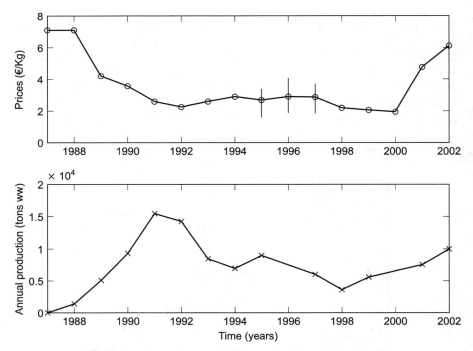

Figure 6.2 Averaged prices for *Tapes philippinarum* in the northern Adriatic (Bencivelli, private communication and Solidoro et al., 2000) and time evolution of estimated clams annual production in Sacca di Goro (Bencivelli, private communication).

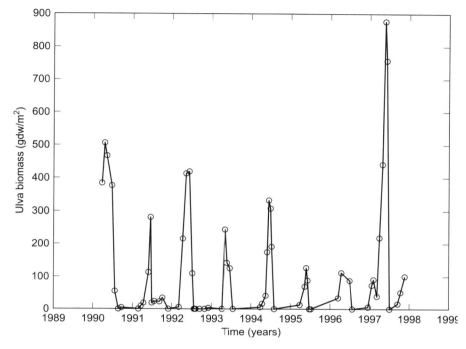

Figure 6.3   Measured annual trends of *Ulva* biomasses in the water column in the sheltered zone of Sacca di Goro (Viaroli et al., 2001).

shellfish production provides work, directly or indirectly, for 5000 people. The economical annual revenue has been varying during the last few years around €100 million.

Water quality is a major problem due to: (1) the large supply of nutrients, organic matter, and sediments that arrive from the freshwater inflows; (2) the limited water circulation due to little water exchange with the sea (total water exchange time is between 2 to 4 days); and (3) the intensive shellfish production. In fact from 1987 to 1992 the Sacca di Goro experienced an abnormal proliferation of macroalgae (*Ulva* sp.), which gradually replaced phytoplankton populations (Viaroli et al., 1992) (see Figure 6.3). This was a clear symptom of the rapid degradation of environmental conditions and of an increase in the eutrophication of this ecosystem.

The decomposition of *Ulva* in summer (at temperatures of 25 to 30°C) produces the depletion of oxygen (Figure 6.4) that can lead to anoxia in the water column. In the beginning of August 1992, after a particularly severe anoxic event that resulted in a high mortality of farmed populations of mussels and clams, a 300- to 400-m-wide, 2-m-deep channel was cut through the sand bank to allow an increase in the sea water inflow and the water renewal in the Valle di Gorino. This measure temporarily solved the situation — during the following years a reduction of the *Ulva* cover (Viaroli et al., 1995) and a clear

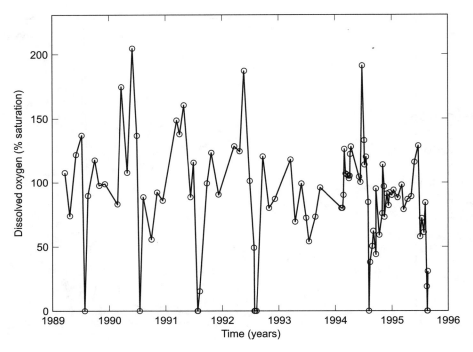

Figure 6.4 Experimental annual trends of dissolved oxygen saturation concentrations in the water column in the sheltered zone of Sacca di Goro (Viaroli et al., 2001).

increase in phytoplankton biomass values were observed (Sei et al., 1996). However, in 1997 another anoxic event took place when an estimated *Ulva* biomass of 100,000 to 150,000 metric tons (enough to cover half of the lagoon) started to decompose. The economic losses due to mortality of the farmed clam populations were estimated at around €7.5–10 million (Bencivelli, 1998).

6.3 SIMULATION MODELS

6.3.1 Biogeochemical Model

A model of the Sacca di Goro ecosystem has been developed and partially validated with field data from 1989 to 1998 (Zaldívar et al., 2003a). The model considers the nutrient cycles in the water column and in the sediments as schematically shown in Figure 6.5. Nitrogen (nitrates plus nitrites and ammonium) and phosphorous have been included into the model, since these two nutrients are involved in phytoplankton growth in coastal areas. Silicate has been introduced to distinguish between diatoms and flagellates, whereas consideration of the dissolved oxygen was necessary in order to study the evolution of hypoxia and the anoxic events that have occurred in the Sacca di Goro during the past few years.

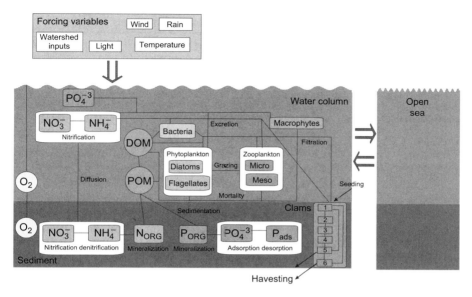

Figure 6.5 General schema of the biogeochemical model for Sacca di Goro.

With regard to the biology, the model considers two types of phytoplankton and zooplankton communities. The phytoplankton model, based on the Aquaphy model (Lancelot et al., 2002), explicitly distinguishes between photosynthesis (directly dependent on irradiance and temperature) and phytoplankton growth (dependent on both nutrients and energy availability). The microbial loop includes the release of dissolved and particulate organic matter with two different classes of biodegradability into the water (Lancelot et al., 2002). Detrital particulate organic matter undergoes sedimentation. Furthermore, the evolution of bacteria biomass is explicitly taken into account.

In shallow lagoons, sediments play an important role in biogeochemical cycles (Chapelle et al., 2000). The sediments have several roles: they act as sinks of organic detritus material through sedimentation and they consume oxygen and supply nutrients through bacterial mineralization, nitrification and benthic fauna respiration. Indeed, depending on the dissolved oxygen concentration, nitrification or denitrification takes place in sediments, and for the phosphorous the sediments usually act as a buffer through adsorption and desorption processes. For all these reasons, the model considers the sediments to be dynamic.

Ulva sp. has become an important component of the ecosystem in Sacca di Goro. The massive presence of this macroalgae has heavily affected the lagoon ecosystem and has prompted several interventions aimed at removing its biomass in order to avoid anoxic crises, especially during the summer. In this case, *Ulva* biomass as well as the nitrogen concentration in macroalgae tissues are considered as other state variables (Solidoro et al., 1997).

Table 6.1 State variables used and units in the biogeochemical model

Variable name	Unit	Variable name	Unit
Inorganic nutrients, water column		Biological variables, Water column	
Nitrate	mmol NO_3^-/m^3	Micro-phytopk (20–200 μm):	
Ammonium	mmol NH_4^+/m^3	Diatoms	mg C/m^3
Reactive phosphorous	mmol PO_4^{3-}/m^3	Flagellates	mg C/m^3
Silicate	mmol $Si(OH)_4$/m^3	Micro-zoopk. (40–200 μm)	mg C/m^3
Dissolved oxygen	g O_2/m^3	Meso-zoopk. (>200 μm)	mg C/m^3
		Bacteria	mg C/m^3
Organic matter (OM), water column		*Ulva*	g dw**/l
Monomeric dissolved OM (C)	mg C/m^3	Nitrogen in *Ulva* tissue	mg N/g dw
Monomeric dissolved OM (N)	mmol N/m^3	Sediments (i. w. = interstitial waters)	
Detrital biogenic silica	mmol Si/m^3	Ammonium (i. w.)	mmol/m^3
		Nitrate (i. w.)	mmol/m^3
High biodegradability:		Phosphorous (i. w.)	mmol/m^3
Dissolved polymers (C)	mg C/m^3	Inorganic adsorbed phosphor.	μg P/g PS***
Dissolved polymers (N, P)	mmol N, P/m^3	Dissolved oxygen (i. w.)	g O_2/m^3
Particulate OM (C)	mg C/m^3	Organic particulate phosphor.	μg P/g PS
Particulate OM (N, P)	mmol N, P/m^3	Organic particulate nitrogen	μg N/g PS
Low biodegradability:			
Dissolved polymers (C)	mg C/m^3		
Dissolved polymers (N, P)	mmol N, P/m^3		
Particulate OM (C)	mg C/m^3		
Particulate OM (N, P)	mmol N, P/m^3		

**g dw is gram-dry-weight, ***PS stands for Particulate Sediment — i.e., dry sediment.

The state space of dynamical variables considered is summarized in Table 6.1. We consider 38 state variables: there are 5 for nutrients in the water column and 5 in the sediments; organic matter is represented by 15 state variables in the water column and 2 in the sediments; 11 state variables represent the biological variables: 6 for phytoplankton, 2 for zooplankton, 1 for bacteria and 2 for *Ulva*.

6.3.2 Discrete Stage-Based Model of *Tapes Philippinarum*

Knowing the importance of *Tapes philippinarum* in the Sacca di Goro ecosystem, it is clear that a trophic model that takes into account the effect of shellfish farming activities in the lagoon is necessary. For this reason a discrete stage-based model has been developed (Zaldívar et al., 2003b). The model considers six stage-based classes (see Figure 6.5). The first one corresponds to typical seeding sizes whereas the last two correspond to the marketable sizes "medium" (37 mm) and "large" (40 mm) according to Solidoro et al. (2000). The growth of *Tapes philippinarum* is based on the continuous growth model from Solidoro et al. (2000) that depends on the temperature and phytoplankton in the water column. This model has been transformed into a variable stage duration for each class in the discrete stage-based model. Furthermore, the effects of harvesting as well as the mortality due to anoxic crisis are taken into account by appropriate functions, as well as the evolution of cultivable area and the seeding and harvesting strategies in use in Sacca di Goro.

6.3.3 Ulva's Harvesting Model

In order to model the *Ulva* biomass harvested by one vessel per unit of time, we followed the model developed by De Leo et al. (2002) assuming that the vessel harvesting capacity, q, is 1.3×0^{-5} g dry weight per 1 (gdw/l) per hour, which corresponds approximately to 100 metric tons of wet weight of *Ulva* per day. Therefore, we have incorporated into the *Ulva*'s model a term that takes this into account:

$$H(U, E) = \begin{cases} q \times E \times R(U) & \text{if} \quad U(t) \geq U_{th} \\ 0 & \text{if} \quad U(t) < U_{th} \end{cases} \qquad (6.1)$$

where E is the number of vessels, U is the *Ulva* biomass (gdw/l) and U_{th} is the threshold density above which the vessels start to operate. R is a function developed by Cellina et al. (2003) to take into account that the harvesting efficiency of vessels decreases when algal density is low. R was defined as:

$$R(U) = \frac{U^2}{U^2 + \delta} \qquad (6.2)$$

where δ is the semisaturation constant set to 2.014×10^{-4} (gdw^2/l^2) according to Cellina et al. (2003).

The function $H(U, E)$ acts as another mortality factor in the *Ulva* equation, with the difference that the resulting organic matter is not pumped into the microbial loop but is removed from the lagoon. The removal of this organic matter decreases the severity and number of anoxic crises in the lagoon and therefore reduces mortality in the clam population.

6.3.4 Cost/Benefit Model

The direct costs of *Ulva* harvesting have been evaluated to be €1000 per vessel per day including fuel, wages, and insurance whereas the costs of biomass disposal are in the range of 150 €/metric ton of *Ulva* wet weight (De Leo et al., 2002). Damage to shellfish production caused by *Ulva* is due to oxygen depletion and the subsequent mortality increase in the clam population. To take into account this factor we have evaluated the total benefits obtained from simulating the biomass increase using the averaged prices for *Tapes philippinarum* in the northern Adriatic (Figure 6.2). Therefore an increase in clam biomass harvested from the lagoon will result in an increase in benefits. The total value obtained (**CB = Costs −**
Benefits) is the difference between the costs associated with the operation of

the vessels as well as the disposal of the harvested *Ulva* biomass minus the profits obtained by selling the shellfish biomass harvested in Sacca di Goro.

6.3.5 Exergy Calculation

The definitions and calculations of exergy and structural exergy (or specific exergy) are discussed in chapter 2.

The Sacca di Goro model considers several state variables for which the exergy should be computed. These are: organic matter (detritus), phytoplankton (diatoms and flagellates), zooplankton (micro- and meso-), bacteria, macroalgae (*Ulva* sp.) and shellfish (*Tapes philippinarum*). The exergy was calculated using the data from Table 6.2 on genetic information content and all biomasses were reduced to gdw/l using the parameters in Table 6.3.

Table 6.2 Parameters used to evaluate the genetic information content, from Jørgensen (2000)

Ecosystem component	Number of information genes	Conversion factor (W_i)
Detritus	0	1
Bacteria	600	2.7 (2)
Flagellates	850	3.4 (25)
Diatoms	850	3.4
Micro-zooplankton	10000	29.0
Meso-zooplankton	15000	43.0
Ulva sp.	2000[*]	6.6
Shellfish (Bivalves)	—	287[†]

[*]Coffaro *et al.* (1997), [†]Marques et al. (1997), Fonseca et al. (2000).

Table 6.3 Parameters used for the calculation of the exergy for the Sacca di Goro lagoon model

	C:dw (gC/gdw)	$-\ln P_i$
Detritus	—	7.5×10^5
Bacteria	0.4	12.6×10^5
Diatoms	0.22	17.8×10^5
Flagellates	0.22	17.8×10^5
Micro-zooplankton	0.45	209.7×10^5
Meso-zooplankton	0.45	314.6×10^5
Ulva	—	41.9×10^5
Shellfish	—	2145×10^5

6.4 RESULTS AND DISCUSSION

6.4.1 The Existing Situation

Sacca di Goro has been suffering from anoxic crises during the warm season. Such crises are responsible for considerable damage to the aquaculture

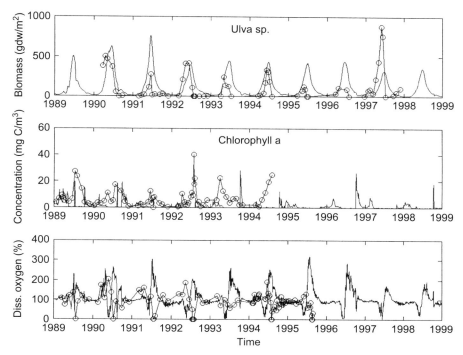

Figure 6.6 Experimental and simulated *Ulva* biomasses; Chlorophyl-a and oxygen concentration in Sacca di Goro.

industry and to the ecosystem functioning. In order to individuate the most effective way to avoid such crises, it is important to understand the processes leading to anoxia in the lagoon. Figure 6.6a shows the experimental and simulated *Ulva* biomasses. The model is able to predict the *Ulva* peaks and for some years their magnitude. For comparing experimental and simulated results we have assumed a constant area in the lagoon of $16.5\,km^2$. As has been observed in Viaroli et al. (2001), the rapid growth of *Ulva* sp. in spring is followed by a decomposition process, usually starting from mid-June. This decomposition stimulates microbial growth. The combination of organic matter decomposition and microbial respiration produces anoxia in the water column, mostly in the bottom water. This is followed by a peak of soluble reactive phosphorous that is liberated from the sediments.

Oxygen evolution in the water column is highly influenced by the *Ulva* dynamics. In fact, high concentrations are simulated in corresponding high algal biomass growth rates. Furthermore, when the *Ulva* biomass starts to decompose the oxygen starts to deplete. Experimental and simulated data are shown in Figure 6.6c. As can be seen, anoxic crises have occurred practically every year in the lagoon.

Figure 6.7 shows the comparison between the estimated and simulated total clam biomass in Sacca di Goro. It can be seen that there is a general agreement between experimental and estimated values. Oxygen also has a strong influence

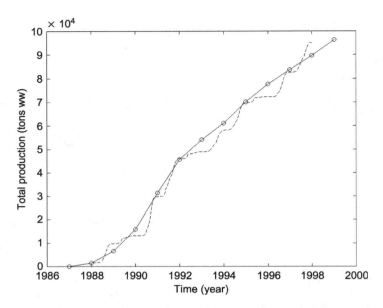

Figure 6.7 Estimated (from Bencivelli, personal communication; continuous line) and simulated (discontinuous line) total production of *Tapes philippinarum*.

on *Tapes philippinarum* dynamics since anoxic crises are responsible for high mortality in the simulated total population (see Figure 6.8). Furthermore, population dynamics in the first stages is controlled by the seeding strategy performed in the lagoon. According to Castaldelli (private communication) there are two one-month seeding periods. The first begins in March; the second from mid-October to mid-November. The dynamics in Class 5 and 6, which correspond to marketable sizes, are controlled by harvesting, since in the model they are harvested all year with an efficiency of 90% and 40%, respectively.

Figure 6.9 shows the values calculated for the exergy and specific exergy. It can be seen that the calculations do not show the annual cycles one should expect in the lagoon, with low exergy during the winter and autumn accompanied by an increase during spring and summer. This is due to the fact that the exergy is practically controlled by shellfish biomass. This can be in Figure 6.10, where the contribution to the exergy of the different variables in the model is plotted as a percentage. Concerning specific exergy there is less variation. The changes are due to the effects of anoxic crises that affect the biomass distribution. As can be seen in Figure 6.10 there are localized peaks of *Ulva* in correspondence with the decrease in *Tapes philippinarum* biomass due to an increase in mortality during anoxic episodes.

6.4.2 Harvesting *Ulva* Biomass

A measure that has been taken in Sacca di Goro to control macroalgal blooms consists of harvesting vessels that remove *Ulva* in zones where clam

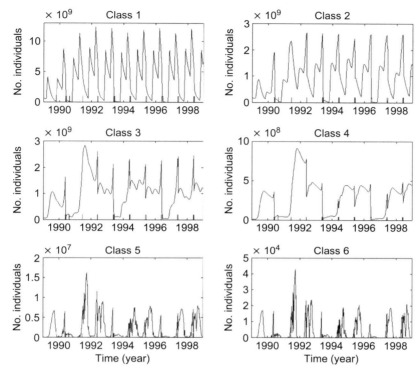

Figure 6.8 Simulated *Tapes philippinarum* population dynamics in Sacca di Goro. The simulated anoxic periods (oxygen concentration below 2 mg/l) are indicated by small bars.

fishery is located. However, it was not clear how the vessels should operate to reduce their costs and obtain the maximum benefit for the shellfish industry. In a series of recent studies, De Leo et al. (2002) and Cellina et al. (2003) developed a stochastic model that allowed the assessment of harvesting policies in terms of cost-effectiveness — that is, the number of vessels and the *Ulva* biomass threshold at which the harvesting should start.

In this study, we have inserted their cost model in the coupled continuous biogeochemical model and discrete stage-based *Tapes philippinarum* population models. Furthermore, no specific functions for evaluating the effects of anoxic crises on *Ulva* and clam dynamics have been introduced. Benefits are calculated as a function of the number of harvested clams in the lagoon and their selling price (see Figure 6.2b).

Several hundreds of simulation runs from 1989 to 1994, using the same initial conditions and forcing functions, have been carried out in order to estimate the optimum solution in terms of costs and benefits, number of operational vessels (from 0 to 20 vessels) and ecosystem (specific exergy) improvement at different *Ulva* biomass thresholds (0.01 gdw/l to 0.16 gdw/l, which corresponds approximately to 20 gdw/m² and 380 gdw/m², respectively).

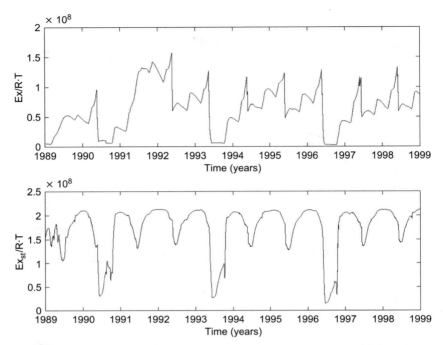

Figure 6.9 Computed exergy (g/l) and specific exergy for the Sacca di Goro model, from 1989 to 1998. Parameters used for the calculation of the genetic information content are given in Table 2.

The results are summarized in Figure 6.11 and Figure 6.12, which show how the relative estimated costs and benefits, $(Cb_i - CB_0)/CB_0$, and specific exergy improvement Ex_{st}^i/Ex_{st}^0, where 0 refers to the existing situation and i to the specific number of vessels and *Ulva* biomass threshold change as a function of the number of boats and different *Ulva* biomass thresholds. The optimum solution would be the one with lower costs and higher specific exergy improvement.

As can be seen from Figure 6.11, there is an optimal solution concerning the costs and benefits: work at low *Ulva* biomass thresholds (0.02 to 0.03 gdw/l (50 to 70 gdw/m^2)) with 10 to 12 vessels — that is, 0.6 to 0.7 vessels/km^2 operating in the lagoon. These values are in agreement with previous studies. De Leo et al. (2002) obtained around 0.5 vessels/km^2 and *Ulva* threshold between 70 to 90 gdw/m^2, whereas Cellina et al. (2003) found values between 50 to 75 gdw/m^2 for 6 to 10 vessels operating in the lagoon.

For the case of relative specific exergy (see Figure 6.12), there is not a global maximum since relative specific exergy continues to increase as we increase the number of vessels operating in the lagoon at low *Ulva* biomass thresholds. However, the optimal solution from the cost/benefit analysis would improve the specific exergy by approximately 21% in comparison with the "do nothing" strategy. The maximum improvement calculated is around 25%.

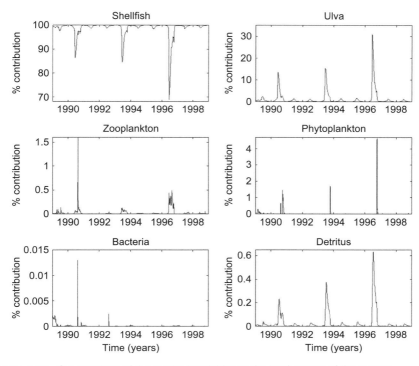

Figure 6.10   Contributions of the models' variables to the total exergy of the system.

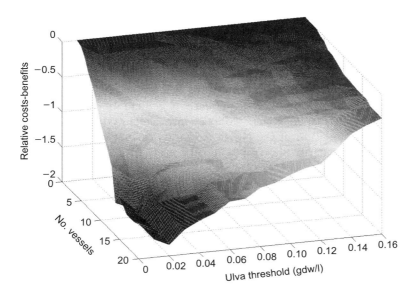

Figure 6.11   Simulated results in terms of relative costs and benefits in Sacca di Goro by changing the number of vessels and the *Ulva* biomass threshold at which they start to operate.

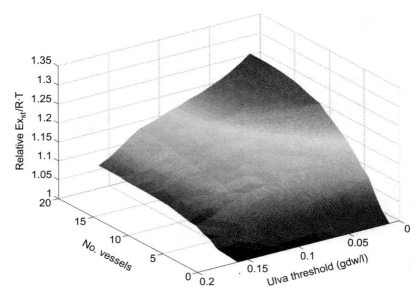

Figure 6.12 Simulated results in terms of relative specific exergy improvement in Sacca di Goro by changing the number of vessels and the *Ulva* biomass threshold at which they start to operate.

6.4.3 Reduction in Nutrient Inputs

Another possible measure to improve the ecosystem functioning would be to reduce the nutrient loads in Sacca di Goro. For this study, we established a scenario that considers the reduction in nutrient loads arriving from Po di Volano, Canale Bianco and Po di Goro compared to the maximum values established by national Italian legislation (based on EU Nitrate Directive) for Case III (poor quality, polluted (NH_4^+ < 0.78 mg N/l, NO_3^- < 5.64 mg N/l, PO_4^{3-} < 0.17 mg P/l)). Furthermore, we have not considered the improvement that the Adriatic Sea should experience if reduction in nutrient loads is accomplished in the Po River. To take into account these effects a three-dimensional simulation of the North Adriatic Sea that considers the nutrient load reduction scenarios should be carried out in order to properly account for these effects in our model.

Figure 6.13 and Figure 6.14 present the evolution of exergy and specific exergy under the two proposed scenarios: *Ulva* removal and nutrient load reduction, in comparison with the "do nothing" alternative. As can be seen, the exergy and specific exergy of both scenarios increase. This is due to the fact that in our model both functions are dominated by clam biomass. This implies that the biomass of *Tapes philippinarum* in Sacca di Goro would have increased whatever the scenario used. This can be seen in Figure 6.15, where the optimal solution in terms of operating vessels would have been multiplied by approximately a factor of three the harvested *Tapes philippinarum* biomasses with the subsequent economic benefits.

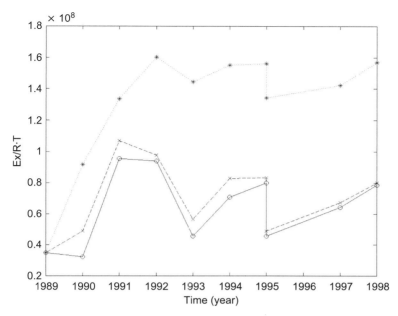

Figure 6.13   Exergy mean annual values: (a) present scenario (continuous line); (b) removal of *Ulva*, optimal strategy from the cost/benefit point of view (dotted line); (c) nutrient load reduction from watershed (dashed line).

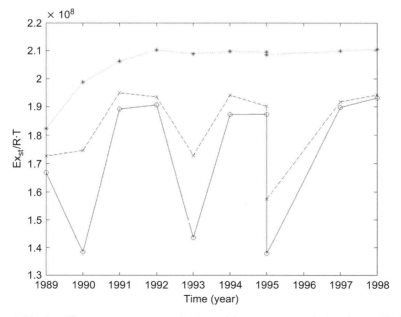

Figure 6.14   Specific exergy mean annual values: (a) present scenario (continuous line); (b) removal of *Ulva*, optimal strategy from the cost/benefit point of view (dotted line); (c) nutrient load reduction from watershed (dashed line).

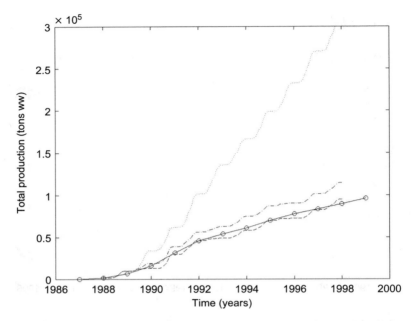

Figure 6.15   Estimated (bencivelli, personal communication) and simulated: (a) total production of *Tapes philippinarum* present scenario (dashed line); (b) removal of *Ulva*, optimal strategy from the cost/benefit point of view (dotted line); (c) nutrient load reduction from watershed (dashed/dotted line).

An evaluation of the costs associated with a reduction in nutrient load is beyond the scope of this paper. However, this evaluation should be carried out when the Water Framework Directive (WFD enters in force, but taking into account the dimensions and importance of the Po River the costs will certainly be higher than the removal of *Ulva* by vessels.

6.5 CONCLUSIONS

The results of the model are in general in good agreement with the stochastic models developed by De Leo et al. (2002) and Cellina et al. (2003). All of these results point towards starting macroalgae removal earlier, when *Ulva* biomasses are relatively low. At higher biomasses, due to the high growth rates of *Ulva* and the nutrient availability in Sacca di Goro, it is more difficult to prevent the anoxic crises. From the point of view of improving specific exergy in the Goro lagoon the best approach would consist of using the maximum amount of vessels operating at thresholds as low as possible. However, the optimal result from the cost/benefit analysis will considerably improve the ecological status of the lagoon in terms of specific exergy. The nutrient reduction scenario considers a small reduction and other more realistic scenarios will be implemented after the first results from the application of the WFD to Italian watershed will become available (Cinnirella et al., 2003).

The assessment of the health of an ecosystem is not an easy task and it may be necessary to apply several indicators simultaneously to obtain a proper estimation. Several researchers have proposed different indicators that cover different aspects of the ecosystem health, but it seems clear that only a coherent application of them would lead us to have a correct indication of the analyzed ecosystem. Between these indicators, exergy expresses the biomass of the system and the genetic information that this biomass is carrying, and specific exergy will tell us how rich on information the system is. These indicators are able to cover a considerable amount of ecosystem characteristics and it has been shown that they are correlated with several important parameters as respiration, biomass, etc. However, it has been found (Jørgensen, 2000) that exergy is not related to biodiversity, and, for example, a very eutrophic system often has a low biodiversity but high exergy.

It seems also clear that both values would give a considerable amount of information when analyzing the ecological status of inland and marine waters as requested by the WFD. However, there is still work to be done in two areas. The first consists on standardizing the genetic information content for the species occurring in EU waters and hence allowing a uniform calculation of exergy, which will allow a useful comparison between studied sites. The second are consists of developing a methodology that would allow the calculation of exergy from monitoring data, already considered in Annex V of the WFD. Unfortunately, ecological data in terms of biomasses of important elements in an ecosystem are not normally available and therefore an important aspect would be to study how to use the physico-chemical parameters (normally the values for which most historical data is currently available) for the calculation of the exergy and specific exergy of a system.

Finally, in order to transform the concept of exergy into an operational tool as an ecological indicator on inland and marine waters, it is necessary to develop a methodology that allows its calculation when models are not available. Of course, if one has enough data on the ecosystem composition one can always calculate the exergy. However, the data that one has available consists mainly of nutrients data and phytoplankton data in the form of chlorophyll-a, which cannot directly provide a good estimation of the exergy of an ecosystem. It is clear that both aspects are related: nutrients allow the growth and development of the ecosystem and their change has a direct effect on the exergy values of our system (see Figure 6.13 and Figure 6.14). But how do we convert these monitoring parameters into a formulation that allows the calculation of exergy? It is still not clear, and the range of validity of such calculation procedure that should be tested in different ecosystems is still an open question.

When managers are confronted to select between different alternatives it is difficult to evaluate, from an ecological point of view, which the optimal solution is. As exergy and specific exergy are global parameters of the ecosystem, they give an idea of the benefits that a measure will produce.

The use of biogeochemical modelling, ecological indicators and cost/benefit analysis seems an adequate combination for developing integrated tools able to build up strategies for sustainable ecosystem management, including ecosystem restoration or rehabilitation.

ACKNOWLEDGMENTS

This research has been partially supported by the EU funded project DITTY (Development of Information Technology Tools for the management of European Southern lagoons under the influence of river-basin runoff, EVK3-CT-2002-00084) in the Energy, Environment and Sustainable Development programme of the European Commission.

REFERENCES

Bacher, C., Bioteau, H., and Chapelle, A. Modelling the impact of a cultivated oyster population on the nitrogen dynamics: the Thau lagoon case (France). *Ophelia* 42, 29–54, 1995.

Bencivelli, S. La Sacca di Goro: La situazzione di emergenza dell'estate 1997. In *Lo Stato Dell'Ambiente Nella Provincia di Ferrara*. Amministrazione Provinciale di Ferrara. Servizio Ambiente, 1998, pp. 61–66.

Cavalier-Smith, T. *The Evolution of the Genome Size*. Wiley, Chichester, 1985.

Cellina, F., De Leo, G.A., Rizzoli, A. E., Viaroli, P., and Bartoli, M. Economic modelling as a tool to support macroalgal bloom management: a case study (Sacca di Goro, Po river delta). *Oceanologica Acta* 26, 139–147, 2003.

Chapelle, A., Lazuer, P., and Souchu, P. Modelisation numerique des crises anoxiques (malaigues) dans la lagune de Thau (France). *Oceanologica Acta* 24 (Supp. 1), s99–s112, 2001.

Chapelle, A., Ménesguen, A., Deslous-Paoli, J. M., Souchu, P., Mazouni, N., Vaquer, A., and Millet, B. Modelling nitrogen, primary production and oxygen in a Mediterranean lagoon. Impact of oysters farming and inputs from the watershed. *Ecol. Model.* 127, 161–181, 2000.

Cinnirella, S., Trombino, G., and Pirrone, N. A multi-layer database for the po catchment-adriatic costal zone continuum structured for the DPSIR framework developed for the implementation of the Water Framework Directive. *Proceedings of the European Conference on Coastal Zone Research: an ELOISE Approach*, Gdansk, Poland, 24–27 March 2003. 116 p.

Coffaro, G., Bocci, M., and Bendoricchio, G. Application of structural dynamical approach to estimate space variability of primary producers in shallow marine waters. *Ecol. Model.* 102, 97–114, 1997.

Colombo, G., Bisceglia, R., Zaccaria, V., and Gaiani, V. "Variazioni spaziali e temporali delle caratteristiche fisicochimiche delle acque e della biomassa fitoplantonica della Sacca di Goro nel quadriennio 1988–1991," in *Sacca di Goro: Studio Integrato Sull'Ecologia*, Bencivelli, S, Castaldi, N., and Finessi, D., Eds. Provincia di Ferrara, FrancoAngeli, Milano, 1994, pp. 9–82.

De Leo, G., Bartoli, M., Naldi, M., and Viaroli, P. A first generation stochastic bioeconomic analysis of algal bloom control in a coastal lagoon (Sacca di Goro, Po river Delta). *Mar. Ecol.* 410, 92–100, 2002.

Fonseca, J.C., Marques, J.C., Paiva, A.A., Freitas, A.M., Madeira, V.M.C., and Jørgensen, S.E. Nuclear DNA in the determination of weighing factors to estimate exergy from organisms' biomass. *Ecol. Model.* 126, 179–189, 2000.

Jørgensen, S.E. *Integration of Ecosystem Theories: A Pattern*, 2nd ed. Kluwer, Dordrecht, 1997.

Jørgensen, S.E. Application of exergy and specific exergy as ecological indicators of coastal areas. *Aquatic Ecosyst. Health Manage.* 3, 419–430, 2000.

Jørgensen, S.E., Nielsen, S.N., and Mejer, H. Energy, environment, exergy and ecological modelling. *Ecol. Model.* 77, 99–109, 1995.

Lancelot, C., Staneva, J., Van Eeckhout, D., Beckers, J.M., and Stanev, E. Modelling the Danube-influenced north-western continental shelf of the Black Sea. II. Ecosystem response to changes in nutrient delivery by the Danube river after its damming in 1972. *Est. Coast. Shelf Sci.* 54, 473–499, 2002.

Levin, B. *Genes V.* Oxford University Press, London, 1994.

Li, W. H. and Grau, D. *Fundamentals of Molecular Evolution.* Sinauer Associates, Sunderland, MA, 1991.

Marques, J.C., Pardal, M.A., Nielsen, S.N., and Jørgensen, S.E. Analysis of the properties of exergy and biodiversity along an estuarine gradient of eutrophication. *Ecol. Model.* 102, 155–167, 1997.

O'Kane, J.P., Suppo, M., Todini, E., and Turner, J. Physical intervention in the lagoon Sacca di Goro. An examination using a 3-D numerical model. *Sci. Total Envir.* (Suppl.), 489–509, 1992.

Rapport, D.J. Ecosystem health: exploring the territory. *Ecosyst. Health* 1, 5–13, 1995.

Sei, S., Rossetti, G., Villa, F., and Ferrari, I. Zooplankton variability related to environmental changes in a eutrophic coastal lagoon in the Po Delta. *Hydrobiologia* 329, 45–55, 1996.

Solidoro, C., Pastres, R., Melaku Canu, D., Pellizzato, M., and Rossi, R. Modelling the growth of Tapes philippinarum in Northern Adriatic lagoons. *Mar. Ecol. Prog. Ser.* 199, 137–148, 2000.

Solidoro, C., Brando, V. E., Dejak, C., Franco, D., Pastres, R., and Pecenik, G. Long term simulations of population dynamics of *Ulva r.* in the lagoon of Venice. *Ecol. Model.* 102, 259–272, 1997.

Viaroli, P., Azzoni, R., Martoli, M., Giordani, G., and Tajé, L. "Evolution of the trophic conditions and dystrophic outbreaks in the Sacca di Goro lagoon (Northern Adriatic Sea)," in *Structure and Processes in the Mediterranean Ecosystems*, Faranda, F.M., Guglielmo, L., and Spezie, G., Eds., Springer-Verlag Italia, Milano, 2001, chapter 56, pp. 443–451.

Viaroli, P., Bartoli, M., Bondavalli, C., and Naldi, M. Oxygen fluxes and dystrophy in a coastal lagoon colonized by *Ulva rigida* (Sacca di Goro, Po River Delta, Northern Italy). *Fresenius Envir. Bull.* 4, 381–386, 1995.

Viaroli, P., Pugnetti, A., and Ferrari, I. "*Ulva rigida* growth and decomposition processes and related effects on nitrogen and phosphorus cycles in a coastal lagoon (Sacca di Goro, Po River Delta)," in *Marine Eutrophication and Population Dynamics*, Colombo, G., Ferrari, I., Ceccherelli, V.U., and Rossi, R., Eds. Olsen & Olsen, Fredensborg, 1992, pp. 77–84.

Water Framework Directive, Directive 2000/60/EC of the European Parliament and of the Council of 23 October 2000 establishing a framework for Community action in the field of water policy. *Offical Journal L* 327, 22/12/2000 p. 0001–0073. Available at: http://europa.eu.int/comm/environment/water/water-framework/index_en.html

Xu, F.L., Jørgensen, S.E., and Tao, S. Ecological indicators for assessing freshwater ecosystem health. *Ecol. Model.* 116, 77–106, 1999.

Zaldívar, J.M., Cattaneo, E., Plus, M., Murray, C.N., Giordani, G., and Viaroli, P. Long-term simulation in coastal lagoons: Sacca di Goro, (1989–1998). *Continent. Shelf Res.* 2003a.

Zaldívar, J.M., Catteneo, E., Plus, M., Murray, C.N., Giordani, G., and Viaroli, P. Long-term simulation of main biogeochemical events in a coastal lagoon: Sacca di Goro (Northern Adriatic Coast, Italy). *Continental Shelf Research*, 23, 1847–1876, 2003.

Zaldívar, J.M., Plus, M., Giordani, G. and Viaroli, P., 2003. Modelling the impact of clams in the biogeochemical cycles of a Mediterranean lagoon. Proceedings of the Sixth International Conference on the Mediterranean Coastal Environment. MEDCOAST 03, E. Ozhan (Editor), 7-11 October 2003, Ravenna, Italy, 2003, pp. 1291–1302.

CHAPTER 7

Application of Ecological and Thermodynamic Indicators for the Assessment of the Ecosystem Health of Coastal Areas

S.E. Jørgensen

Six ecological indicators, taken from Odum's attributes (1969, 1971), and 5 thermodynamic indicators were studied in 12 coastal ecosystems. The correlations among the 11 indicators were examined, and the extent to which the 5 thermodynamic indicators: exergy, exergy destruction, exergy production, exergy destruction/exergy, and specific exergy, can be applied to assess ecosystem health was discussed. It can be concluded that the thermodynamic indicators cover a range of important properties of ecosystems and correlate well with several of Odum's attributes, which are widely applied as ecological indicators. To give a sufficiently comprehensive assessment of ecosystem health for environmental management, however, will probably require other indicators in addition to the thermodynamic indicators.

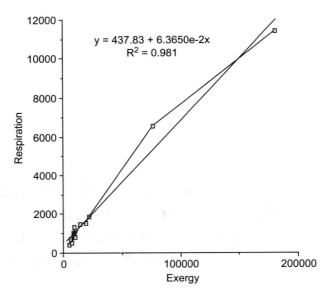

Figure 7.1   Flow diagram of the fish community of Tamiahua Lagoon, Mexico. The diagram indicates the results by use of ECOPATH II in g/m^2 for biomasses and in $g/m^2/year$ for rates.

7.1 INTRODUCTION

This chapter presents the results of a study in which several ecological indicators (including recently proposed thermodynamic indicators, exergy, exergy destruction, exergy destruction/exergy, exergy production increase in exergy, and specific exergy) were applied to 12 coastal ecosystems (described in detail by Christensen and Pauly, 1993). The extent to which the ecological indicators are correlated with the thermodynamic indicators will be explained. A recommendation on the application of thermodynamic indicators for the assessment of ecosystem health in environmental management can probably be derived from these results. Figure 7.1 gives an example of the steady-state models available for all 12 case studies.

7.2 RESULTS

The 12 ecosystems are:

1. Tamaihua, a Coastal Lagoon in Mexico
2. Celestun Lagoon, southern Gulf of Mexico
3. A coastal fish community in southwestern Gulf of Mexico
4. The Campeche Bank, Mexico
5. The Maputo Bay, Mozambique
6. A Mediterranean lagoon: Etang de Tahu, France
7. Pangasinan coral reef, Philippines

8. Caribbean coral reef
9. Yucatan Shelf Ecosystem, Mexico
10. Continental Shelf Ecosystem. Mexico
11. Shelf Ecosystem, Venezuela
12. Brunei Darassulak, South China Sea.

The following ecological indicators were determined for all 12 ecosystems:

- Biomass (g dry weight/m^2)
- Respiration (g dry weight/m^2 y)
- Exergy (kJ/m^2)
- Exergy destruction (kJ/m^2/year)
- Diversity as number of species included in the model (—)
- Connectivity as number of connections relatively to the total number of possible connections (—)
- Complexity expressed as "diversity" times "connectivity" (—)
- Respiration/biomass = B/A (year − 1)
- Exergy destruction/exergy = D/C (year − 1) Jørgensen (2002)
- Exergy production (kJ/m^2/year) Jørgensen (2002)
- Specific exergy (kJ/g) Jørgensen (2002).

Using a correlation matrix it was found that only the following of the 11 indicators were correlated with a correlation coefficient ≥ 0.64:

- Exergy production to exergy, $R^2 = 0.93$, see Figure 7.2.
- Respiration to exergy, $R^2 = 0.98$, see Figure 7.3.
- Respiration to biomass, $R^2 = 0.68$, see Figure 7.4. Notice in this context that respiration is considerably better correlated to exergy than to biomass.
- Respiration to exergy production, $R^2 = 0.855$, see Figure 7.5.

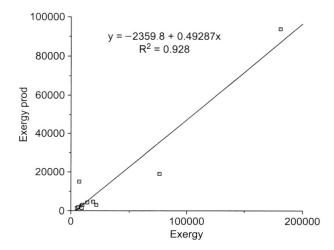

Figure 7.2 Exergy production plotted against exergy.

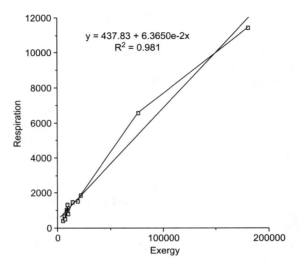

Figure 7.3   Respiration plotted against exergy.

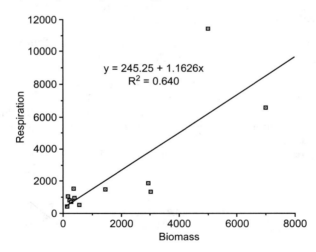

Figure 7.4   Respiration plotted against biomass. Notice that the correlation in Figure 7.3 is considerably better than this correlation.

- Exergy destruction to respiration, $R^2 = 0.87$ see Figure 7.6.
- Respiration/biomass to specific exergy, $R^2 = 0.86$, see Figure 7.7.
- Respiratory to exergy dissipation or destruction, $R^2 = 0.86$, see Figure 7.8.

7.3 DISCUSSION

Higher exergy levels are, at least for the examined marine ecosystems, associated with higher rates of exergy production which is consistent with the

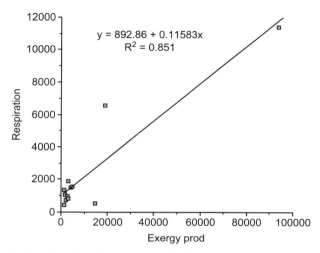

$$y = 892.86 + 0.11583x$$
$$R^2 = 0.851$$

Figure 7.5 Respiration plotted against exergy production (increase in exergy storage).

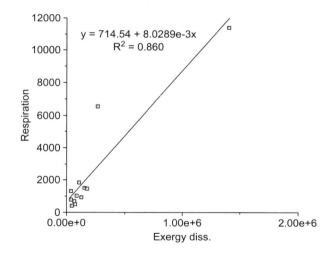

$$y = 714.54 + 8.0289e\text{-}3x$$
$$R^2 = 0.860$$

Figure 7.6 Exergy dissipation plotted against respiration.

translation of Darwin's theory to thermodynamics by the use of exergy. As an ecosystem develops, its biomass increases, and when all the inorganic matter is used to build biomass, a reallocation of the matter in form of species with more information may take place. Increased information gives increased possibility to build even more exergy (information).

The respiration levels for the various examined ecosystems are considerably better correlated with the exergy levels than with the amount of biomass in the examined ecosystems, although there is a tendency slope respiration/exergy to decrease as exergy increases (as shown in Figure 7.3). This tendency cannot be

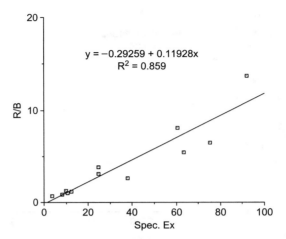

Figure 7.7   Respiration/Biomass plotted against specific exergy.

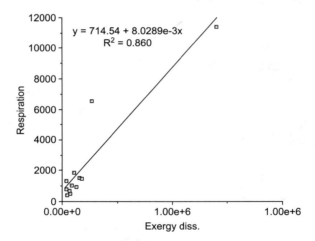

Figure 7.8   Respiration is plotted versus exergy dissipation (or destruction).

shown to be statistically significant as it would require information from more marine ecosystems.

Biomass includes plants (algae) that have relatively low exergy and also lower respiration. It explains why exergy with high weighting factors for fish and other higher organisms is better correlated with respiration than biomass (see Figures 7.3 and 7.4). The relationship is not surprising, as more stored exergy means that the ecosystem becomes more complex and more developed, which implies that it also requires more energy (exergy) for maintenance. More developed ecosystems also mean that bigger and more complex organisms become more dominant. As bigger organisms have less respiration relative to the biomass (according to the allometric principles), it is not surprising that the ratio of respiration to exergy (the slope of the plot in Figure 7.3) decreases with increasing exergy. These results are also inconsistent with Figure 7.5, where respiration is well correlated with exergy production. A high respiration level is associated with higher organisms with more information, which again gives the opportunity to increase the information further.

The correlation between the respiration level and the rate of exergy destruction in Figure 7.6 is not surprising, as the exergy destruction is caused by respiration. It is just two sides of the same coin.

Figure 7.7 indicates that the specific exergy for the examined ecosystems — higher specific exergy means more dominance of higher organisms — is well correlated with the ratio of respiration to biomass, which is also consistent with the results presented in Figure 7.5.

The results are consistent with the discussion in chapter 2, where the concepts of exergy-specific exergy were presented and associated with ecosystem health characteristics:

1. Exergy measures the distance from thermodynamic equilibrium. Svirezhev (1992) has shown that exergy measures the amount of energy needed to break down the ecosystem. Exergy is therefore a reasonably good measure of the following (compare with Costanza, 1992).

   a. Absence of disease (may be measured by the growth potential).
   b. Stability or resilience (destruction of the ecosystem is more difficult the more exergy the ecosystem has).
   c. Vigor or scope for growth (notice in this context that Figure 7.2 shows a good correlation between exergy and exergy production [growth]).

2. Specific exergy measures the organization in the sense that more developed organisms correspond to higher specific exergy. More developed organisms usually represent higher trophic levels. It implies a more complicated food web. Specific exergy is a therefore a reasonably good measure of:

   a. Homeostasis (more feed back is present in a more complicated food web)
   b. Diversity or complexity
   c. Balance between system components — the ecosystem is not dominated by the first trophic levels as this is usually for ecosystems at an early stage.

Notice that exergy or specific exergy is not correlated to diversity or complexity as determined by the connectivity. Complexity of ecosystems has several dimensions as illustrated by this chapter: complexity due to: (1) the presence of more complex organisms; (2) diversity; and (3) a more complex network. These three complexities increase independently of each other.

7.4 CONCLUSIONS

Eleven ecological and thermodynamic indicators were examined for 12 marine ecosystems. The results showed that a good correlation could only be obtained for the following pairs: exergy/exergy production, exergy/respiration, biomass/respiration, exergy production/respiration, respiration/exergy dissipation, specific exergy/respiration/biomass.

It was discussed that exergy and specific exergy and the other three thermodynamic indicators together cover the properties normally associated with ecosystem health (Costanza, 1992). It is probably not possible to assess the health of such a complex system as an ecosystem by means of only two to five indicators, which is also consistent with the lack of correlation between these two concepts and the other attributes included in this examination. It can, however, be assessed that exergy is a good measure of the ability of the ecosystem to grow (see Figure 7.2). Exergy is also a good measure of the energy (exergy) required for maintenance — better than biomass on its own — as more stored exergy and higher exergy production mean that more exergy is also needed for maintenance (see Figure 7.3, Figure 7.5 and Figure 7.6). Exergy or specific exergy is not well correlated with diversity (expressed simply as the number of state variables in the model) or complexity (measured simply as the product of number of state variables in the model and the connectivity), which is consistent with the results of several other chapters in this volume. On the other hand, specific exergy is a good expression for the presence of more developed organisms (and a more complex ecosystem).

The two concepts of exergy and specific exergy cover a certain range of properties that we generally associate with ecosystem health. They should, however, be supplemented by other indicators in most practical management situations, as they are not strictly correlated to other important attributes.

REFERENCES

Christensen, V. and Pauly, D. *Trophic Models of Aquatic Ecosystems*. ICLARM, Manila, Philippines. 1993, 390 p.

Costanza, R. "Toward an operational definition of ecosystem health," in *Ecosystem Health, New Goals for Environmental Management*, Costanza, R., Norton, B.G., and Haskell, B.D., Eds. Island Press. Washington, D.C., 1992, pp. 239–256.

Jørgensen, S.E., Nielsen, S.N., and Mejer, H. Emergy, environ, exergy and ecological modelling. *Ecol. Mod.* 77, 99–109, 1995.

Jørgensen, S.E. *Integration of Ecosystem Theories: A Pattern* 3. Kluwer, Dordrecht, 428 p.

Odum, E.P. The strategy of ecosystem development. *Science* 164, 262–270, 1069 1969.

Odum, E.P. *Fundamentals of Ecology*. W.B. Saunders, Philadelphia, 1971.

Svirezhev, Y. Exergy as a measure of the energy needed to decompose an ecosystem. Presented as a poster at ISEM's International Conference on State-of-the-Art of Ecological Modelling, 28, 1992, Kiel.

CHAPTER 8

Application of Ecological Indicators for Assessing Health of Marine Ecosystems

Villy Christensen and Philippe Cury

8.1 INTRODUCTION

"Roll on, thou deep and dark blue ocean — roll! Ten thousand fleets sweep over thee in vain; Man marks the Earth with ruin — his control stops with the shore," Lord Byron wrote two hundred years ago. Much has happened since, and humans now impact the marine environment to an extent far greater than thought possible centuries or even decades ago.

The impact comes through a variety of channels and forcing factors. Eutrophication and pollution are examples, and while locally they may be important, they constitute less of a direct threat at the global scale. A related issue, global warming and how it may impact marine ecosystem may be of more concern in the foreseeable future. This is, however, presently being evaluated as part of the "Millennium Ecosystem Assessment,"[69] to which we will refer for further information.

1-56670-665-3/05/$0.00 + $1.50
© 2005 by CRC Press

Habitat modification, especially of coastal and shelf systems, is of growing concern for marine ecosystems. Mangroves are being cleared at an alarming rate for aquaculture, removing essential habitat for juvenile fishes and invertebrates; coastal population density is exerting growing influence on coastal systems; and bottom trawls perform clear-cutting of marine habitat, drastically altering ecosystem form and functioning. The looming overall threat to the health of marine ecosystems, however, is the effect of overfishing,[2] and this will be the focus of the present contribution.

We have in recent years witnessed a move from the perception that fisheries resources need to be developed by expanding the fishing fleet toward an understanding that the way we exploit the marine environment is bringing havoc to marine resources globally, endangering the very resources on which a large part of the human population rely for nutrition. Perhaps most alarming in this development is that the global fisheries production appears to have been declining steadily since 1990,[3] the larger predatory fish stocks are being rapidly depleted,[4,5] while ecosystem structure and habitats are being altered through intense fishing pressure.[1,6,7]

In order to evaluate how fisheries impact marine ecosystem health, we have to expand the toolbox traditionally applied by fisheries researchers. Fisheries management builds on assessments of fish populations. Over the years, a variety of tools for management have been developed, and a variety of population-level indicators have seen common use.[8] While such indicators serve and will continue to serve an important role for evaluating best practices for management of fish populations, the scope of fisheries research has widened. This is due to a growing understanding that where fish populations are exploited, their dynamics must be considered as integral components of ecosystem function, rather than as epiphenomena that operate independently of their environment. Internationally, there has been wide recognition of the need to move toward an ecosystem approach to fisheries (EAF), a development strengthened by the Food and Agricultural Organization of the United Nations (FAO) through the Reykjavik Declaration of 2001,[9] and reinforced at the 2002 World Summit of Sustainable Development in Johannesburg, which requires nations to base policies for exploitation of marine resources on an EAF. Guidelines for how this can be implemented are developed through the FAO Code of Conduct for Responsible Fisheries.[10] The move is widely supported by regional and national institutions as well as academia, nongovernmental organizations and the public at large, and is mandated by the U.S. National Oceanic and Atmospheric Administration.[11]

Internationally, the first major initiative related to the use of ecosystem indicators for evaluating sustainable fisheries development was taken by the Australian government in cooperation with the FAO, through a consultation in Sydney, January 1999, involving 26 experts from 13 countries.[12] The consultation resulted in "Technical Guidelines No. 8 for the FAO Code of Conduct for Responsible Fisheries: Indicators for Sustainable Development of Marine Capture Fisheries."[13] These guidelines were produced to support the implementation of the code of conduct, and deal mainly with the development of frameworks, setting the stage for using indicators as part of the management decision process.

The guidelines do not discuss properties of indicators, nor how they are used and tested in practice. This instead became the task of an international working group, established jointly by the Scientific Committee on Oceanic Research (SCOR) and the Intergovernmental Oceanographic Committee (IOC) of UNESCO. SCOR/IOC Working Group 119 entitled "Quantitative Ecosystem Indicators for Fisheries Management" was established in 2001 with 32 members drawn internationally. The working group's aim was defined as to support the scientific aspects of using indicators for an ecosystem approach to fisheries, to review existing knowledge in the field, to demonstrate the utility and perspectives for new indicators reflecting the exploitation and state of marine ecosystems, as well as to consider frameworks for their implementation. The current overview article is influenced by the work of the SCOR/IOC Working Group 119, while prepared prior to the conclusion of the working group.

We see the key aspects of ecosystem health as a question of maintaining biodiversity and ecosystem integrity, in line with current definitions of the term. What actually constitutes a "healthy" ecosystem is a debatable topic. This debate includes the way we can promote reconciliation between conservation and exploitation interests. It also includes the recognition and understanding of system states to minimize the risk for loss of integrity when limits are exceeded.[14] From a practical perspective we assume here that we can define appropriate indicators of ecosystem health and evaluate how far these are from a reference state considered representative of a healthy ecosystem. We will illustrate this describing indicators in common use as well as the reference state they refer to.

8.2 INDICATORS

A vast array of indicators have been described and used for characterizing aspects of marine ecosystem health; a non-exhaustive review found upwards of two hundred related indicators.[15] On this background it is clear that the task we are faced with is not so much one of developing new indicators, but rather one of setting criteria for selecting indicators and evaluating the combination of indicators that may best be used to evaluate the health of marine ecosystems. Indeed, the key aspects of using indicators for management of ecosystems is centered on defining reference states and on development of indicator frameworks, as discussed above.[16] However, here we will focus on a more practical aspect: What are the indicators that have actually been applied to evaluate the health status of marine ecosystems?

8.2.1 Environmental and Habitat Indicators

Human health is impacted by climate; many diseases break out during the colder winter months in higher latitudes or during the monsoon in the lower.

We do not expect to see a similar, clear impact when discussing the marine environment, given that seasonal variability tends to be quite limited in the oceans. We do, however, see longer-term climate trends impacting ocean systems, typically over a timescale of decades, and often referred to as regime shifts.[17,18] Climate changes especially become important when ecosystem indicators signal change — is a change caused by human impact through, for example, fishing pressure, or are we merely observing the results of a change in, for example, temperature? Understanding variability in environmental indicators is thus of fundamental importance for evaluating changes in the status of marine ecosystems. This conclusion is very appropriately supported by the first recommendation of the U.S. Ecosystem Principles Advisory Panel on developing a fisheries ecosystem plan: "[T]he first step in using an ecosystem approach to management must be to identify and bound the ecosystem. Hydrography, bathymetry, productivity and trophic structure must be considered; as well as how climate influences the physical, chemical and biological oceanography of the ecosystem; and how, in turn, the food web structure and dynamics are affected."[11]

A variety of environmental indicators are in common use, including atmospheric, (wind, pressure, circulation), oceanographic (chemical composition, nutrients/eutrophication, temperature and salinity), combined (upwelling, mixed layer depth), and indicators of the effect of environmental conditions for, for example, primary productivity, plankton patterns, and fish distribution.[19]

Habitat impacts of fisheries have received increasing attention in recent years, focusing on biogenic habitats such as coral reefs, benthic structure, seagrass beds and kelp forests, which are particularly vulnerable to mechanical damage from bottom trawl and dredging fisheries.[20] The trawling impact on marine habitats has been compared to forest clear-cutting and estimated to annually impact a major part of the oceans shelfs.[21] While habitat destruction has direct consequences for species that rely on benthic habitats for protection (as is the case for juveniles of many fish species),[22] it is less clear how even intensive trawling impact benthic productivity.[20,23] A recent study found though that the productivity of the benthic megafauna increased by an order of magnitude in study sites where trawling had ceased, compared to control sites with continued trawling.[24]

Habitat indicators for ecosystem health are in other ecosystems typically focused on describing communities and community change over time. As marine ecosystems are generally less accessible for direct studies, habitats descriptions are mostly lacking. Indeed, for many ecosystems the only informative source may be charts, which traditionally include descriptions of bottom type as an aid to navigation. In recent years critical habitats has, however, received increased focus, and aided by improved capabilities for linking geopositioning and underwater video surveys, habitat mapping projects are now becoming widespread activities, providing data material that in a foreseeable future will be useful for deriving indicators of ecosystem health.

As indicators for human impact on marine habitats proxies such as, for example, proportion of the seabed trawled annually, the ratio of bottom-dwelling and demersal fish abundance, and proportion of seabed area set aside for marine protected areas have been used.[21]

8.2.2 Species-Based Indicators

Indicators of the level of exploitation is central to management of fisheries, focusing on estimating population size and exploitation level of target species.[25] Such applications of indicators are, however, of limited use for describing fisheries' impact on ecosystem health if they only consider target species. Instead the aim for this is to identify species that may serve as indicators of ecosystem-level trends. For example, the breeding success and feeding conditions of marine mammals and birds may as serve as indicators of ecosystem conditions.[26]

Another approach is to examine community-level effects of fishing, and indications are that indicators for which the direction of change brought about by fishing can be predicted may serve as useful indicators of ecosystem status.[27] Examples of potential indicators may be the average length of fishes or proportion of high-trophic-level species in the catch.

Most studies dealing with community-aspects related to species in an ecosystem describes species diversity, be it as richness or evenness measures.[28] A variety of diversity indices have been proposed, with selection of appropriate indices very much related to the type of forcing function that is influencing ecosystem health. However, it is often a challenge when interpreting such indices to describe the reference states for "healthy" ecosystems.[29,30]

Using indicators to monitor individual species is of special interest where there are legal or other obligations; for example, for threatened species. From an ecological perspective, special interest has focused on keystone species due to their capability to strengthen ecosystem resilience and thus positively impact ecosystem health.[31] Keystone species are defined as strongly interacting species that have a large impact on their ecosystems relative to their abundance. Who are they, and what are their roles in the ecosystem? The classical example from the marine realm is one of sea otters keeping a favorite prey, sea urchins in check, allowing kelp forests to abound.[32] Eradication of sea otters has a cascading effect on sea urchin, which in turn deplete the kelp forests. Identification of keystone species is currently the focus of considerable research efforts, reflecting that protection of such species is especially crucial for ecosystem health. Surprisingly, few examples of keystone species in marine systems have been published so far.

8.2.3 Size-Based Indicators

It was demonstrated more than thirty years ago that the size distribution of pelagic communities could be described as a linear relationship between (log) abundance and size.[33] It is commonly observed that there will be a decreasing

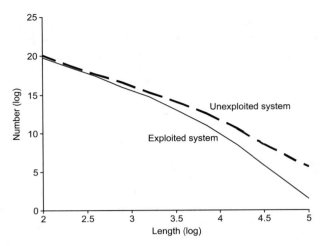

Figure 8.1   Particle size distribution curves for an ecosystem in unexploited and exploited states. Data are binned in size classes and logarithmic abundance (usually of numbers, occasionally of biomass) is presented. Exploitation is assumed to mainly reduce abundance of larger-sized organisms, while cascading may cause increase of intermediate sized (not shown here).

relationship between the log abundance and size. The intercept of the size distribution curve will be a function of ecosystem productivity, while the slope is due to differential productivity with size. Forcing functions, such as fisheries, are expected to impact notably the slope of the size distribution curves, with increasing pressure associated with increased slopes as larger-sized organisms will be relatively scarce in an exploited system (Figure 8.1). The properties of size distribution curves and how they are impacted by fishing are well understood,[15,29,34,35] while there is some controversy around the possibility of detecting signals from changes in exploitation patterns based on empirical data sets.[30] Still, size distribution curves have been widely used to describe ecosystem effects of fishing, and studies have indeed shown promising results, as demonstrated in one of the main contributions to the 1999 International Symposium on Ecosystem Effects of Fishing.[36]

Fisheries impact fish populations by selectively removing larger individuals (see also section 8.5 below), and thus by removing the faster-growing, large size-reaching part of the populations. It is widely assumed that if such phenotypic variability has a genetic basic, then exploitation will result in a selective loss in the gene pool with potentially drastic consequences.[37] There is, however, limited empirical evidence of such loss of genetic diversity and genetic drift, but this may well be because the area so far hasn't been the subject of much research. New studies indicate that it may be a real phenomenon.[38]

8.2.4   Trophodynamic Indicators

Fish eat fish, and the main interaction between fish may well be through such means,[39] indeed a large proportion of the world's catches are of piscivorous fishes.[40] There has, for this reason, been considerable attention for development of trophic models of marine ecosystems over the past decades,[41,42] and this has led to such modeling reaching a state of maturity where it is both widely applied and of use for ecosystem-based fisheries management.[43,44] When extracting and examining results from ecosystem models it becomes a key issue to select indicators to describe ecosystem status and health, we describe aspects of this in the next sections.

8.3 NETWORK ANALYSIS

One consequence of the current move toward ecosystem approaches to management of marine resources is that representations of key parameters and processes easily get really messy. When working with a single species it is fairly straightforward to present information in a simple fashion. But what do you do at the ecosystem level when dealing with a multitude of functional groups? One favored approach for addressing this question is network analysis, which has identification of ecosystem-level indicators at its root.

Network analysis is widely used in ecology (as discussed in several other contributions in this volume), and also in marine ecology.[45] In marine ecosystem applications, interest has focused on using network analysis to describe ecosystem development, notably through the work of R.E. Ulanowicz, centered around the concept of ecosystem ascendancy.[46,47] Related analyses have seen widespread application in fisheries-related ecosystem modeling where it is of interest to describe how humans impact the state of ecosystems.[48,49] Focus for many of the fisheries-related modeling has been on ranking ecosystems after maturity *sensu Odum*.[50] The key aspect of these approaches is linked to quantification of a selection of the 24 attributes of ecosystem maturity described by E.P. Odum, using rank correlation to derive an overall measure of ecosystem maturity.[51]

8.4 PRIMARY PRODUCTION REQUIRED TO SUSTAIN FISHERIES

How much do we impact marine ecosystems? This may be difficult to quantify, but the probable first global quantification that went beyond summing up catches, and incorporated an ecological perspective estimated that human appropriation of primary production through fisheries around 1990 globally amounted to around 6% of the total aquatic primary production, while the appropriation where human impact was the biggest reached much higher levels: for upwelling ecosystems, 22%; for tropical shelves, 20%; for nontropical shelves, 26%; and for rivers and lakes, 23%.[52] These coastal system levels are thus comparable to those estimated for terrestrial systems, where humans appropriate 35 to 40% of the global primary production, be it directly, indirectly or foregone.[53]

In order to estimate the primary production required (PPR) to sustain fisheries, we use an updated version of the approach used for the global estimates reported above. Global, spatial estimates of fisheries catches are now available for any period from 1950, along with estimates of trophic levels for all catch categories.[54,55] We estimate the PPR for any catch category as follows,

$$PPR = C_y \left(\frac{1}{TE} \right)^{TL} \qquad (8.1)$$

where C_y is the catch in year y for a given category with trophic level TL, while TE is the trophic transfer efficiency for the ecosystem. We use a trophic transfer efficiency of 10% per trophic level throughout based on a meta-analysis,[52] and sum over all catch categories to obtain system-level PPR.

We obtained estimates of total primary production from Nicolas Hoepffner from the Institute for Environment and Sustainability, based on SeaWiFS chlorophyll data for 1998 and the model of Platt and Sathyendranath.[56]

8.5 FISHING DOWN THE FOOD WEB

Fishing tales form part of local folklore throughout the world. I caught a big fish. What a big fish is, is however a moving target as we all tend to judge based on our own experience, making us part of a shifting-baseline syndrome.[57] As fishing impact intensifies, the largest species on top of the food web become scarcer, and fishing will gradually shift toward more abundant, smaller-prey species. This form part of a process, termed "fishing down the food web"[7] in which successive depletion results in initially increasing catches as the fishery expands spatially and starts targeting low-trophic-level prey species rather than high-trophic-level predatory species, followed by a steady phase, and often a decreasing phase caused by overexploitation, possible combined with shift in the ecological functioning of the ecosystems (see Figure 8.2).[7]

A series of publications based on detailed catch statistics and trophic-level estimates typically from FishBase have demonstrated that "fishing down the food web" is a globally occurring phenomenon.[58-60] Indeed, there seems to be a general trend that the more detailed catch statistics that are available for the analysis, the more pronounced the phenomenon.[60]

8.6 FISHING IN BALANCE

An important aspect of "fishing down the food web" is that we would expect to get higher catches of the more productive, lower-trophic-level catches of prey fishes in return for the loss of less productive, higher-trophic-level

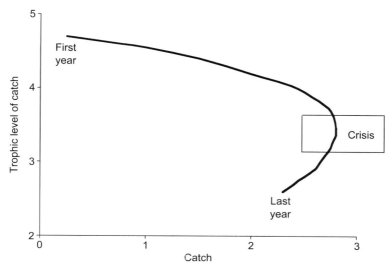

Figure 8.2   Illustration of "fishing down the food web" in which fisheries initially target high-trophic-level species with low catch rates. As fishing intensity increases catches shift toward lower-trophic-level species. At high fishing intensity it has often been observed that catches will tend to decrease along with the trophic level of the catch (backward-bending part of curve, starting where "crisis" is indicated).

catches of predatory fishes. With average trophic transfer efficiencies of 10% between trophic levels in marine systems,[52] we should indeed expect, at least theoretically, a ten-fold increase in catches if we could fully eliminate predatory species and replace them with catches of their prey species.

To quantify this aspect of "fishing down the food web" an index, termed "fishing in balance" (FiB) has been introduced.[61] The index is calculated based on the calculation of the PPR index (see Equation 8.1):

$$\text{FiB} = \log\left(\left[C_y \times \left(\frac{1}{TE}\right)^{TL_y}\right] \Big/ \left[C_1 \times \left(\frac{1}{TE}\right)^{TL_1}\right]\right) \tag{8.2}$$

where, C_y and C_1 are the catches in year y and the first year of a time series, respectively, and TL_y and TL_1 are the corresponding trophic levels of the catches; TE is the trophic transfer efficiency (10%). The index will start at unity for the first year of a time series, and typically increase as fishing increases (due to a combination of spatial expansion and "fishing down the food web"), and then often show a stagnant phase followed by a decreasing trend. During the stagnant phase where the FiB index is constant, the effect of lower-trophic-level of catches will be balanced by a corresponding increase in catches level. A decrease of 0.1 in the trophic level of the catches will as an example be balanced by a $10^{0.1}$ (25%) increase in catch level. There has so far been few applications of the FiB index,[62] but indications are that the index has some potential by virtue of being dimensionless, sensitive, and easy to interpret.

8.7 APPLICATION OF INDICATORS

We illustrate the application of indicators by presenting accessible information for the North Atlantic Ocean, defined as comprising FAO Statistical Areas 21 and 27. The North Atlantic was the initial focus area for the Sea Around Us project through which information about ecosystem exploitation and resource status has been derived for the period since 1950.[4,63–65] During the second half of the twentieth century, the catches increased from an already substantial level of 7 million metric tonnes per year to reach double this level by the 1970s, but it has since declined gradually (Figure 8.3). Catch composition changed over the period from being dominated by herring and large demersals to lower-trophic-level groups, with high landings of fish for fish meat and oil. The biomass of higher-trophic-level fish in the North Atlantic has been estimated to have decreased by two-thirds over the past half century.[4]

8.7.1 Environmental and Habitat Indicators

There are indications, notably from the continuous plankton recorder surveys, of decadal changes linked to the atmospheric North Atlantic Oscillation Index, causing marked changes in productivity patterns as well as zooplankton composition.[66] Overall, the changes do not have consequences for ecosystem health, but they change the background at which to evaluate health, and as such should be considered.

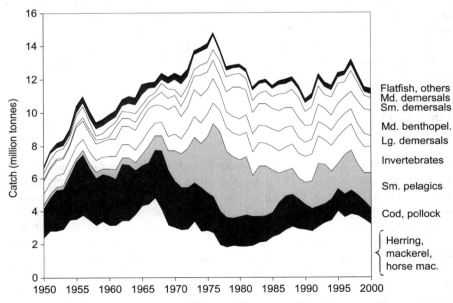

Figure 8.3   Total catches and catch composition for the North Atlantic (FAO Areas 21 and 27) estimated based on information from FAO, ICES, NAFO and national sources. Source: http://www.seaaroundus.org.

Fishing pressure, notably by habitat-damaging bottom trawls, increased drastically during the second half of the twentieth century, where low-powered fleets of gill-netters, Danish seines, and other small-scale fisheries were largely replaced with larger-scale boats dominated by trawlers. The consequence of this has been widespread habitat damage, as illustrated by a large cold-water coral reef area south of Norway, where trawling was impossible until the 1990s when beam-trawlers had grown powerful enough to exploit and completely level the area within a few years.

It is unfortunately characteristic for fisheries science in the second half of the twentieth century that emphasis has been on fish population dynamics, and very little information is available about the effort exerted to exploit the resources, and of the consequences the exploitation has had on habitats. It is thus not possible at present to produce indices of habitat impact at the North Atlantic scale (or of any larger part of the area for that matter).

8.7.2   Size-Based Indicators

Particle size distributions have been constructed for several areas of the North Atlantic illustrating how fisheries have reduced the abundance of larger fish.[34,58] We do not yet, however, have access to abundance information at the North Atlantic level that makes construction of particle size distributions possible at this scale. If we instead examine how the average of the maximum standard length of species caught in the North Atlantic has developed over the last fifty years we obtain the picture in Figure 8.4. This illustrates a gradual erosion of fish capable of reaching large sizes, with the average maximum size decreasing from 120 to 70 cm over the period. This finding links to what is

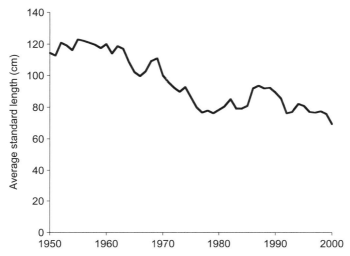

Figure 8.4 Average maximum standard length for all catches of the North Atlantic. Source: http://www.seaaroundus.org.

presented below on trophodynamic indicators as size and tropic level are correlated measures.[67]

8.7.3 Trophodynamic Indicators

Network indicators covering the North Atlantic are not available as no model has been constructed for the overall area. There are a large number of models for various North Atlantic ecosystems, including some that cover the time period of interest here. We have, however, not been able to identify any network indicators that could be used to describe aspects of ecosystem health based on the available models. Instead we focus on other trophodynamic indicators that can be estimated from catch statistics.[68]

We estimate the primary production required (PPR) to sustain the North Atlantic fisheries varied from 9% of the primary production in 1950 to nearly 16% in the late 1960s. It then gradually declined to 11% (Figure 8.5), a level around which it has been since. The appropriation is thus in between the 6% and 26% estimated globally for open oceans and nontropical shelves, respectively.[52] Since, the vast majority of the North Atlantic area is oceanic, the PPR is relatively high compared to other areas. Examining the trend in PPR is by itself not very meaningful for drawing inferences about ecosystem status or health; it is more telling when including information about trends in trophic and catch levels in the considerations as demonstrated below.

The North Atlantic has been exploited for centuries, and has seen its fair share of devastation from the demise of northern right whales and to more recent fisheries collapses throughout the area.[69] Reflective of the changes within the fish populations is the "fishing down the food web" index, which for

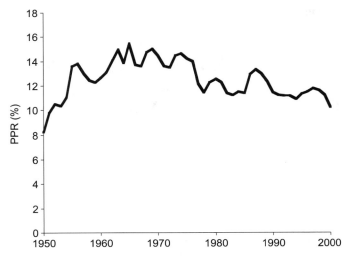

Figure 8.5 Primary production required to sustain the fisheries of the North Atlantic, expressed as percentage of the total primary production for the area.

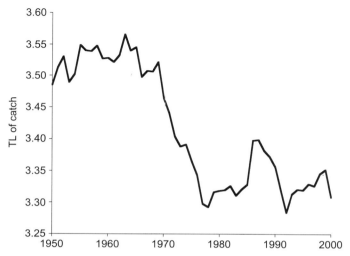

Figure 8.6   "Fishing down the food web" in the North Atlantic as demonstrated by the trend in the average trophic level of the catches during the second half of the twentieth century.

the North Atlantic takes the shape presented in Figure 8.6. In the 1950s the average trophic level of the catches hovered around 3.50 to 3.55, before decreasing sharply during the 1960s and 1970s, reaching a level of around 3.3, where it has remained since.

The decrease in trophic level that occurred during the 1960s and 1970s was associated with an increase in catches as one may have expected, see Figure 8.7. The catches increased up to the mid-1960s without any impact on the average

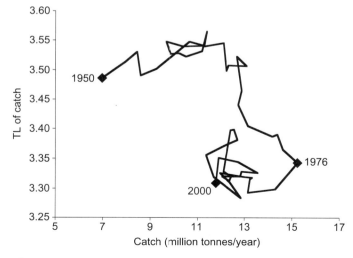

Figure 8.7   Phase plot of catches versus the average trophic levels of catches in the North Atlantic, 1950–2000.

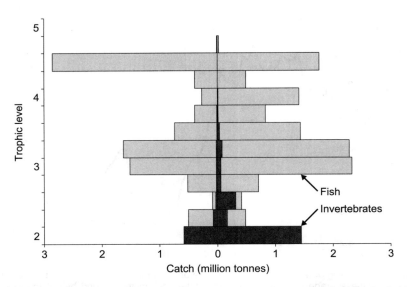

Figure 8.8   Catch composition of fish (light-shaded bars) and invertebrates (dark-shaded bars) by trophic level in the 1950s and the 1990s for the North Atlantic (FAO areas 21 and 27). Source: FishBase.

trophic level, indicating that the fisheries during this period were in a spatial expansion phase. Through the 1960s up to the mid 1970s the fisheries catches continued to increase but this was now associated with a marked decrease in trophic level of the catches. This in turn is indicative of a "fishing down the food web" effect, where higher-trophic-level species are replaced with more productive lower-trophic-level species (Figure 8.8). From the mid-1970s the catches have been decreasing, while remaining at a low trophic level, and without any sign of a return to increased importance of high-trophic-level species. This backward-bending part of the catch–trophic level phase plot (Figure 8.7) seems to be a fairly common phenomenon, and may be associated with a breakdown of ecosystem functioning or increased nonreported discarding.[7]

A closer examination of the catch composition for the North Atlantic in the 1950s compared to the 1990s shows that the more recent, lower trophic levels of the catches are indeed associated with lower catches of the highest-trophic-level species and higher catches of lower-trophic-level fish species as well as of invertebrates (Figure 8.8). The catch of the uppermost trophic level category was nearly halved over the period.

As discussed, we would expect that a reduction in the trophic level of the catches should be associated with a corresponding increase in catches (as indeed observed in the 1960s), with the amount being a function of the trophic transfer efficiencies in the system. For the North Atlantic we estimate the corresponding FiB index as presented in Figure 8.9. As expected, the FiB index increased from its 1950 level up to the mid-1960s — that is, through the period

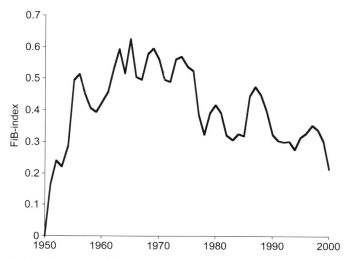

Figure 8.9   "Fishing-in-balance index" for the North Atlantic, 1950–2000, estimated based on catches and the average trophic level of the catches.

characterized by spatial expansion and relatively low resource utilization. From the mid-1960s the index is stable for a decade — that is, the fishing was "in balance." This was, however, followed with steady erosion from the mid-1970 through the century, where the index shows a clear decline, indicating that the reduction in the average trophic level of the catches is no longer compensated for by a corresponding increase in overall catch levels. The major conclusion that can be drawn from this is that the fisheries of the North Atlantic are unsustainable.

8.8 CONCLUSION

Ecosystem-based indicators have only recently become a central focus for the scientific community working on marine ecosystems. However, there exists a range of potential indicators that can provide useful information on ecological changes at the ecosystem level, and can help us move towards implementation of an ecosystem approach to fisheries.

We have used the North Atlantic Ocean as a case study to demonstrate the use of indicators for describing aspects of ecosystem status and health. The North Atlantic has been exploited for hundreds of years for some species, even in a sustainable manner up to a few decades ago. Recent trends, however, are far from encouraging, and the indicators we have selected largely indicate that the fisheries of the North Atlantic are of a rather unsustainable nature.

If other aspects of the way we impact the North Atlantic are included it doesn't improve the picture. This is clear from the detailed study of the fisheries

Table 8.1 "Report Card" for the health status of the North Atlantic Ocean[65]

Name:	**North Atlantic Ocean**
Class:	**Health Status**

Subjects:	Grade:
Long-term productivity of fisheries	F
Economic efficiency of the fisheries	C–
Energy efficiency of the fisheries	D–
Ecosystem status	F
Effects of fisheries on marine mammals	D

and ecosystems of the North Atlantic presented by Pauly and Maclean, who concluded by presenting a "report card" for the North Atlantic Ocean where a "failing grade" was passed for its health status and the way we exploit it (Table 8.1).[6]

There are no comparable report cards for other areas to facilitate drawing inferences at the global level; it is clear, however, that there are problems globally with the exploitation status of marine ecosystems. The North Atlantic is no special case, indicating that the way the world's fisheries are being conducted is in general far from sustainable.[2]

There is, worldwide, much effort being directed toward improving the exploitation status for marine ecosystems as discussed earlier, and we need to consider how we track the success of such efforts, should there be any. This question is very much related to how we assess ecosystem health, and we have here attempted to highlight some related, current research.

The indicators we have presented all relate to the composite ecosystem level, and we note that they all have maintenance of larger-sized, long-lived species as an integral component. We think that maintenance of such species in an ecosystem is important for ecosystem health status.[40] This is in accordance with E.P. Odum's maturity measures;[50] if large-size predators are depleted and marine ecosystems drastically altered through overfishing, the risk of radical changes in ecosystem status increases drastically; for example, through shifts from demersal to pelagic fish-dominated ecosystems or through outbreaks of jellies or red tide. At the decadal-level, ecosystems may experience alternate semi-stable states, with potential drastic consequences for food supply, the current problems with cod populations across the North Atlantic serving as a case in point. The safe approach for maintaining healthy, productive ecosystems involves maintaining reproductive stocks of marine organisms at all trophic levels.

REFERENCES

1. Pauly, D., Alder, J., Bennett, E., Christensen, V., Tyedmers, P., and Watson, R. The future for fisheries. *Science* 302, 1359–1361, 2003.

2. Pauly, D., Christensen, V., Guénette, S., Pitcher, T.J., Sumaila, U.R., Walters, C.J., Watson, R., and Zeller, D. Towards sustainability in world fisheries, *Nature* 418, 689–695, 2002.
3. Watson, R. and Pauly, D. Systematic distortions in world fisheries catch trends. *Nature* 414, 534–536, 2001.
4. Christensen, V., Guénette, S., Heymans, J.J., Walters, C.J., Watson, R., Zeller, D., and Pauly, D. Hundred-year decline of North Atlantic predatory fishes. *Fish and Fisheries* 4, 1–24, 2003.
5. Myers, R.A. and Worm, B. Rapid worldwide depletion of predatory fish communities. *Nature* 423, 280–283, 2003.
6. Hall, S.J., *The Effects of Fisheries on Ecosystems and Communities*. Blackwell, Oxford, 1998.
7. Pauly, D., Christensen, V., Dalsgaard, J., Froese, R., and Torres, F., Jr. Fishing down marine food webs. *Science* 279, 860–863, 1998.
8. Collie, J.S. and Gislason, H., Biological reference points for fish stocks in a multispecies context. *Canadian Journal of Fisheries and Aquatic Sciences* 58, 2167–2176, 2001.
9. FAO. Report of the Reykjavik Conference on Responsible Fisheries in the Marine Ecosystem — Reykjavik, Iceland, 1–4 October 2001. FAO Fisheries Report No. 658, 2001.
10. FAO. The Ecosystem Approach to Fisheries. FAO Technical Guidelines for Responsible Fisheries. No. 4, Suppl. 2. FAO Fisheries Department, Rome, 2003, 112 p.
11. National Marine Fisheries Service. Ecosystem-based Fishery Management. *A report to Congress by the Ecosystems Principles Advisory Panel*. National Marine Fisheries Service, U.S. Department of Commerce, Silver Spring, MD, 1999.
12. Anon. Sustainability indicators in marine capture fisheries: introduction to the special issue. *Marine and Freshwater Research* 51, 381–384, 2000.
13. FAO. Indicators for Sustainable Development of Marine Capture Fisheries. FAO Technical Guidelines for Responsible Fisheries No. 8, FAO, Rome, 1999.
14. Fowler, C.W. and Hobbs, L. Limits to natural variation: implications for systemic management. *Animal Biodiversity and Conservation* 25, 7–45, 2002.
15. Rice, J.C. Evaluating fishery impacts using metrics of community structure. *ICES Journal of Marine Science* 57, 682–688, 2000.
16. Hall, S.J. Managing fisheries within ecosystems: can the role of reference points be expanded? *Aquatic Conservation-Marine and Freshwater Ecosystems* 9, 579–583, 1999.
17. Beamish, R.J., Noakes, D.J., McFarlane, G.A., Klyashtorin, L., Ivanov, V.V., and Kurashov, V. The regime concept and natural trends in the production of Pacific salmon. *Canadian Journal of Fisheries and Aquatic Sciences* 56, 516–526, 1999.
18. Scheffer, M., Carpenter, S., Foley, J.A., Folke, C., and Walker, B., Catastrophic shifts in ecosystems, *Nature* 413, 591–596, 2001.
19. Brander, K. Ecosystem indicators in a varying environment. *ICES Journal of Marine Science* (in press).
20. Moore, P.G. Fisheries exploitation and marine habitat conservation: a strategy for rational coexistence. *Aquatic Conservation-Marine and Freshwater Ecosystems* 9, 585–591, 1999.
21. Watling, L. and Norse, E.A. Disturbance of the seabed by mobile fishing gear: a comparison to forest clearcutting. *Conservation Biology* 12, 1180–1197, 1998.

22. Sainsbury, K.J., Campbell, R.A., and Whitelaw, W.W. "Effects of trawling on the marine habitat on the North West Shelf of Australia and implications for sustainable fisheries management," in *Sustainable Fisheries through Sustainable Habitat*, Hancock, D. A., Ed. Bureau of Rural Sciences Proceedings, AGPS, Canberra, 1993, pp. 137–145.

23. Schratzberger, M., Dinmore, T.A., and Jennings, S. Impacts of trawling on the diversity, biomass and structure of meiofauna assemblages. *Marine Biology* 140, 83–93, 2002.

24. Hermsen, J.M., Collie, J.S., and Valentine, P.C. Mobile fishing gear reduces benthic megafaunal production on Georges Bank. *Marine Ecology Progress Series* 260, 97–108, 2003.

25. Jennings, S. and Kaiser, M.J. The effects of fishing on marine ecosystems. *Advances in Marine Biology* 34, 201–351, 1998.

26. Carscadden, J.E., Montevecchi, W.A., Davoren, G.K., and Nakashima, B.S. Trophic relationships among capelin (Mallotus villosus) and seabirds in a changing ecosystem. *ICES Journal of Marine Science* 59(5), 1027–1033, 2002.

27. Trenkel, V.M. and Rochet, M.J. Performance of indicators derived from abundance estimates for detecting the impact of fishing on a fish community. *Canadian Journal of Fisheries and Aquatic Sciences* 60, 67–85, 2003.

28. Rice, J. Environmental health indicators. *Ocean & Coastal Management* 46, 235–259, 2003.

29. Gislason, H. and Rice, J. Modelling the response of size and diversity spectra of fish assemblages to changes in exploitation. *ICES Journal of Marine Science* 55, 362–370, 1998.

30. Rochet, M.J. and Trenkel, V.M. Which community indicators can measure the impact of fishing? A review and proposals. *Canadian Journal of Fisheries and Aquatic Sciences* 60, 86–99, 2003.

31. McClanahan, T., Polunin, N., and Done, T. Ecological states and the resilience of coral reefs. *Conservation Ecology* 6(2), 18, 2002.

32. Kvitek, R.G., Oliver, J.S., Degange, A.R., and Anderson, B.S. Changes in Alaskan soft-bottom prey communities along a gradient in sea otter predation. *Ecology* 73, 413–428, 1992.

33. Sheldon, R.W., Prahask, A., and Sutcliffe, Jr., W.H.J. The size distribution of particles in the ocean. *Limnology and Oceanography* 17, 327–340, 1972.

34. Rice, J. and Gislason, H. Patterns of change in the size spectra of numbers and diversity of the North Sea fish assemblage, as reflected in surveys and models. *ICES Journal of Marine Science* 53, 1214–1225, 1996.

35. Shin, Y.-J. and Cury, P. Using an individual-based model of fish assemblages to study the response of size spectra to changes in fishing. *Canadian Journal of Fisheries and Aquatic Sciences* 61(3), 414–431, 2004.

36. Bianchi, G., Gislason, H., Graham, K., Hill, L., Jin, X., Koranteng, K., Manickchand-Heileman, S., Paya, I., Sainsbury, K., Sanchez, F., and Zwanenburg, K. Impact of fishing on size composition and diversity of demersal fish communities. *ICES Journal of Marine Science* 57, 558–571, 2000.

37. Heino, M. and Godo, O.R. Fisheries-induced selection pressures in the context of sustainable fisheries. *Bulletin of Marine Science* 70, 639–656, 2002.

38. Hutchinson, W.F., van Oosterhout, C., Rogers, S.I., and Carvalho, G.R. Temporal analysis of archived samples indicates marked genetic changes in declining North Sea cod (Gadus morhua). *Proceedings of the Royal Society of London Series B-Biological Sciences* 270, 2125–2132, 2003.

39. Bax, N.J. The significance and prediction of predation in marine fisheries. *ICES Journal of Marine Science* 55, 997–1030, 1998.
40. Christensen, V. Managing fisheries involving predator and prey species. *Reviews in Fish Biology and Fisheries* 6, 417–442, 1996.
41. Andersen, K.P. and Ursin, E. A multispecies extension to the Beverton and Holt theory of fishing, with accounts of phosphorus circulation and primary production. *Meddelelser fra Danmarks Fiskeri og Havundersogelser* 7, 319–435, 1977.
42. Polovina, J.J. Model of a coral reef ecosystems I. The ECOPATH model and its application to French Frigate Shoals. *Coral Reefs* 3, 1–11, 1984.
43. Latour, R.J., Brush, M.J., and Bonzek, C.F. Toward ecosystem-based fisheries management: strategies for multispecies modeling and associated data requirements. *Fisheries* 28, 10–22, 2003.
44. Christensen, V. and Walters, C.J. Ecopath with Ecosim: methods, capabilities and limitations. *Ecological Modelling* 172, 109–139, 2004.
45. Wulff, F., Field, J.G., and Mann, K.H. *Network Analysis in Marine Ecology.* Springer-Verlag, Berlin, 1989.
46. Ulanowicz, R.E. *Growth and Development: Ecosystem Phenomenology.* Springer-Verlag (reprinted by iUniverse, 2000), New York, 1986.
47. Ulanowicz, R.E. *Ecology, the Ascendent Perspective.* Columbia University Press, Columbia, 1997.
48. Christensen, V. and Pauly, D. Trophic models of aquatic ecosystems. *ICLARM Conference Proceedings, Manila*, 26, 1993, 390 p.
49. Christensen, V. and Pauly, D. Placing fisheries in their ecosystem context, an introduction. *Ecological Modelling* 172, 103–107, 2004.
50. Odum, E.P. The strategy of ecosystem development. *Science* 104, 262–270, 1969.
51. Christensen, V. Ecosystem maturity — towards quantification. *Ecological Modelling* 77, 3–32, 1995.
52. Pauly, D. and Christensen, V. Primary production required to sustain global fisheries. *Nature* 374, 255–257, 1995.
53. Vitousek, P.M., Ehrlich, P.R., and Ehrlich, A.H. Human appropriation of the products of photosynthesis. *BioScience* 36, 368–373, 1986.
54. Froese, R. and D. Pauly, Editors. FishBase. World Wide Web electronic publication. www.fishbase.org, 2004.
55. Watson, R., Alder, J., Christensen, V., and Pauly, D. "Probing the depths: reverse engineering of fisheries landings statistics," in *Place Matters: Geospatial Tools for Marine Science, Conservation, and Management in the Pacific Northwest*, Wright, D.J. and Scholz, A.J., Eds. Oregon State University Press, Corvallis, OR, (in review).
56. Platt, T. and Sathyendranath, S. Oceanic primary production: estimation by remote sensing at local and regional scales. *Science* 241, 1613–1620, 1988. See http://www.seaaroundus.org.
57. Pauly, D. Anecdotes and the shifting baseline syndrome of fisheries. *Trends in Ecology & Evolution* 10, 430, 1995.
58. Jennings, S., Greenstreet, S.P.R., Hill, L., Piet, G.J., Pinnegar, J.K., and Warr, K.J. Long-term trends in the trophic structure of the North Sea fish community: evidence from stable-isotope analysis, size-spectra and community metrics. *Marine Biology* 141, 1085–1097, 2002.
59. Heymans, J.J., Shannon, L.J., and Jarre, A. Changes in the northern Benguela ecosystem over three decades: 1970s, 1980s, and 1990s. *Ecological Modelling* 172, 175–195, 2004.

60. Pauly, D. and Palomares, M.L.D. Fishing down marine food web: it is far more pervasive than we thought. *Bulletin of Marine Science* (in press).
61. Pauly, D., Christensen, V., and Walters, C. Ecopath, Ecosim, and Ecospace as tools for evaluating ecosystem impact of fisheries. *ICES Journal of Marine Science* 57, 697–706, 2000.
62. Christensen, V. Indicators for marine ecosystems affected by fisheries. *Marine and Freshwater Research* 51, 447–450, 2000.
63. Guénette, S., Christensen, V., and Pauly, D. Fisheries impacts on North Atlantic ecosystems: models and analyses. *Fisheries Centre Research Reports* 9(4), 344, 2001.
64. Zeller, D., Watson, R., and Pauly, D. Fisheries impact on North Atlantic marine ecosystems: catch, effort and national and regional data sets. *Fisheries Centre Research Reports* 9(3), 254, 2001.
65. Pauly, D. and Maclean, J.L. *In a Perfect Ocean: The State of Fisheries and Ecosystems in the North Atlantic Ocean.* Island Press, Washington, D.C., 2003.
66. Beaugrand, G., Reid, P.C., Ibanez, F., Lindley, J.A., and Edwards, M. Reorganization of North Atlantic marine copepod biodiversity and climate. *Science* 296, 1692–1694, 2002.
67. Jennings, S., Pinnegar, J.K., Polunin, N.V.C., and Warr, K.J. Linking size-based and trophic analyses of benthic community structure. *Marine Ecology Progress Series* 226, 77–85, 2002.
68. Mowat, F. *Sea of Slaughter.* Monthly Press, Boston, 438 p.
69. www.millenniumassessment.org

CHAPTER **9**

Using Ecological Indicators in a Whole-Ecosystem Wetland Experiment

W.J. Mitsch, N. Wang, L. Zhang, R. Deal, X. Wu, and A. Zuwerink

Indicators were used to estimate wetland divergence and convergence in a whole ecosystem experiment in central Ohio. Two 1-ha flow-through-created wetland basins with similar geomorphology were maintained with similar inflow and water depth for six years. One basin was planted with 2,500 individual rootstocks of 13 species of macrophytes at the beginning of the six-year study; the second basin was left unplanted. Both basins were then subjected to natural additional colonization of plants, algae, microbes, and animals. Macrophyte community diversity was estimated by a new community diversity index (CDI) and this was treated as the independent variable. By the sixth year, CDI diverged in the two wetlands with the "planted" basin supporting several macrophyte communities of mostly introduced species and the "unplanted" or "naturally colonizing" basin dominated by an invasive *Typha* sp. monoculture. With this difference in community diversity came a divergence in ecosystem structure and function. Sixteen indicators of wetland function (dependent variables) were observed annually and relative differences between the wetland basins were determined by a nonparametric similarity index. The basins were ecologically similar in years 3 to 5, but by the sixth year of the experiment, the basins diverged in function with only 44% similarity after similarity between 75% and 87% for years 3 to 5. The macrophyte-diverse wetland that resulted from planting had a higher water-column productivity, water temperature, dissolved oxygen, and pH. The naturally colonizing *Typha* wetland had a higher macrophyte productivity, benthic invertebrate diversity, outflow suspended sediments (turbidity), and dissolved ions (conductivity). Different bird, amphibian, and fish use are also hypothesized as having resulted from the planting and differential colonization. Our large-scale, long-term, whole-ecosystem findings dispute some findings of small-scale, short-term, replicated mesocosm experiments.

9.1 INTRODUCTION

Few studies have investigated how macrophyte diversity affects ecosystem function in created and restored wetlands, despite the frequent use of macrophyte cover and diversity as determinants of legal and ecological success of these wetlands in mitigating wetland loss, particularly in the U.S.[1-5] Engelhardt and Ritchie[6] manipulated 70 1.5-m-diameter wading pools with one, two, and three species of submersed pondweed (*Potamogeton* sp.) and found that higher algal biomass and higher phosphorus uptake occurred in the pools with highest macrophyte species richness. They concluded that higher species richness created up to 25% higher algal biomass that caused 30% more phosphorus uptake and thus would support more wildlife and fish. They further concluded that a wetland with high richness or diversity due to disturbance might better "sustain ecosystem functioning and promote the services of those wetlands to humans."[6]

Alternatives to the replicated small-scale mesocosms for wetland study are large-scale, long-term whole ecosystem studies that include more components of the ecosystem. Whole-ecosystem experiments, which have been carried out for terrestrial systems,[7-10] lakes,[11-14] and wetlands,[2,15-16] are often criticized because the size, cost, and logistics alone do not allow for much, if any, replication. Some researchers suggest that there is no single optimum scale for ecosystem experimentation but state that it is easier to apply statistical methods successfully to many small replicated systems.[17-18]

This chapter presents a six-year, whole-ecosystem wetland experiment that illustrates: (1) the effect of macrophyte introduction and subsequent macrophyte community development on ecosystem function; and (2) the use of simple, easy-to-measure, indicators for assessing large-scale, long-term, whole-ecosystem experiments in wetland ecology. Our study investigates the relationship between macrophyte community diversity and ecosystem function in light of current theories on biodiversity and ecosystem function. Results of this study follow those from the first three years of this study that were previously published.[2] Those early results illustrated that marsh functions in the experimental wetlands diverged and then converged in concert with divergence and convergence of macrophyte development. After six years, some of those findings were validated, while our conclusion on the time over which the introduction of plant diversity has a measurable effect on ecosystem function has been determined to be longer than we originally thought.

9.2 METHODS

9.2.1 Site History

Two 1-ha experimental wetlands and a river water delivery system were constructed in 1993 to 1994 at the Olentangy River Wetland Research Park, a 10-ha site on the campus of The Ohio State University in Columbus (Figure 9.1a). Over 2,400 plant propagules (mostly root stock and rhizomes) representing 13 species typical of midwestern U.S. marshes were planted in one wetland (termed W1) in May 1994. The second area of wetland (termed W2)

(a)

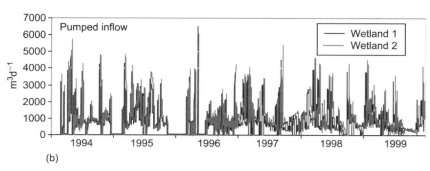

(b)

Figure 9.1   (a) Paired 1-ha experimental wetlands at the Olentangy River Wetland Research Park at Ohio State University (photo in August 1999). The planted wetland basin (W1) is on the right; the basin planted by nature (W2) is on the left; (b) pumped inflow to planted (W1) and naturally colonizing (W2) wetland basins for the six years described here. Water levels and water budgets were essentially identical over the six-year period discussed here.

remained unplanted. Both wetlands received the same amount and quality of pumped river water and both have had essentially identical hydroperiods for the entire six-year study period (Figure 9.1b). Pumped river water generally flows into the wetlands continuously, day and night, except for planned drawdowns, and occasional short-term unscheduled pump failures. After start-up trials in 1994, a pumping protocol was developed that involves changing the pumping rate manually two or three times per week according to a formula that calls for more pumping when river discharge is high and less pumping when river flow is low. On average, pumped inflow to each wetland has been 20 to 40 m per year. Water depths in the major portions of the wetland are generally 20 to 40 cm in the shallow areas where most of the emergent macrophytes grow, and 50 to 80 cm in the deepwater areas that were constructed in the wetland to allow over wintering of nekton and additional sediment storage. Five river flooding events occurred during the study period. During each of these floods, water from the river spilled into the wetlands in approximately equal amounts.

9.2.2 Macrophyte Community Index

Macrophyte coverage by dominant community is estimated each year from aerial color photography taken at the period of peak biomass (late August), coupled with ground truth surveys. Ground surveys involved mapping plants along 500 m of seven transects (shown in Figure 9.1a) in each wetland. Transect permanent walkways are about 1.5 m above the wetland soil, thus giving a good perspective even with 3-m-tall plants. A 10 m × 10 m grid system marked with permanent numbered white poles is used to identify the location of plant communities in each wetland. Maps for each year are normalized to the same size basin map utilizing geographic information system software.

We developed a macrophyte community diversity index (CDI) to quantify the diversity in the wetland basins. The index used relative areas of macrophyte community cover from the maps derived above and the mathematics of the Shannon–Weaver diversity index, with the area of each community instead of number of individuals of each species used. It is expressed as:

$$\text{CDI} = -\sum_{i=1}^{N} (C_i \ln C_i)$$

where C_i is the percentage cover of wetland community i (0 to 1) in the wetland basin and N is the number of wetland communities.

Overall, there were seven different communities identified by our combination of aerial photography and subsequent ground surveys during the six-year study. They were named for the dominant species in community:

- *Schoenoplectus tabernaemontani*
- *Typha sp.*
- *Scirpus fluviatilis*

- *Nelumbo lutea*
- *Sparganium eurycarpum*
- *Spartina pectinata*
- Open water/submersed aquatics.

9.2.3 Field Indicators

Above-ground biomass during August was used as an estimate of above-ground net primary productivity (ANPP). It was estimated directly beginning in 1997 by direct above-ground harvesting of 16 1-m^2 plots in each wetland along sampling boardwalks and from aerial photography and fewer sample plots before that. Algae are sampled several times each year at inflow, middle, and outflow areas in both wetlands with a plankton net tossed 5 m and retrieved with a cord. Samples representative of metaphyton such as attached and benthic algae are taken by hand. Algal species are identified by microscope at magnifications of 100× and 400× and relative abundance of each genus is estimated. Daily dawn–dusk–dawn readings of dissolved oxygen from manually taken measurements at the outflows are used to estimate aquatic productivity of the water column.[19] More than 60 such paired dawn–dusk samples were available each year. Benthic invertebrates are sampled in late October until early November annually with Hester–Dendy plates (11 cm^2) placed at nine stations in each wetland. Sometimes this sampling has been supplemented with dip net collections and bottle collections. Invertebrates are sorted to lowest recognizable taxa and Shannon–Weaver diversity indices (see chapter 2) are estimated with these taxa counts. Taxa diversity measures with and without pollution-tolerant organisms (oligochaetes, tubificids, and chironomids) were used as relative indicators in the two wetlands of aquatic community diversity.

Manual sampling of water temperature, dissolved oxygen, pH, conductivity, and redox has been done twice per day (dawn and dusk) with Hydrolab H20G or YSI 6000 water quality sondes at the inflow of the wetlands and outflows of both wetland basins. Samples of 100 ml were also taken dawn and dusk each day at the inflow and two outflows for turbidity analyses in the laboratory with a Hach ratio turbidimeter. In addition to the twice-per-day manual sampling, weekly water samples were taken at inflows and outflows of the wetlands for nutrients which were determined by standard methods.[20,21]

Basins were observed for avian activity in the early years through exact walking paths by experienced observers several times during the year. Comparison of avian use of the basins in 1999 was made through frequent visits to the basins in spring. Bird presence is noted by both songs and sightings.

9.2.4 Similarity Index

Basins were compared yearly by using 16 indicators, listed in Table 9.1 and described above, that represent the structure and function of the wetland

Table 9.1   Indicators of ecosystem structure and function used to compare experimental 1-ha wetlands at Olentangy River Wetland Research Park, 1994 to 1999

Indicator	Ecosystem structure or function
I. Macrophyte community function	
1. Net above-ground primary productivity	Macrophyte community organic production
II. Aquatic community development	
2. Algal species richness	Water column diversity
3. Aquatic metabolism	Water column organic production
4. Macroinvertebrate diversity	Benthic biodiversity
5. "Clean water" species richness	Balance of P and R
III. Bio-geochemistry	
6. Temperature change	Shading of water column
7. Turbidity change	Sedimentation/resuspension
8. Dissolved oxygen change	Oxidation/reduction
9. pH change	CO_2 uptake/release in water
10. Specific conductance change	Chemical precipitation/absorption
11. Redox change	Oxidation/reduction balance
IV. Nutrient dynamics	
12. Total phosphorus change	Phosphorus retention
13. Soluble reactive phosphorus change	Phosphorus microbial uptake
14. Nitrate + nitrite change	Denitrification/nitrogen retention
V. Avian use	
15. Abundance	Food source/habitat abundance
16. Species richness	Food source/habitat richness

ecosystems. We used one indicator of macrophyte function (above-ground net primary productivity), four indices of aquatic community development (two macroinvertebrate diversity indices, aquatic GPP, and algal species richness), six indicators of wetland bio-geochemistry (outflow concentrations of pH, conductivity, redox, turbidity, dissolved oxygen, and temperature), three indicators of nutrient retention (SRP, total P, and nitrate-nitrogen retention), and two indicators of avian use (bird abundance and species richness) in each wetland basin. The nonparametric similarity of each basin was estimated with a similarity index (SI, see chapter 2) calculated as:

$$SI = [I_s/I] \times 100$$

where I_s is the number of similar functional indicators in a given year and I is the total number of functional indicators (16).

9.3 RESULTS

9.3.1 Macrophyte Community Diversity

Emergent macrophyte communities developed in both wetlands to the point where almost all of the available shallow water area was vegetated by the end of the fourth growing season (Table 9.2). Vegetation appeared to converge by

Table 9.2 Coverage (m^2) in each basin by dominant vegetation communities, 1994 to 1999

Zone	1994 W1	1994 W2	1995 W1	1995 W2	1996 W1	1996 W2	1997 W1	1997 W2	1998 W1	1998 W2	1999 W1	1999 W2
Total basin	8903	8672	8903	8672	8903	8672	8903	8672	8903	8672	8903	8672
Community												
Open water	1451	2567	7746	8672	5333	5498	3579	3035	3490	2567	3276	2914
Algal mat	7452	6105										
Schoenoplectus tabernaemontani			1157		3205	3018	4149	4163	3668	2333	1914	876
Typha sp.					365	165	445	1440	98	3772	792	4882
Scirpus fluviatilis							205		392		258	
Nelumbo lutea/Potamogeton sp.							107	35	89			
Sparganium eurycarpum						418	9		1166		2261	
Sagittaria latifolia												
Spartina pectinata											401	
Total-rooted macrophytes	0	0	1157	0	3570	3661	4915	5638	5413	6105	5626	5758
% macrophyte cover	0	0	13.0	0	40.1	42.2	55.2	65.0	60.8	70.4	63.2	66.4

the third year with each basin dominated by *Schoenoplectus tabernaemontani* but with some presence of naturally colonizing *Typha* in each basin. *Typha* sp., a clonal dominant, was not planted but was seen in both wetlands approximately three months after the 1994 planting. In the first year of convergence, 1996, *Typha* actually had slightly greater cover in the planted W1 (Figure 9.2a).

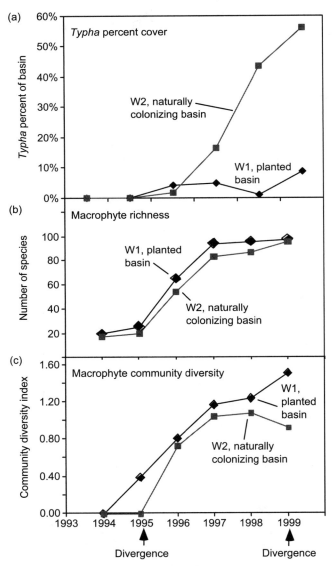

Figure 9.2 Indicators of macrophyte diversity in the two experimental wetlands, 1994 to 1999: (a) Percent of cover that is *Typha* sp.; (b) macrophyte species richness; (c) macrophyte community diversity index (CDI). Note divergence of macrophyte community diversity in 1995 (year 2) and in 1999 (year 6).

By 1997 (fourth year) it began to develop at a more rapid rate in W2 and by 1999 it has increased to 56% in basin W2, while remaining at only 9% cover in basin W1. At this point all shallow areas in W2 were almost completely dominated by *Typha* sp. Similar areas in W1 in 1999 were dominated by communities in the following order from the most to the least dominant: *Sparganium eurycarpum > Schoenoplectus tabernaemontani > Typha* sp. > *Scirpus fluviatilis*.

Macrophyte species richness increased dramatically in years 3 and 4 to almost 100 species in both basins (Figure 9.2b). Species richness did not prove to be a useful metric for comparing the two basins. When the wetlands are viewed in terms of plant cover using our community diversity index (CDI), which includes indications of evenness of plant cover as well as number of dominant communities, a different conclusion about macrophyte community diversity is reached (Figure 9.2c). The data show two years of basin divergence. The first year of divergence was in year 2 when W1, the planted basin, had 13% plant cover but W2 had no macrophyte development. The second macrophyte community divergence occurred in the sixth year (1999) when, after three years of similar plant cover in the two basins, different spatial community diversity developed. The CDI in W2 dropped as *Typha* formed close to a monoculture while it increased in W1 as a good balance among five communities developed, with four of those communities dominated by plants introduced in the planting (see Figure 9.2c and Table 9.2).

9.3.2 Macrophyte Productivity

Productivity ranged from 657 to $729 \, \text{g m}^{-2} \, \text{year}^{-1}$ in W1 and 756 to $1127 \, \text{g m}^{-2} \, \text{year}^{-1}$ in W2. In the first year that macrophyte above-ground net primary productivity (ANPP) was measured by our standard techniques (1997; fourth year of the study) it was statistically similar in both basins (Figure 9.3a). ANPP was statistically higher ($\alpha = 0.05$) in W2 in both year 5 (1998) and year 6 (1999) due to the dominance of the highly productive *Typha* that covered 43% and 56% of W2 in those two years respectively.

9.3.3 Algal Development

Dense benthic and floating metaphyton were significant in both wetlands throughout the study because of more than adequate nutrient concentrations in the inflowing river water. In the first year, large metaphyton mats developed in both basins. These mats were composed of *Hydrodictyon reticulatum* (L) Lag. and *Rhizoclonium* sp. along with extensive epiphytes of several species of Chlorophyta and Chrysophyta.[22] In the second year, the planted wetland (W1) carried an average of 80% of all of the genera identified while the unplanted wetland (W2) supported less (70% of those genera). By the third year (1996), that statistic was 79% for W1 and 74% for W2, illustrating some convergence. The apparent increase in algal diversity in W2 in the third year correlated with the natural colonization of macrophyte cover.[2] Macrophytes may have

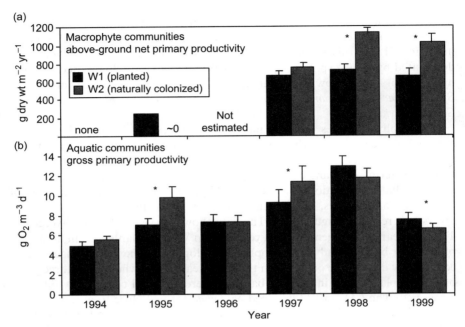

Figure 9.3   (a) Macrophyte net above-ground primary productivity (NAPP) in grams dry weight; (b) aquatic community gross primary productivity (GPP) in grams O_2 of two experimental wetlands. * Indicates significant difference between wetlands ($\alpha = 0.05$). Macrophyte NAPP was not estimated during the first three years by harvesting because of the feared impact that would have on the experiment when vegetation was just starting. Vegetation was different in 1995 because there were essentially no macrophytes in W2 in that year while the planted macrophytes covered 13% of W1.

increased habitats for the microphytes. By the fifth year (1998), the deep-water areas in the two wetlands started to become dominated by *Lemna minor*, causing dramatic decreases in metaphyton during some periods. Although there have been some differences in the two basins in 1998 and 1999, it appears that there has been general convergence of algal species since year 3 (1996).

Gross primary productivity (GPP) in the water column, as a functional measure of algae and submersed aquatic communities, was generally inversely related to macrophyte productivity (Figure 9.3b). When productivity of macrophytes was different between the two wetland basins, as was the case in the second (1995) and fifth (1998) and sixth (1999) years, GPP in the water column compensated by being higher in the basin with lower macrophyte productivity. When there were considerable macrophytes in the planted wetland (W1) but not in the unplanted wetland (W2) in the second year, water column GPP was higher in W2 by 38%. The situation reversed in 1999 when statistically higher water column GPP occurred in the planted wetland (W1) than in the naturally colonizing wetland (W2), consistent with the greater macrophyte biomass cover in W2.

Table 9.3 Benthic invertebrate diversity in two experimental wetlands, 1994 to 1999

Year	Total count diversity index		"Clean water" diversity index	
	W1	W2	W1	W2
1994 (planting)	0.63	0.69	0.45	0.60
1995 (divergence)	0.50	0.62	0.98	0.51
1996 (convergence)	0.88	0.83	0.88	0.86
1997 (convergence)	1.34±0.02	1.41±0.18	1.23±0.28	0.96±0.17
1998 (convergence)	0.58±0.43	0.82±0.56	0.58±0.43	0.73±0.45
1999 (divergence)	0.63±0.05*	0.91±0.12*	0.49±0.13	0.61±0.17

Indices are Shannon–Weaver.
W1 = planted wetland; W2 = natually colonizing wetland.
"Clean water" = all taxa execpt chironomids, oligochaetes, and tubificids.
*Significant differences ($\alpha = 0.10$) between wetlands from last three years for paired samples. Earlier years' data indicate overall diversity indices for entire wetland basin.

9.3.4 Macroinvertebrate Diversity

Macroinvertebrate taxa diversity remained statistically similar from 1996 through 1998 when plant communities were well developed but similar in both wetlands (Table 9.3). Taxa diversity diverged in 1999 when *Typha*-dominated W2 had a statistically higher benthic invertebrate diversity. Except for the first two years of wetland development, clean water species richness and diversity (count and diversity of all taxa minus chironomids, oligochaetes, and tubificids) were generally similar in both wetlands (Table 9.3).

9.3.5 Water Chemistry

Water chemistry changes as the water flowed through the wetlands showed several differences between the two wetlands in certain years and for certain parameters (Figure 9.4). In the sixth year, seven of nine parameters were different from inflow to outflow in each wetland ($\alpha = 0.05$) and the wetlands were different from one another ($\alpha = 0.05$) in five of nine water-quality parameters. Changes also occurred from year to year as macrophyte cover developed and began to influence water quality. For example, temperature increased through the wetlands on average in each of the first five years, but the increase was less each year (Figure 9.4a). By the sixth year (1999), temperature actually decreased on average as water flowed through the wetlands and was also statistically different in the two basins after two years of convergence.

Average dissolved oxygen (daily average of dawn and dusk readings; Figure 9.4b) was initially different in the two basins (1994 and 1995) and increased by 30% to 45% through W2, which did not have any macrophytes shading the water during this period. When macrophytes developed in the basins, as they did from 1995 in W1 and from 1996 in W2, dissolved oxygen increased by 25% or less. The dissolved oxygen increase was higher in the planted W1 in 1999 after three years of similarity.

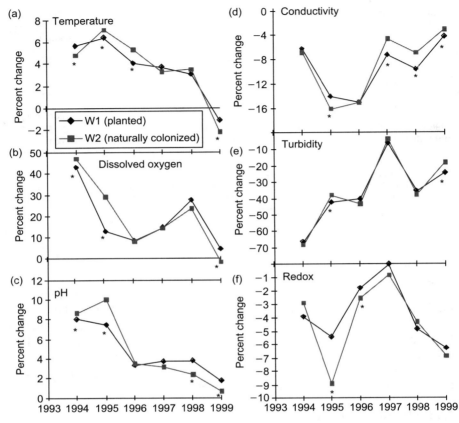

Figure 9.4 Water quality changes through experimental wetlands for 1994 to 1999: (a) temperature; (b) dissolved oxygen; (c) pH; (d) conductivity; (e) turbidity; and (f) redox potential. data points indicate overall percent change from inflow to outflow of averages of all inflow and outflow concentrations during that year. Each data point represents hundreds of samples per year. * Indicates significant difference between the wetlands as determined by t-test comparing outflow concentrations ($\alpha = 0.05$).

pH (Figure 9.4c) was significantly different in the two wetlands during four of the six years. It increased more in the then-unvegetated W2 than in the planted W1 in the early years (1994 and 1995). This pattern reversed in later years when W1 had significantly higher pH than the naturally colonizing wetland W2 in years 5 and 6 (1998 and 1999).

Conductivity changes (Figure 9.4d) were significantly different between the two wetlands during the second year and in the final three years. Conductivity decreased more in the then-unvegetated W2 in 1995. Then the pattern switched, with conductivity decreasing more in the planted W1 than in W2 in 1997 through 1999. The pattern also showed that, except for the first year, there was generally less change in conductivity through the wetlands with each succeeding each year as macrophyte vegetation developed.

Turbidity, as a measure of suspended materials that include allochthonous clay and silt particles and autochthonous algal cells, decreased through the wetland every year (Figure 9.4e). Changes through the two wetlands were different in the second (1995) and sixth (1999) years, the same two years in which the macrophyte diversity (CDI) diverged. In 1995 turbidity decreased more in the planted W1 than in the still-unvegetated W2. In 1999, turbidity also decreased more in the diverse wetland W1 than in the *Typha* monoculture W2.

Redox potential (Figure 9.4f) in the outflow water has not differed much in the two wetlands except early in the study. It also appears that redox potential is decreasing more each year in both wetlands' outflows. In 1997, redox potential was essentially unchanged from inflow to outflow. In 1999 it decreased on average between 6% and 7%.

9.3.6 Nutrient Retention

Nutrient (phosphorus and nitrogen) reduction was consistent and significant in both wetlands throughout the six years for the three nutrient species analyzed (Figure 9.5). Total phosphorus concentration annual reductions ranged from 18% to 73% per wetland and there appears to be a trend of less total phosphorus retention each year. Percentage reduction of soluble reactive phosphorus (SRP) through the wetland basins has been higher (50% to 90% removal) and more consistent than that of total phosphorus. Percentage reduction of nitrate plus nitrite–nitrogen has remained consistent from year to year (generally between 20% and 50% reduction) in each wetland. Decreases from inflows to outflow have been significantly different for both wetlands and all years ($\alpha = 0.05$). There were few statistically significant differences in nutrient retention between the wetlands over the six-year study with only two differences out of a possible 18 parameter years.

9.3.7 Avian Use

A total of 150 bird species were identified at the Olentangy River Wetland Research Park from 1992 to 1999, with a 20% increase in species richness in the first year after wetland construction, another 8% increase during the second year, and an additional 5% increase in the third year. The creation of the wetlands resulted in the addition of about 35 wetland-specific bird species to the site. Because of their proximity, it has been generally difficult to compare avian use of the two wetlands. Nevertheless, surveys in the second year and reported previously did find that the planted W1 consistently supported a greater number of species (nesting and migratory) and more individuals than the unplanted W2.[2] Two species in particular, the sora (*Porzana carolina*) and marsh wren (*Cistothorus palustris*), were found exclusively in the planted wetland in the second year. By the third year, with the development of vegetation cover in the unplanted W2, differential bird use between the two wetlands declined and similar numbers and richness of species were found in each. With the development of different macrophyte communities in 1999 in the two wetlands,

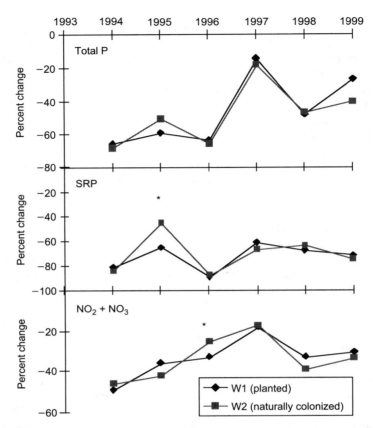

Figure 9.5 Nutrient retention in the two experimental wetlands for 1994 to 1999: (a) total phosphorus (total P); (b) soluble reactive phosphorus (SRP); and (c) nitrite plus nitrate–nitrogen ($NO_2 + NO_3$). Data points indicate overall percent change from inflow to outflow of averages of all inflow and outflow concentrations. Each data point represents overall results of weekly readings over a year. * Indicates significant difference between the wetlands as determined by t-test comparing outflows ($\alpha = 0.05$).

differences in bird use between the basins were observed (Figure 9.6). There were significantly greater numbers of red-winged blackbirds (*Agelaius phoeniceus*) in W2 and significantly greater numbers of song sparrows (*Melospiza melodia*) in W1. Red-winged blackbirds have a great affinity for *Typha* while song sparrows favor the less dense vegetation in wetland W1.

9.3.8 Basin Similarity

According to our community diversity index (CDI) and similarity index (SI) of wetland structure and function, there were two years out of six where the two wetlands diverged (see Table 9.4 and Figure 9.7). In the second year (1995) after wetland construction, substantial macrophyte cover developed only in the planted wetland W1 as expected and none was present in W2; therefore

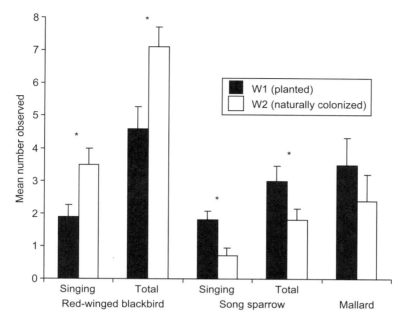

Figure 9.6 Bird observations comparing the two experimental wetlands in 1999. * Indicates significant difference between wetlands ($\alpha = 0.05$).

the CDI index was different in the two wetlands (it was 0.0 for the unplanted wetland). Only 13% of the ecosystem indicators were similar (SI = 13%).

In years 3, 4, and 5, the macrophyte CDI was similar between the two wetlands and the ecosystem similarity converged to 75% to 87%. *Typha* invasion into the unplanted wetland in the sixth year (1999) caused a second divergence in the CDI between the two basins (Figure 9.3c) and wetland function diverged a second time (Figure 9.7). During this year the similarity of the two wetlands dropped to 44%.

The similarity index between the two wetlands for each year is plotted versus the difference in the community diversity index (CDI) between the basins for the same year (Figure 9.8). The regression ($r^2 = 0.53$) suggests that when the two basins were different in community diversity (as in 1995 and 1999) the two wetlands were dissimilar in structure and function (SI = 13% to 43%). When the basins were similar in community diversity, the two basins were similar in structure and function (SI = 70% to 88%).

9.4 DISCUSSION

9.4.1 Community Diversity and Ecosystem Function

Our study suggests that structure and function are related to macrophyte community diversity in wetlands. In years when macrophyte community

Table 9.4 Summary of 16 indices comparing Wetland 1 (W1) and Wetland 2 (W2) for 6 years, 1994 to 1999

	1994	1995	1996	1997	1998	1999
Macrophyte diversity, CDI	W1 = W2	W1 > W2	W1 = W2	W1 = W2	W1 = W2	W1 > W2
I. Macrophyte community function						
NAPP	W1 = W2	W1 > W2	W1 = W2	W1 = W2	W1 < W2	W1 < W2
II. Aquatic community development						
Algal species richness	W1 = W2	W1 > W2	W1 > W2	W1 = W2	W1 = W2	W1 = W2
Aquatic metabolism	W1 = W2	W1 < W2	W1 = W2	W1 < W2	W1 = W2	W1 > W2
Benthic invertebrate diversity	W1 = W2	W1 < W2	W1 = W2	W1 = W2	W1 = W2	W1 < W2
Clean water species richness	W1 < W2	W1 > W2	W1 = W2	W1 = W2	W1 = W2	W1 = W2
III. Bio-geochemistry						
Temperature	W1 > W2	W1 < W2	W1 < W2	W1 = W2	W1 = W2	W1 > W2
Turbidity	W1 = W2	W1 < W2	W1 = W2	W1 = W2	W1 = W2	W1 < W2
Dissolved oxygen	W1 < W2	W1 < W2	W1 = W2	W1 = W2	W1 = W2	W1 > W2
pH	W1 < W2	W1 < W2	W1 = W2	W1 = W2	W1 > W2	W1 > W2
Conductivity	W1 = W2	W1 > W2	W1 = W2	W1 > W2	W1 < W2	W1 < W2
Redox	W1 = W2	W1 > W2	W1 > W2	W1 = W2	W1 = W2	W1 = W2
IV. Nutrient dynamics						
Total P	W1 < W2	W1 = W2	W1 = W2	W1 = W2	W1 = W2	W1 = W2
SRP	W1 = W2	W1 > W2	W1 = W2	W1 = W2	W1 = W2	W1 = W2
$NO_3 + NO_2$	W1 = W2	W1 = W2	W1 > W2	W1 = W2	W1 = W2	W1 = W2
V. Avian Use						
Bird abundance	W1 = W2	W1 > W2	W1 = W2	W1 = W2	W1 = W2	W1 < W2
Bird species richness	W1 = W2	W1 > W2	W1 = W2	W1 = W2	W1 = W2	W1 = W2
Similarity index: %	69	13	75	87	81	44

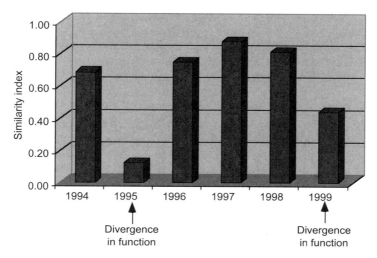

Figure 9.7 Indicators or similarity between two experimental wetland basins, 1994 to 1999.

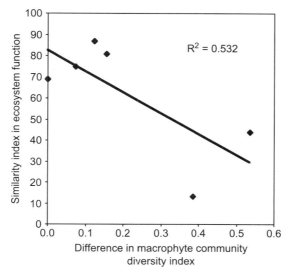

Figure 9.8 Plot of ecosystem function similarity index (SI) between the two wetlands versus difference in macrophyte community diversity indices (CDI) for the same years. Data illustrate that when the two wetlands were different in CDI, they were also dissimilar in ecosystem function. When wetlands had similar CDI, their ecosystem function was similar.

diversity was similar between our two wetlands, ecosystem indicators were also similar. When wetlands had different macrophyte community, diversity, structure, and function were different. Our macrophyte community diversity index (CDI) is not a traditional species diversity of small, managed plots but rather is a measure of the spatial diversity that can be seen from good aerial

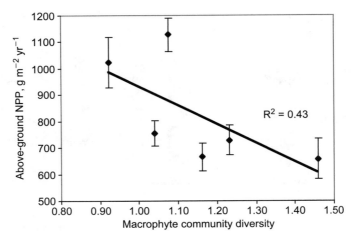

Figure 9.9   Relationship of above-ground net primary productivity (ANPP) to macrophyte community diversity index (CDI). Each data point is one of two 1-ha basins for years 1997 to 1999.

photography. It includes both the richness and evenness of community patterns.

When above-ground net primary productivity is plotted versus CDI for six wetland-years, a negative relationship can be observed (see Figure 9.9) when the macrophyte community is diverse in terms of spatial patterns. While our study uses a measure of spatial community diversity (the diversity of pixels in a color map) and other studies count individual plants or stems, some comparison can be made between our general findings and those suggested by others on the relationship between productivity and diversity. Plot and small-scale terrestrial research often illustrates that plant diversity has a positive effect on primary productivity. While this may hold true for managed diversity of plants on small plots, our results show exactly the opposite for large-scale wetlands. Higher diversity on a community scale (that is, how many different macrophyte communities are present and how even their distribution is) not only does not enhance productivity, our study suggests that it reduces it. This tempers the universality of conclusions suggested by studies in grasslands,[27,28] and more recently for mesocosm wetlands,[6] that diversity enhances productivity. These studies were carried out on smaller scales than ours and were also subject to some manipulation to maintain certain diversities.[23,24]

There was no significant relationship seen in our study between community diversity index (CDI) and gross primary productivity of algal communities ($r^2 = 0.004$) or nutrient and sediment retention (phosphorus: $r^2 = 0.017$; nitrates: $r^2 = 0.004$; turbidity: $r^2 = 0.05$). This last lack of a significant relationship is in direct conflict with the conclusion by Engelhardt and Ritchie[6] that "higher vascular plant richness in wetlands may potentially yield up to 25% more algal biomass ... and retaining (sic) up to 30% more potentially polluting nutrients, such as P." While these conclusions may be applicable to small-scale experiments they were not supported by our full-scale wetland ecosystem study.

9.4.2 Productivity as the Independent Variable

We suspect that asking what the effect of diversity on productivity is, may be the wrong question to ask as ultimately, physical and chemical factors affect productivity, which in turn often determine biodiversity which feedbacks and affects the physics and chemistry. If we turn the cause and effect around by rotating Figure 9.9 90 degrees, we have illustrated what is already fairly well established in the wetland literature:[25–27] species richness and diversity in freshwater marshes and lakes are inversely related to biomass and productivity. Productivity in turn is related to factors such as nutrients, sunlight, and flooding. When primary productivity is viewed in this way as the independent variable, then several ecosystem functions, which do not correlate well with macrophyte community diversity, correlate much better with marsh net primary productivity. Total phosphorus, nitrate–nitrogen, and to a lesser degree, turbidity reductions are positively related to ANPP (Figure 9.10). Although long suspected, there has been no study to the author's knowledge that has confirmed this relationship at a full-ecosystem scale. Tanner[28] found a relationship between biomass (productivity) and nitrogen uptake similar to the one we found between ANPP and nitrogen retention. His New Zealand study involved gravel-bed mesocosms fed by high-concentration dairy wastes. Therefore, nitrogen concentrations were substantially greater in his study than ours. Experiments in Norway[29] involving four small wetlands showed that suspended sediment retention increased with greater vegetation cover, not so much because of any increase in sedimentation but because of a reduction in resuspension. Factors such as this are a more likely explanation for the Figure 9.10 patterns than any direct uptake by the macrophytes.

9.4.3 Diversity at Different Levels

It is often assumed that if one trophic level of an ecosystem is diverse, that diversity will "spill over" to other parts of the ecosystem. This is a "structure affecting structure" argument and is partially the theory of consumer control of species diversity.[30] The effect of macrophyte community diversity on benthic invertebrate diversity was compared for 1997 to 1999 (three wetland-years; Figure 9.11) in the same fashion as the macrophyte comparison above. There was a weaker inverse relationship ($r^2 = 0.23$) than that for the effect of macrophyte productivity on invertebrate diversity ($r^2 = 0.43$). Macrophyte community diversity does not appear to necessarily increase benthic diversity; in fact our data suggest that in some cases it reduced it.

9.4.4 Aquatic Consumers

Aquatic consumers were sampled in the wetlands immediately after the year 6 divergence in macrophyte community diversity. There were more *Rana catesbeiana* (bullfrog tadpoles), *Nerodia sipedon* (northern water snakes) and

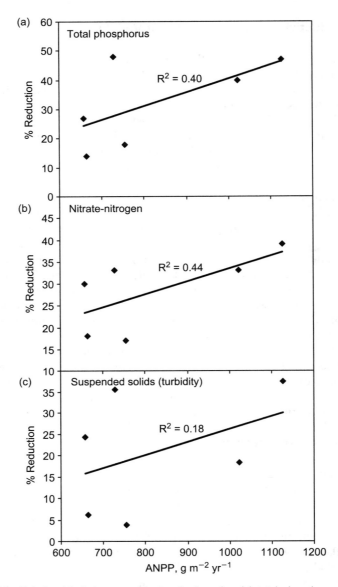

Figure 9.10 Relationship between nutrient reductions for: (a) total phosphorus; (b) nitrate–nitrogen; and (c) suspended solids (measured as turbidity) as a function of above-ground net primary productivity (ANPP) of macrophytes.

Lepomis cyanellus (green sunfish) in the naturally colonizing wetland (W2) than in the planted wetland (W1) in the early 2000 sampling (Table 9.5).

The greater abundance of tadpoles (which consume detritus and small insects) and snakes (which consume tadpoles) suggests a more powerful detrital food chain in W2 in early 2000, which is supported by a macrophyte net

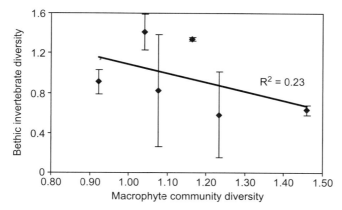

Figure 9.11   Relationship of benthic invertebrate diversity to macrophyte community diversity index (CDI). Each data point is one of two 1-ha basins for years 1997 to 1999.

Table 9.5   Comparison of amphibians, reptiles, and fish caught in 20 traps in two wetlands in spring of 2000. Numbers are organism caught per trap-day

Species	Wetland 1 (mean±SE)	Wetland 2 (mean±SE)
Rana catesbeiana	0.015±0.006	0.14±0.024*
Rana clamitans	0.03±0.01	0.02±0.01
Nerodia sipedon	0.006±0.004	0.023±0.008*
Lepomis cyanellus	21±2	34±3*

* Indicates significantly higher number compared to wetland 1 ($\alpha = 0.05$)[31].

primary productivity that was almost 50% higher there in 1999 than in W1. Green sunfish, a species common in wetlands because of its tolerance for warm summer temperatures, may also have been more abundant in W2 because of slightly cooler temperatures (water temperature was significantly cooler in W2 in 1999) caused by the higher macrophyte productivity which, in turn, provides the water with more shade. Greater macrophyte productivity also provides more locations for hiding from predators, including wading birds. The comparison between the wetland basins on fish populations should be made with caution as mark and recapture studies carried out later in 2000[31] showed significantly higher fish populations in W1 after vegetation in both wetlands was lost by a muskrat eatout.

9.4.5   Replication and Experimental Scale

Whole ecosystem studies, such as the one being conducted here, can provide useful comparisons of ecosystem functions when performed over a long period, even when replication is not possible due to the large size of these systems. We believed that 1 ha was of a sufficient size to provide development of all potential ecosystem engineers including ducks, geese, frogs, snakes, and muskrats. C. Korner (conference communication) has made the case for more understanding and acceptance of large-scale experiments and observations in the literature, if for no other reason than that they are a necessary check on theories being published from smaller-scale studies chosen primarily because of elegant replications and statistics. Experiments at small scales allow us to have confidence in the ecological function of small-scale systems. We consider the conclusions of research such as those by Engelhardt and Ritchie's wading pool study[6] on wetland macrophyte richness effects on wetland function as a major overreach, given the limitations of small experiments representing full-scale ecosystem function. Using the appropriate experimental scale for extrapolating to the scale of ecosystem function is essential.[32] The large size of the experimental ecosystems and the long period of the study compensates to some extent for the lack of replication. The insights of diversity and ecosystem function would not have manifested themselves in a short-time, small-scale experiment. Many of our findings from this large-scale experiment illustrate that cause and effect on the role of diversity and function may be the reversed of what is postulated in many previous studies.

Carpenter et al.[18] suggest that rather than having an unreplicated experiment, one could actually have two or more similar ecosystems, each with different management or experimental schemes. In our case, we have two large wetland basins, each with a different means of plant propagule introduction — one by humans and nature and the other by nature only. In effect there is not a control where propagules are weeded or otherwise prevented from entering the wetlands. When one wetland became dominated by *Typha* and the other was not, we were in a good position to see the effects of *Typha* invasions and natural reduction in plant diversity. We believe that use of many physical, chemical, and biological indicators gives strength to our arguments that the basins are similar or dissimilar in any given year.

Specific to this wetland experiment, we compared one of the full-scale 1-ha wetlands with ten 1-m^2 mesocosms in similar hydrologic conditions.[33] While the mesocosm results had statistical rigor because of replication, they had so many scale effects and ecosystem simplifications as to prevent conclusions from being extended to large-scale wetlands without verification with full-scale wetlands. Mesocosm-scale wetlands could not duplicate hydrodynamic features and lacked important aspects of full-scale wetlands such as wind mixing. Also, mesocosm studies of wetlands do not include proportional scales of ducks, geese, muskrats, beavers, and wading birds, all of which can be important "ecosystem engineers" of wetland function.[26,34-35]

We believe that our 1-ha wetlands, with the large area for all processes to manifest themselves in time and space, are inherently less variable and thus require far fewer replications. As suggested by Carpenter et al.[18] we believe that there is no optimal scale for ecosystem experiments. But if we desire a system that can grow macrophytes, algae and invertebrates equally in the water column, and at the same time allow for a complete food web including ducks, wading birds, muskrats, and other important parts of any natural wetland, an experiment at 1-ha scale is much more likely to yield true ecosystem functions than hundreds of 1-m² mesocosms. The use of easily measured multiple indicators allowed us to have more confidence in our relative comparison of these large-scale experimental units than if we relied on only a few indices.

ACKNOWLEDGMENTS

We were assisted by many others in this multi-year study including Changwoo Ahn, Virginie Bouchard, Charles Boucher, Randy Bruins, Lesley Carmichael, Melanie Ford, Scott Frazier, Amie Gifford, Sarah Harter, Cheri Higgins, Megan Hunter, Anne Johnson, David L. Johnson, Sharon Johnson, John Kantz, Katie Kleber, Michael Liptak, Kathy Metzker, Tony Minamyer, Holly Montgomery, Robert Nairn, Uygar Özesmi, Jeff Pearcy, Doug Spieles, Lisa Svengsouk, Rachael Thiet, Paul Weihe, and Renée Wilson. Also appreciated are a number of other students and volunteers who helped with the twice daily and weekly sampling and laboratory work. This research was made possible by funding from Mid-American Waste Systems, Inc. Ohio Agricultural Research and Development Center, and several other private and public sources. Olentangy River Wetland Research Park Publication Number 2004–05.

REFERENCES

1. Mitsch, W.J. and Wilson, R.F. Improving the success of wetland creation and restoration with know-how, time, and self-design. *Ecol. Appl.* 6, 77–83, 1996.
2. Mitsch, W.J. et al. Creating and restoring wetlands: a whole-ecosystem experiment in self-design. *BioScience* 48, 1019–1030, 1998.
3. Mitsch, W.J. et al. To plant or not to plant — response. *BioScience* 50, 188–191, 2000.
4. Streever, B. and Zedler, J.B. To plant or not to plant. *BioScience* 50, 188–191, 2000.
5. National Research Council. Compensating for wetland losses under the Clean Water Act. Committee on Mitigating Wetland Losses, National Academy Press, Washington, D.C., 2001.
6. Engelhardt, K.A.M. and Ritchie, M.E. Effects of macrophyte species richness on ecosystem functioning and services. *Nature* 411, 687–689, 2001.
7. Odum, H.T. and Pigeon, P., Eds. *A Tropical Rain Forest.* AEC Division of Technical Information, Oak Ridge, TN, 1970.
8. Likens, G.E. et al. *Biogeochemistry of a Forested Ecosystem.* Springer-Verlag, New York, 1977.
9. Sullivan, T.J. Whole-ecosystem nitrogen effects research in Europe. *Env. Sci. Tech.* 27, 1482–1486, 1993.
10. Beier, C. and Rasmussen, L. Effects of whole-ecosystem manipulations on ecosystem internal processes. *Trends Ecol. Evol.* 9, 218–223, 1994.
11. Schindler, D.W. Evolution of phosphorus limitation in lakes. *Science* 195, 260–262, 1977.
12. Schindler, D.E. et al. Influence of food web structure on carbon exchange between lakes and the atmosphere. *Science* 227, 248–251, 1997.

13. Carpenter, S.R. et al. Chlorophyll variability, nutrient input and grazing: evidence from whole-lake experiments. *Ecology* 77, 725–735, 1996.
14. Carpenter, S.R. et al. Impact of dissolved organic carbon, phophorus and grazing on phytoplankton biomass and production in experimental lakes. *Limnol Oceanogr.* 43, 73–80, 1998.
15. Odum, H.T. et al. "Recycling treated sewage through cypress wetlands," in *Wastewater Renovation and Reuse*, D'Itri, F.M., Ed. Marcel Dekker, New York, 1977, pp. 35–67.
16. Mitsch, W.J. et al. Phosphorus retention in constructed freshwater riparian marshes. *Ecol. Appl.* 5, 830–845, 1995.
17. Carpenter, S.R. "The need for large-scale experiments to assess and predict the response of ecosystems to perturbation," in *Successes, Limitations, and Frontiers of Ecosystem Science*, Pace, M.L. and Groffman, P.M., Eds. Springer-Verlag, New York, 1998, pp. 287–312.
18. Carpenter, S.R. et al. Evaluating alternative explanations in ecosystem experiments. *Ecosystems* 1, 335–344, 1998.
19. Mitsch, W.J. and Kaltenborn, K.S. Effects of copper sulfate application on diel dissolved oxygen and metabolism in the Fox Chain of Lakes. *Trans. Ill. State Acad. Sci.* 73, 55–64, 1980.
20. U.S. Environmental Protection Agency. *Handbook for Methods in Water and Wastewater Analysis.* U.S. Environmental Protection Agency, Cincinnati, OH, 1983.
21. American Public Health Association. *Standard Methods for the Analysis of Wastewater*, 17th ed. APHA, Washington, D.C., 1989.
22. Wu, X. and Mitsch, W.J. Spatial and temporal patterns of alga in newly constructed freshwater wetlands. *Wetlands* 18, 9–20, 1998.
23. Tilman, D., Wedin, D., and Knops, J. Productivity and sustainability influenced by biodiversity in grassland ecosystems. *Nature* 379, 718–720, 1996.
24. Tilman, D., Knops, J., Wedin, D., Reich, P., Ritchie, M., and Siemann, E. The influence of functional diversity and composition on ecosystem processes. *Science* 277, 1300–1302, 1998.
25. Moore, D.R.J. et al. Conservation of wetlands: do infertile wetlands deserve a high priority. *Biological Conservation* 47, 203–217, 1989.
26. Mitsch, W.J. and Gosselink, J.G. *Wetlands*, 3rd ed. Wiley, New York, 2000.
27. Mitsch, W.J. and Jørgensen, S.E. *Ecological Engineering and Ecosystem Restoration.* Wiley, New York, 2004.
28. Tanner, C.C. Plants for constructed wetland treatment systems — a comparison of the growth and nutrient uptake of eight emergent species. *Ecol. Eng.* 7, 59–83, 1996.
29. Braskerud, B.C. The influence of vegetation on sedimentation and resuspension of soil particles in small constructed wetlands. *J. Environ. Qual.* 30, 1447–1457, 2001.
30. Worm, B. et al. Consumer versus resource control of species diversity and ecosystem functioning. *Nature* 417, 848–851, 2002.
31. Gifford, A.M. The Effect of Macrophyte Planting on Amphibian and Fish Community use of Two Created Wetland Ecosystems in Central Ohio. MSc Thesis, Ohio State University, Columbus, OH, 2002.
32. Pace, M.L. "Getting it right and wrong: Extrapolations across experimental scales," in *Scaling Relations in Experimental Ecology*, Gardner, R.H., Kem, W.M., Kennedy, V.S., and Petersen, J., Eds. Columbia University Press, New York, 2001, pp. 157–177.

33. Ahn, C. and Mitsch, W.J. Scaling considerations of mesocosm wetlands in simulating large created freshwater marshes. *Ecol. Eng.* 18, 327–342, 2002.
34. Jones, C.A., Lawton, J.H., and Shachak, M. Organisms as ecosystem engineers. *Oikos* 69, 373–386, 1994.
35. Jones, C.A., Lawton, J.H., and Shachak, M. Positive and negative effects of organisms as physical ecosystem engineers. *Ecology* 78, 1946–1957.

CHAPTER 10

The Joint Use of Exergy and Emergy as Indicators of Ecosystems Performances

S. Bastianoni, N. Marchettini, F.M. Pulselli, and M. Rosini

Orientors have been introduced at the interface between ecology and thermodynamics. Two have been chosen to compare the characteristics of ecological systems: (1) exergy, which is related to the degree of organization of a system and represents the bio-geochemical energy of a system; and (2) emergy, which is defined as the total amount of solar energy directly or indirectly required to generate a product or a service. They represent two complementary aspects of a system: the actual state and the past work needed to reach that state. The ratio of exergy to the emergy flow indicates the efficiency of an ecosystem in producing or maintaining its organization. The ratio of the variations of exergy and emergy flow over time gives a general definition of "pollutant" and "nutrient" of a system.

10.1 INTRODUCTION

Ecology has given many examples of numeraires that can be used as indicators of performances in ecosystem analysis. With the same scope, thermodynamics and general system theory has developed functions that have been widely used as holistic indicators (see Von Bertallanffy, 1968; Odum, 1983, 1988; and Prigogine, 1955, for example). In the intention of those who invented or adapted these concepts, these functions are "orientors" because they show tendencies in complex, adaptive, hierarchical systems, either towards a maximization of the emergy flow (according to Odum, 1983, 1988); or towards a maximization of exergy content (according to Jørgensen, 1992). These two approaches are not necessarily in contrast. On the contrary, they describe the possible behavior of a system at different stages of its development (Patten et al., 2002; Bastianoni, 2002). For a general description of orientors, see Müller and Leupelt (1998). For complex, adaptive, hierarchical systems, see Patten et al. (2002).

We can say that *emergetics* and *exergetics* are two parallel paths to adapting classical thermodynamics to the specific condition of our living biosphere, adopting a diachronic and a synchronic perspective, respectively.

Exergy-oriented researchers root themselves in the terrestrial specificity defining, as a reference state, the mean composition of the Earth's crust, or of the atmosphere, or of a peculiar local context, considered to be in a steady state and without introducing further assumptions about the datum.

Emergy-oriented scientists, on the other hand, base their descriptions on previous knowledge of the biosphere, which has a very general tendency in concentrating energy in more and more condensed forms through trophic chains, metabolism of organisms as well as through bio-geochemical cycles.

With exergy we have a measure, surely closer to classical science canons, of a system's distance from thermodynamic equilibrium, with a snapshot of our environment (or of a more restricted, local bulk), identified with its mean values. On the other hand, the emergy description is more dependent on the actual metabolism of the biosphere and its evolutionary history, building its transformities — the coefficients used to express the ecological value of a material, a flux, or a specific good — on that background.

Emergy analysis can provide a budget of solar energy memory necessary (e.g., to produce a university-level book of 200 pages). In the same way we can express the exergetic value of that book, considering the uncompressible information of the text (giving 2.9×10^{-21}) for each bit of information, at room temperature) plus the chemical potential of the book — that is, the energy extractable with a complete combustion of the book itself. But neither exergy nor emergy can say anything about the actual scientific or artistic content of the book — the meaning that it can provide to a reader.

In the same way an exergetic potential, or an emergy storage, could be either a resource or a toxic substance, depending on the specific ecological meaning that it will express when it will be in contact with a specific organism, or an ecological association in the environment.

10.2 EXERGY AND ECOLOGY

Exergy is the maximum work that can be obtained from a system when the system is brought from its present state to the state of thermal, mechanical, and chemical equilibrium with the surrounding environment (see chapter 2).

The basic idea of the application of exergy to ecological systems is that as the exergy stored in raw clay is less than the same amount of clay as bricks, which is less than the same number of bricks organized in a building; the same holds, with larger differences, when biology is involved. Following the same reasoning, the exergy content of a certain number and types of atoms is not the same if they are random atoms, atoms in a protein, in a cell, in a plant or in an animal. Jørgensen and co-workers have developed a theory and a series of formulae to estimate the exergy content in living organisms and ecosystems (Jørgensen, 1992; Jørgensen et al., 1995, 2000 and Fonseca et al., 2000). The result, due to the unavoidable approximations and hypotheses, is more an index related to exergy than the "real" exergy content, meant as work which can be extracted by these organisms or ecosystems. For instance, the application to aquatic ecosystems produced the formula for exergy "density" (J/l):

$$Ex = RT \sum_i \left[c_i \ln\left(\frac{c_i}{c_{i,eq}}\right) + (c_i - c_{i,eq}) \right]$$

where c_i is the concentration of the element under concern in compartment i of the system, $c_{i,eq}$ is the "hypothetical" concentration of the same compartment, but at thermodynamic equilibrium (Jørgensen and Meyer, 1977). Starting from the equation above, following considerations about the relationship between concentration and probability and between probability and information content (see, for example, Bendoricchio and Jørgensen, 1997; Fonseca et al., 2000), the exergy index has been derived as:

$$Ex = \sum_{i=0}^{n} \beta_i c_i$$

where β_i are weighting factors that the various components (i) of the ecosystem possess due to their chemical energy and to the information embodied in the DNA (Bendoricchio and Jørgensen, 1997; Fonseca et al., 2000; Jørgensen et al., 2000).

This procedure has several shortcomings, as recognized by Jørgensen and co-workers themselves, and it is not strictly based on thermodynamics (Fonseca et al., 2000). Nonetheless, the attempt to use thermodynamics, namely exergy, for living systems is in our opinion a goal to pursue, especially if exergy is to be used in a sustainability framework: in this case we have to be able to distinguish, for example, between living and nonliving organisms.

The distribution of the exergy among compartments, which, when summarized, gives Ex_s of the system, is viewed as a result of the fluxes taking place in the system and is thus a result of the system function as a whole.

10.3 EMERGY AND ECOLOGY

In energy transformations, output has less energy but is (usually) of higher quality than the input(s). Many joules of low quality are needed for a few joules of high quality; thus, in many cases, it is not correct to use energy as a measure of a system contribution. "A joule of sunlight, a joule of coal, a joule of human effort are of different quality and represent vastly different convergences of energy in their making" (Odum, 1991). For this reason, to compare all kinds of energy on a common basis, solar transformity (referred to throughout the chapter as transformity) has been defined as the solar energy directly and indirectly required to generate one joule of a product. (Solar) emergy is defined in chapter 2.

We can view emergy as the work that the biosphere has to do in order to maintain a system far from equilibrium or in order to reproduce an item once it has been used. If natural selection has been given time to operate, the higher the emergy flux necessary to sustain a system or a process, the higher their hierarchical level and the usefulness that can be expected from them (the "maximum empower principle," see Odum, 1988). This is often not sufficient when dealing with shorter runs and with systems involving relations between humans and natural systems.

Among emergy-related indices, the empower density is particularly interesting from an ecological viewpoint: it is the emergy flow per unit time and unit area, and is a measure of the spatial and temporal concentration of emergy flow within a system. A high value of this index can signify a high stress on the environment due to large quantities of inputs converging on the system, or of a situation where space is becoming a limiting factor for further development of the system.

Previous work has been done on the relationship between energy and information in the transmission of messages (Tribus and McIrvine, 1971) and the relation between emergy and information in biological systems of different dimensions (Odum, 1988; Keitt, 1991). In his Crafoord prize lecture in Stockholm, Odum stated that the emergy/information ratio is a measure of the information hierarchy: the higher the energy hierarchy of a system, the higher the ratio in sej/bit (Odum, 1988). He also discussed the results of a comparison of four types of system at different levels having the same number of bits of structural information. One thousand bits of molecular glucose, algae, forest and science journal were examined. The production of the same quantity of information on different spatial scales requires quite different energy inputs. This gives a scale factor that cannot be obtained from simple energy analysis. The emergy/information ratio was greatest for the science journal, followed by the forest. Analyzing energy to information ratio (Tribus and McIrvine, 1971), an inverse result was obtained.

10.4 THE RATIO OF EXERGY TO EMERGY FLOW

The relationship between emergy and information used by Odum gives a good indication of general character but has problems related to Shannon's formula. This is why we replaced the measure of information with exergy and introduced a relation between emergy flow and exergy to indicate the solar energy equivalent required by the ecosystem to produce or maintain a unit of organization or structure of a complex system (Bastianoni and Marchettini, 1997). At the beginning, the emergy flow to exergy ratio was used in order to maintain coherence with the definition of transformity and point out the differences: transformity is the emergy that contributes to a production system divided by the energy content of a product. The emergy flow to exergy ratio, on the other hand, represents an emergy flow divided by the exergy of the whole system driven by this emergy flow. The dimensions of this ratio are sej/(J·time). In general the reciprocal is more meaningful since it would present the state of the system (as exergy) in comparison with the inputs (as emergy). Therefore the exergy/empower ratio can be regarded as the efficiency of a system, even though this ratio is not dimensionless, as efficiency usually is, as it has the dimension of time. Svirezhev (1999) found this fact normal, since this concept, in his opinion, resembles that of a relaxation time — that is, the time necessary to recover from disturbances.

This parameter indicates the quantity of external input necessary to maintain a structure far from equilibrium. The higher its value, the higher the efficiency of the system. If the exergy/empower ratio tends to increase (apart from oscillations due to normal biological cycles), it means that natural selection is making the system follow a thermodynamic path that will bring the system to a higher organizational level.

This efficiency index have been applied to several aquatic ecosystems. Two of the water bodies used for comparison are in North Carolina, U.S., and are part of a group of similar systems, constructed to purify urban wastewater. Of the six ponds that compose the system, three are "control" ponds that receive a mixture of estuarine waters and purified waters from the local sewage treatment plant, and three are "waste" ponds that receive estuarine waters mixed with more polluted, or nutrient-rich, wastewater. Plants and animals were introduced to the ponds to create new ecosystems by natural selection. The different conditions have produced quite different ecosystems in the two types of pond, with a prevalence of phytoplankton and crustaceans in the waste ponds and a great abundance of aquatic plants in the control ponds (Odum, 1989; Bastianoni and Marchettini, 1997).

The third water body was the lake of Caprolace in Latium, at the edge of the Circeo National Park. This is an ancient natural formation fed mostly by rainwater, plus an input rich in nitrogen, phosphorus, and potassium that percolates from nearby agricultural land. Human impact is low. A quantity of fish is taken each year, but is not such that the fish population decreases (Bastianoni and Marchettini, 1997).

The fourth ecosystem was Lake Trasimeno in Umbria (Ludovisi, 1998; Ludovisi and Poletti, 2003). Lake Trasimeno is the largest lake in peninsular Italy, but is very shallow, its theoretical water retention time is very high and the accumulation processes are favored. The water level of the lake shows strong fluctuations: under particular meteorological conditions (several years with annual rainfall below 700 mm), hydrological crises may occur.

The fifth system was a fish-farming basin in the central part of a lagoon in Venice. Fish-farming basins consist of peripheral areas of lagoon surrounded by banks in which local species of fish and crustaceans are raised. Saltwater from the sea and freshwater from canals and rivers are regulated by locks and drains. Control of water levels, salt content, and drainage towards the sea are part of an ancient tradition which is an economic and cultural heritage.

The sixth and the seventh ecosystems were two internal lagoons in northern Argentina: Laguna Iberá (Mazzuoli et al., 2003) and Laguna Galarza (Loiselle et al. 2001). The Esteros del Iberá is one of the largest wetland ecosystems in South America that has remained significantly unmodified by man's activities. Laguna Iberá is a large ($54 \, \text{km}^2$) shallow lake on the eastern border of the Ibera wetlands ($12,000 \, \text{km}^2$). This permanent lake has an average depth of 3 meters and a maximum of 4.5 meters, with an annual water level variation of 0.5 m. The lake has two small inlets which drain wetland areas. To the north, a small stream connects the lake to an extensive wetland area, dominated by dense emergent vegetation. To the south a small river connects the lake to a smaller wetland area that is surrounded by cultivated areas.

The Galarza lagoon is $14 \, \text{km}^2$ and averages 2 m in depth. The lagoon is fed by a small stream that originates in the large marsh area directly above the lagoon and feeds into another small stream that leads to another large shallow lagoon.

Table 10.1 shows empower and exergy density values and the ratio of exergy to empower. Densities were used to enable comparison between ecosystems in different areas. It was observed that the natural lake (Caprolace) had a higher exergy/emergy ratio than the control and waste ponds, due to a higher exergy density and a lower emergy density (Bastianoni and Marchettini, 1997). These observations were confirmed by the study of Lake Trasimeno (Ludovisi, 1998). Figheri basin is an artificial ecosystem, but has many characteristics typical of natural systems. This depends partly on the long tradition of fish-farming basins in the Venetian lagoon, which has "selected" the best management strategies (Bastianoni, 2002).

The human contribution at Figheri Basin manifests as a higher emergy density (of the same order of magnitude as that of artificial systems) than in natural systems. However, there is a striking difference in exergy density, with

Table 10.1   Empower density, exergy density and exergy/empower ratio for seven ecosystems

	Control pond	Waste pond	Caprolace Lagoon	Trasimeno Lake	Figheri Basin	Iberá Lagoon	Galarza Lagoon
Empower density (sej/year·l)	20.1×10^8	31.6×10^8	0.9×10^8	0.3×10^8	12.2×10^8	1.0×10^8	1.1×10^8
Exergy density (J/l)	1.6×10^4	0.6×10^4	4.1×10^4	1.0×10^4	71.2×10^4	7.3×10^4	5.5×10^4
Exergy/empower (J·year/sej) ($\times 10^{-5}$)	0.8	0.2	44.3	30.6	58.5	73	50.0

values of a higher order of magnitude than in any of the other systems used for comparison: man and nature are acting in synergy to enhance the performance of the ecosystem. The fact that Figheri can be regarded as a stable ecosystem makes this result even more interesting and significant.

The emergy flow to Iberá Lagoon has been underestimated due to lack of data about the release of nutrients from the surrounding rice farms. In a sense this explains the highest value for exergy to empower ratio, while the ecosystem does not seems to be in an ideal condition (Bastianoni et al., 2004). Nonetheless, the important fact is that all the natural systems that are better protected from human influence show very close figures. It seems that there is a tendency common to different ecosystems in different areas and of different characteristics to evolve towards similar thermodynamic efficiencies.

In general we can say in that natural systems, where selection has acted undisturbed for a long time, the ratio of exergy to empower is higher, and decreases with the introduction of artificial stress factors that increase the emergy flow and lower the exergy content of the ecosystem.

10.5 THE RATIO OF ΔEX TO ΔEM

As shown by Fath et al. (2001), orientors (not only emergy and exergy but also ascendency and others) are consistent with each other and are able to represent different stages of ecosystem development. But what if there is a change in inputs? How would a system respond to this change with regard to its self-organization. This problem can also be seen in emergy flows and exergy terms.

If we consider the emergy flow to a system to vary between two equal and contiguous intervals, these intervals must be significant for the system under study in order to annul the effect of periodic variations like daily and seasonal cycles. In effect, emergy analysis is almost always performed considering an interval of one year during which all the emergy inputs and energy outputs are accounted for in obtaining transformities. We indicate the variation of emergy flow with ΔEm (Bastianoni, 1998).

What will be the change in organization due to the change in emergy input ΔEm? To answer this question we have to be able to calculate the variation of the exergy content of the system, ΔEx. We therefore introduce the quantity

$$\sigma = \frac{\Delta Ex}{\Delta Em}$$

with the dimensions of J·s·sej^{-1}, and representing the change of level of organization (exergy) of the system under study, when it is involved in a change of the emergy flow. It is a quantity that is specific to the inputs that are subtracted or added.

To explain what scenarios are possible we can consider that if σ is positive, the addition of emergy input gives rise to further organization, whereas a lowering of emergy has a negative effect on the system. On the other hand, when σ is negative, a higher emergy flow causes a decrease in organization or a lower quantity of one or more inputs causes increasing organization.

We can say that in both the latter cases the inputs (added or removed) can generally be regarded as pollutants: if we remove them, the system self organizes, if we add them, the system is damaged. We can therefore have a definition of pollution based on two orientors — emergy and exergy — that focus their attention not on particular aspects of a system, but on the system as a whole. The intensity of the "pollution" is proportional to the absolute value of the slope of the segment connecting the origin to the point that describes the system, since a small increase (decrease) in emergy flow produces a large loss (gain) of organization.

The same reasoning can be applied to the cases where σ is positive. The slope of the line connecting the point with the origin represents the benefit that a set of inputs — when added — are able to produce on a system.

The points on the diagram correspond to singular situations that can evolve over time. We have a succession of points, one for each subsequent interval, during which we can calculate the emergy flow. To clarify this point we refer to what we previously said about the differences existing between emergy and exergy from a mathematical viewpoint.

Let us consider $t_0, t_1, \ldots t_{k-1}, t_k, t_{k+1}$, a set of points at the axis of time representing the extreme of the closed intervals on which we calculate the emergy flows to the system, $\mathrm{Em}([t_0,t_1]), \ldots \mathrm{Em}([t_{k-1},t_k]), \mathrm{Em}([t_k,t_{k+1}])$. At each point t_j we can also calculate the exergy $\mathrm{Ex}(t_j)$.

The succession of points of the ratio $\Delta\mathrm{Ex}/\Delta\mathrm{Em}$ can be written as:

$$\sigma_k = \frac{\mathrm{Ex}(t_{k+1}) - \mathrm{Ex}(t_k)}{\mathrm{Em}([t_k,t_{k+1}]) - \mathrm{Em}([t_{k-1},t_k])}$$

where σ_k is the ratio calculated considering the differences between the two flows of emergy during the intervals $[t_k, t_{k+1}]$ and $[t_{k-1}, t_k]$, and the value of the exergy at the right extreme of these two intervals. In this way we have a succession of σ_k points that represent the way the system responds to changing surrounding conditions. We can consider a succession starting from a point δ with a negative σ to a point with a positive value of σ. These would mean that the system is "learning" how to use other available inputs and self-organizes. On the other hand, a pattern of inputs that is initially positive for a system can become negative if there is a longer-term toxic effect.

As an example of the application of these concepts, consider the change in the composition of rain that falls upon a forest. If the rain becomes more acidic, its emergy content rises as does the emergy flow through the forest. On the other hand, the exergy of the forest is likely to decrease because of the loss of biomass density and of the consequent loss of biodiversity. In this case, σ would be negative at least until the acidity of the rain decreases again or the species in the forest learn how to survive in the modified environment or how to use a different input.

This framework has been found helpful in solving some shortcomings of the use of a pure life-cycle assessment (LCA) approach (see Heijungs et al., 1996, for example). As stated by Bakshi (2002), in LCA there is a lack of systematic and quantitative framework that does not allow comparison of the environmental sustainability of processes when we want to consider both the use of resources and the global effects of the outputs of a process. The use of emergy and exergy, and especially a wider use of the ratio of the variations of exergy and empower can be a step towards a thermodynamic foundation of LCA (Bakshi, 2002).

REFERENCES

Bakshi, B.R. A thermodynamic framework for ecologically conscious process systems engineering. *Computers and Chemical Engineering* 26, 269–282, 2002.

Bastianoni, S. A definition of pollution based on thermodynamic goal functions. *Ecological Modelling* 113, 163–166, 1988.

Bastianoni, S. Use of thermodynamic orientors to assess the efficiency of ecosystems: a case study in the lagoon of Venice. *The Scientific World* 2, 255–260, 2002.

Bastianoni, S. and Marchettini, N. Emergy: exergy ratio as a measure of the level of organization of systems. *Ecological Modelling* 99, 33–40, 1997

Bastianoni, S., Focardi, S., Loiselle, S., Rossi, C., and Tiezzi, E. Emergy flows and exergy storages in Iberà and Galarza lagoons. *Ecological Modelling* (in press), 2004.

Bendoricchio, G. and Jørgensen, S.E. Exergy as goal function of ecosystems dynamic. *Ecological Modelling* 102, 5–15, 1997.

Fath, B.D., Patten, B.C., and Choi, J.S. Complementarity of ecological goal functions. *Journal of Theoretical Biology* 4, 493–506, 2001.

Fonseca, J.C., Marques, J.C., Paiva, A.A., Freitas, A.M., Madeira, V.M.C., and Jørgensen., S.E. Nuclear DNA in the determination of weighing factors to estimate exergy from organisms biomass. *Ecological Modelling* 126, 179–189, 2000.

Heijungs, R., Huppes, G., Udo de Haes, H., Van den Berg, N., and Dutlith, C. E. Life cycle assessment, UNEP, 1996.

Jørgensen, S.E. *Integration of Ecosystem Theories: A Pattern.* Kluwer, Dordrecht, 1992.

Jørgensen, S.E. Application of exergy and specific exergy as ecological indicators of coastal areas. *Aquatic Ecosystem Health and Management* 3, 419–430, 2000.

Jørgensen, S.E. and Mejer, H. Ecological buffer capacity. *Ecological Modelling* 3, 39–61, 1977.

Jørgensen, S.E., Nielsen, S.N., and Mejer H. Emergy, environ, exergy and ecological modelling. *Ecological Modelling* 7, 99–109, 1995.

Jørgensen, S.E., Patten, B.C., and Straskraba, M. Ecosystems emerging: 4. growth. *Ecological Modelling* 126, 249–284, 2000.

Jørgensen, S.E. and Costanza, R. Understanding and solving environmental problems in the 21st century. Elsevier, Amsterdam, 2002, 324 p.

Keitt, T.H. Hierarchical organization of energy and information in a tropical rain forest ecosystem. M.S. thesis, University of Florida, 1991.

Loiselle, S., Bracchini, L., Bonechi, C., and Rossi, C. Modelling energy fluxes in remote wetland ecosystems with the help of remote sensing. *Ecological Modelling* 145, 243–261, 2001.

Ludovisi, A. Approcci olistici applicati allo studio degli ecosistemi lacustri. Il caso del Lago Trasimeno. Ph.D dissertation, University of Perugia (Italy), 1998.

Ludovisi, A. and Poletti, A. Use of thermodynamic indices as ecological indicators of the development state of lake ecosystems. 1. Entropy production indices. *Ecological Modelling* 159, 203–222, 2003.

Mazzuoli, S., Bracchini, L., Loiselle, S., and Rossi, C. An analysis of the spatial and temporal variation evolution of humic substances in a shallow lake ecosystem. *Acta Hydrochimica et Hydrobiologica* (in press).

Müller, F. and Leupelt, M., Eds. *Ecotargets, Goal Functions, and Orientors.* Springer-Verlag, Berlin, 1998.

Odum, H.T. *Systems Ecology: An Introduction.* New York, Wiley, 1983.

Odum, H.T. Self-organization, transformity and information. *Science* 242, 1132–1139, 1988.

Odum, H.T. "Experimental study of self-organization in estuarine ponds." in *Ecological Engineering: An Introduction to Ecotechnology*, Mitsch, W.J. and Jørgensen, S.E., Eds. Wiley, New York, 1989.

Odum, H.T. Emergy and biogeochemical cycles. In *Ecological Physical Chemistry*, Rossi, C and Tiezzi, E., Eds. Elsevier Science, Amsterdam, 1991.

Patten, B.C., Fath, B.D., Choi, J.S., Bastianoni, S., Borret, S.R., Brandt-Williams, S., Debeljak, M., Fonseca, J., Grant, W.E., Karnawati, D., Maques, J.C., Moser, A., Müller, F., Pahl-Wostl, C., Seppelt, R., Steinborn, W.H., and Svirezhev, Y.M. *Understanding and Solving Environmental Problems in the 21ˢᵗCentury*, Costanza, R. and Jørgensen, S.E., Eds. Elsevier, Amsterdam, 2002, pp. 41–94.

Prigogine, I. *Thermodynamics of Irreversible Processes.* Wiley, New York, 1955.

Svirezhev, Personal communication, 1999.

Tribus, M. and McIrvine, E.C. Energy and information. *Scientific American* 225, 179–188, 1971.

Von Bertalanffy, L. *General System Theory; Foundations, Development, Applications.* G. Braziller, New York, 1968.

CHAPTER **11**

Application of Thermodynamic Indices to Agro-Ecosystems

Y.M. Svirezhev

A traditional criterion for the evaluation of efficiency of different agricultural technologies was always their crop production in relation to energy, fertilizers, human, and animal labor, spent. In this ratio the first term was preferable. Agro-ecosystems with maximum production or with maximal ratio were considered to be the most efficient. In order to estimate agro-ecosystems from the viewpoint of the latter criterion, D. Pimentel (1973, 1980) has developed the so-called "eco-energetic analysis." However, the agriculture intensification leads to degradation of both the agro-system and its environment. Moreover, today this phenomenon acquires such a scale that in developed countries the conservation of environmental quality becomes the main criterion of efficiency for agriculture. We therefore need a universal index, which could quantitatively estimate the impact of agro-system on the environment. From the physical point of view, any degradation can be associated with an increase in entropy. Therefore if we know how to calculate the entropy balance for agro-ecosystem, then the value of the entropy over-production can be used as a measure (index) of the agro-ecosystem's degradation, caused by the intensification of agriculture. In other words, the entropy measure estimates the load of intensive technology on the environment

1-56670-665-3/05/$0.00 + $1.50
© 2005 by CRC Press

that allows us to support and ensure ecological and sustainable public policies for agricultural land use. For this purpose, a thermodynamic model of ecosystem under the anthropogenic pressure was developed. The pressure shifts the equilibrium of a natural (reference) ecosystem to a new equilibrium corresponding to an agro-ecosystem. In the course of the transition, some quantity of entropy is produced which cannot be balanced by the natural processes. These are the main contents of the "entropy pump" concept. In this case the excess of entropy has to be balanced by the destruction of the agro-ecosystem, particularly due to soil erosion. If the entropy overproduction is equal to zero, then we deal with a sustainable agro-ecosystem that can exist for a long enough time. The condition of sustainability allows the determination of either the maximal energy input for a given yield, or the maximal yield for a given energy input that does not cause the agro-ecosystem's degradation. The developed method was applied to such case studies as Hungarian maize production, agriculture in northern Germany and agriculture in Sachsen-Anhalt (eastern Germany).

11.1 INTRODUCTION

A traditional criterion for the evaluation of efficiency of different agricultural technologies was always their crop production in relation to energy, fertilizers, human, and animal labor spent. In this ratio the first term was preferable. Agro-ecosystems with maximum production or with maximal ratio were considered to be the most efficient.

On the other hand, any agro-ecosystem is an ecosystem in a traditional sense which is under anthropogenic pressure. Following the classic "ecological" tradition originating from Lindeman (1942), an agro-ecosystem is considered as a transformer of the inflows of "natural" and "artificial" energy into the outflow of agricultural production. The ratio of the latter to the expended "artificial" energy is then naturally considered as a measure (coefficient) of efficiency for agro-ecosystems. The use of more effective technologies also requires an increase in inflows of matter and energy into agro-ecosystems.

In the 1970s, Pimentel suggested a method — eco-energetic analysis — for comparative estimation of agro-systems (Pimentel et al., 1973). It is based on: (1) categorizing of material and energy flows that are the most significant for an agro-ecosystem; and (2) determination of energy equivalents for these flows. Thereby, it allowed the determination of the intensity of all inflows and outflows in the same energy units and the calculation of the ratio of outputs to inputs — that is, the efficiency of energy transformation, η, by agro-ecosystem. Pimentel's method has ensured that an eco-energetic efficiency of different agro-ecosystems could be compared with respect to an eco-energetic criterion. By the same token, we could describe the evolution of agro-ecosystem in time and compare different agriculture technologies. The certain advantage of Pimentel's method is the use of standard agricultural statistics. It is necessary to note that in these data, different energy and matter flows are expressed in different units, so that the construction of some universal criterion is connected with the problem of weighting different items in the expression for the total energy flow. The problem is resolved by the introduction of so-called Pimentel's conversion coefficients that are the main contents of Pimentel's book (1980). It allows us to determine all inflows and outflows in the same energetic units.

A situation often occurs where $\eta > 1$ for the crop production systems. Why is this so? The flux of solar energy is two orders of magnitude higher than the flows of different kinds of artificial energy that have approximately the same order. Therefore if we take into account the solar energy in our calculations of the efficiency coefficient then we obtain its very low value, which is also almost insensitive to the structure of artificial energy inflows. Of course, the "solar energy" item might be taken as consisting of the parts of percent of the total solar flux; for example, a part we can consider the fraction that is absorbed by vegetation in the process of photosynthesis (less than 1%). However, problems immediately arise connected with the accuracy of the photosynthetic efficiency measurements. Apparently, for the agro-ecosystem, it would be simpler not to consider the flow of "natural" solar energy as one of the energy inflows. In other words, the eco-energetic analysis disregards the "natural" solar energy.

Traditionally, the utilized energy of anthropogenic origin is divided into two types: direct and indirect ("gray") energy. Direct energy input implies the flows of resources, directly associated with *energetics*: oil, coal, peat, electricity, etc. (Smil et al., 1979). By indirect energy we denote the flows of resources that are not actually energetic but take part in the operation of the system. These flows involve mineral fertilizers, pesticides, machinery, agro-systems infrastructure and some other resources.

The energy content of the flows is estimated by taking into account the *total primary energy* expenses needed for their formation. Thus, for example, the energy equivalent of electricity is the heat equivalent of the fuel burnt at power stations to produce this amount of electricity. The most questionable elements in such approach are relative to the quantification of multiple processes participating in the production flow. For example, by estimating the flow of "human labor," one can account for not only the calories of the nutrition necessary for maintaining the required physical activities, but also for all the other "energetic" expenses to provide an adequate living standard (e.g., using gasoline for private motor-vehicle transport) (Smil et al., 1983).

The Pimentel method is still very popular. However, in the last decades many imperatives and preferences of social development have been changed: the problem of how to minimize the environmental degradation and to support environmental protection has taken the first place. Some international boards (The Club of Rome, The Brundtland Commission, IPCC, etc.) were focused on environmental problems at global and local scales. After these activities there is a general consensus that such conditions for economical development must be created that can cover the current needs of societies and will simultaneously apace needs and aspirations of future generations. This concept is known as *sustainable development.* All these concern such systems that supply the food requirements of society — that is, agro-ecosystems. On the one hand, we have to provide a certain level of food production (for this we have to intensify the process of agriculture production), and, on the other hand, we have to minimize the load of this process on environment. In order to draw it from the domain of only verbal estimation we should find out how to compare different agro-ecosystems in respect to their impact on environment — that is, to have some quantitative indices describing the degree of the impact.

The society of developed countries demands an access to the information about environmental degradation mainly for agricultural areas. This is because the degree of degradation and soil pollution has a vital influence on food quality and is strongly linked with health quality of the whole society. There is therefore a severe need for development of a coherent system of information on "environmental" quality of agricultural products. This information should be somehow parameterized so that it is easily available. Finally, agricultural products should be labeled according to their contribution or noncontribution to sustainable development.

Although this concept is relatively new, many scientific groups work on the creation of some universal quantity index, which could give information about the condition of the agricultural areas, where the food production takes place (quantitative measure of sustainability) (D'Agostini and Schlindwein, 1996; Eulenstein, 1995; Eulenstein et al., 2003ab; Lindenschmidt et al., 2001; Steinborn and Svirezhev, 2000; Svirezhev, 1990, 1998, 2000, 2001). This index, being somehow reversed to environmental degradation (Svirezhev, 1990, 1998; Steinborn and Svirezhev, 2000), is expected to supply information about the quality of agricultural areas, from where the crops come from.

From the physical point of view, the environmental degradation can be identified with the concept of entropy (often used as some degradation index). Thus, one value (entropy) can be a measure (indicator) of two opposing processes, the environmental degradation (or the degradation of agro-ecosystem, if its definition includes both crop field and soil, water, etc., see below) from one side, and its sustainability from the other side.

Let we find out how to describe an agro-ecosystem as an open thermo-dynamic system. We know how to calculate the entropy production within the system and the entropy exchange between the system and its environment, and we can attempt to construct different entropy measure for agro-ecosystem — that is, to calculate the necessary index (or indices).

However, we do not want to "throw out a baby with the bathwater," and for that reason we shall use Pimentel's approach as a part of our method. Pimentel's method is not more than the simple application of the first law of thermodynamics to some concrete problem. However, as Sommerfeld said in his *Thermodynamics* (1952): "The Queen of the World, Energy, has her shadow, and the shadow is Entropy." From the thermodynamics point of view, the conservation energy law is only the first law of thermodynamics. There is the second law, which deals with entropy, and which is not less (and may be more) important than the first law. Note that in physics, entropy is a measure of energy degradation. Therefore, it is natural to assume that the further development of Pimentel's approach has to be connected to the concept of entropy.

It is obvious that Pimentel's method was quite acceptable in the course of extensive period of agriculture development when the increase in crop production remained the principal task. However, this period was over in most developed countries and the impact of intensive agriculture production on environment became an acute problem. It is clear that an increase agricultural production leads to further degradation of environment. Agricultural production with a minimal impact on environment is now beginning to be regarded as more effective and sustainable. Pimentel's approach was evidently insufficient for these conditions, because the estimation of impact of agro-ecosystems on the environment was not foreseen. The problem arises of how to estimate agriculture pressure and the degree of environmental degradation. We suggest the following criterion: On the set of ecosystems with the same eco-energetic efficiency, an ecosystem with minimal impact on the environment is the most effective one. This is equivalent to the statement that the total entropy production should be minimal in this optimal (preferable) system.

Certainly, the estimation of impact of ecosystem on the environment could be performed in various ways and comparison criteria could differ as well. For instance, such kind of criteria could be the degrees of chemical pollution or soil erosion, the loss in soil fertility, and so on. However, all these criteria are determined in different units, and the construction of universal criteria demands the reduction of different particular processes of degradation to the same unit. This is one more argument in favor of using the same entropy unit. We could compare different agro-ecosystems in respect to the intensity of their degradation processes and, finally, in respect to a degree of their negative impact on environment.

11.2 SIMPLIFIED ENERGY AND ENTROPY BALANCES IN AN ECOSYSTEM

Since an agro-ecosystem is nothing more than a natural ecosystem under anthropogenic pressure, we start our consideration from the analysis of energy and entropy flows in a natural ecosystem (see also Jørgensen and Svirezhev, 2004). From the viewpoint of thermodynamics, any ecosystem is an open thermodynamic system. An ecosystem being in a "climax" state corresponds to a dynamic equilibrium, when the entropy production within the system is balanced by the entropy outflow to its environment.

Let us consider a single unit of the Earth's surface, which is occupied by some natural ecosystem (i.e., meadow, steppe, forest, etc.) and is maintained in a climax state. Since the main component of any terrestrial ecosystem is vegetation, we assume that the area is covered by a layer, including both dense enough vegetation and the upper layer of soil with litter, where dead organic matter is decomposed. The natural periodicity in such a system is equal to one year; therefore all processes are averaged over a one-year interval.

Since the energy and matter exchanges between the system and its environment are almost completely determined by the first autotrophic level — that is, vegetation — then it is considered as the system, whereas its environment is the atmosphere and soil. We assume that all exchange flows of matter as well as energy and entropy are vertical — that is, we neglect all horizontal flows and exchanges between ecosystems located at different geographic points.

The simplified equation of the annual energy balance for a vegetation layer is $R = \gamma_w q_w + q_h + \gamma_c q_c$, where R is the radiation balance at this point, $\gamma_w q_w$ is a latent heat flux, $\gamma_c q_c$ is the gross primary production (GPP), and q_h q_h is a sensible heat flux (i.e., a turbulent flux transporting heat from the surface layer into the atmosphere). Since in our case there is an additional income of heat from biomass oxidation (respiration and decomposition of the dead organic matter), then the left side of the balance has to be represented as $R + q_{ox}$ (instead of R) where $q_{ox} = Q_{met} + Q_{dec}$. Here Q_{met} is a metabolic heat, and Q_{dec} is heat released in the process of decomposition. The equation of energy balance is written as:

$$[R - \gamma_w q_w - q_h] + [Q_{met} + Q_{dec} - GPP] = 0 \tag{11.1}$$

In such a form of representation, all items of the energy balance are arranged into two groups (in square brackets), members of these groups differ from each other by the orders of magnitude. For instance, the characteristic energies of the processes of a new biomass formation and its decomposition (the second brackets) are lower by a few orders of magnitude than the radiation balance and the energies that are typical for evapotranspiration and turbulent transfer (the first brackets). In the standard expression for energy balance, $R = \gamma_w q_w + q_h$, the terms within the second brackets are usually omitted (e.g., see Budyko, 1977). However, we shall not apply a so-called "asymptotic splitting." Note that such kind of method is widely used in the theory of climate: a so-called "quasi-geostrophic approximation" (Pedlosky, 1979). Thus if we follow the logic of asymptotic splitting, then instead of an exact balance for Equation 11.1, we get the two asymptotic equalities:

$$R - \gamma_w q_w - q_h \approx 0 \tag{11.2a}$$

and

$$Q_{met} + Q_{dec} - GPP \approx 0 \tag{11.2b}$$

We assume that fulfillment of the first equality provides the existence of some "thermostat," which should be called the "environment." The thermostat maintains the constancy of temperature and pressure within the environment. The fulfillment of the second equality is then determined by a consistency of the processes of production on the one hand, and metabolism of vegetation and decomposition of dead organic matter in litter and soil on the other.

In accordance with Glansdorff and Prigogine (1971), the entropy production within the system is equal to $d_iS/dt \approx q_{ox}/T$, where T is the temperature and q_{ox} is the thermal (heat) production of the system. As shown above, the total heat production is a result of two processes: metabolism or respiration, Q_{met}, and decomposition of dead organic matter, Q_{dec}. Since these processes can be considered as a burning of corresponding amount of organic matter then the values of Q_{met} and Q_{dec} can be also expressed in enthalpy units:

$$\frac{d_iS}{dt} = \frac{1}{T}(Q_{met} + Q_{dec}) \tag{11.3}$$

The value of T is the mean air temperature in the given area averaged over the entire year, or the vegetation period, or the period when the decomposition is occurring. In the latter case, all other values also have to be averaged along this period. Since the second equality of Equation 11.2a has to hold, then:

$$\frac{d_iS}{dt} = \frac{GPP}{T} \tag{11.4}$$

At the dynamic equilibrium the internal entropy production has to be compensated by the entropy export from the system, so that:

$$\frac{d_iS}{dt} = \left|\frac{d_eS}{dt}\right| = \frac{GPP}{T} \tag{11.5}$$

The last equality is a quantitative formulation of so-called "entropy pump" concept (Svirezhev, 1990, 1998, 2001; Svirezhev and Svirejeva-Hopkins, 1998; Steinborn and Svirezhev, 2000). We assume that the entropy pump "sucks" entropy produced by the ecosystem. As a result, natural ecosystems do not accumulate entropy, and as a result it can exist during a sufficiently long time period. The power of entropy pump, $|d_eS/dt|$, depends on the GPP of the natural ecosystem located in situ and the local temperature (see Equation 11.5). We assume that the local climatic, hydrological, soil, and other environmental conditions are adjusted in such a way that only a natural ecosystem corresponding specifically to these local conditions can exist at this site and be here in a steady state (dynamic equilibrium). All these statements constitute the main contents of the "entropy pump" concept.

There is never any "overproduction of entropy" in natural ecosystems because the "entropy pump" sucks the entire entropy out of ecosystems, by the same token preparing them for a new one-year period. The picture may be clearer if we consider the process of ecosystem functioning as a cyclic process with one-year natural periodicity. At the initial point of the cycle, the ecosystem is in thermodynamic equilibrium with its environment. Then, as a result of work done by the environment on the system, it performs a *forced* transition to a new *dynamic* equilibrium. The transition is accompanied by the creation of new biomass, and the ecosystem entropy decreases. After this, the reversible *spontaneous* process is started, and the system moves to the initial point producing the entropy in the course of this path. The processes that accompany the transition are metabolism of vegetation and the decomposition of the dead organic matter in litter and soil. If the cyclic process is reversible — that is, the cycle can be repeated infinitely, then the total production of entropy by the system has to be equal to its decrease at the first stage. Substantively, these both transitions take place simultaneously, so that the entropy is produced within the system, and at the same time is "sucked out" by the environment. At the equilibrium state, the annual quantities of entropy produced by the system and the decrease of entropy caused by the work of environment on the system, are equivalent.

Let us imagine that this balance was disturbed (this is a typical situation for agro-ecosystems). Under the impact of new energy and matter inflows, the system moves towards a new state, which differs from the dynamic equilibrium of a natural ecosystem. As a rule, the entropy produced by the system along the reversible path to the initial point cannot be compensated by its decrease at the first stage of the cycle. We obtain a typical situation of the *entropy overproduction*. The further fates of this overproduced entropy could be different. The entropy can be accumulated within the system. As a result, it degrades and after a while, deteriorates (the first fate). The second fate is that the entropy can be "sucked out" by the environment, and the equilibrium will be re-established. In turn, this can be realized in two ways: (1) importing an additional low-entropy energy, which can be used for the system restoration, or (2) environmental degradation.

11.3 ENTROPY OVERPRODUCTION AS A CRITERION OF THE DEGRADATION OF NATURAL ECOSYSTEMS UNDER ANTHROPOGENIC PRESSURE

Let us assume that the considered area is influenced by anthropogenic pressure — that is, there is an inflow of artificial energy (W) to the system. In this notion ("the inflow of artificial energy") we include both the direct energy inflow (fossil fuels, electricity, etc.) and the inflow of chemical substances (pollution, fertilizers, etc.). The anthropogenic pressure can be described by the vector of direct energy inflow, $\mathbf{W}_f = \{W_1^f, W_2^f, \ldots\}$, and the vector of anthropogenic chemical inflows, $\mathbf{q} = \{q_1, q_2, \ldots\}$, to the ecosystem. A state of the "anthropogenic" ecosystem can be described by the vector of concentrations of chemical substances, $\mathbf{C} = \{C_1, C_2, \ldots\}$, and such a macroscopic variable as the mean gross primary production of ecosystem, GPP. Undisturbed state of the corresponding natural ecosystem in the absence of anthropogenic pressure is considered as a *reference state* (see below) and is denoted by $\mathbf{C}_0 = \{C_1^0, C_2^0, \ldots\}$ and GPP_0.

We assume that the first inflow is dissipated inside the system when transformed directly into heat and, moreover, modifies the plant productivity. The second inflow, changing the chemical state of environment, also modifies the plant productivity. In other words, there is a link between the input variables, $\mathbf{W_f}$ and the state variables, $\mathbf{C}$, on the one hand, and the macroscopic variable, GPP, on the other. It is given by the function $GPP = GPP(\mathbf{C}, \mathbf{W_f})$. Obviously, if we deal with contamination that inhibits plants' productivity, then this function must be monotonously decreasing in respect to its arguments. On the contrary, if the anthropogenic inflows stimulate plants (as fertilizers) then the function increases. The typical "dose–effect" curves belong to such functional class.

By formalizing the previous arguments we can represent the entropy production within this "disturbed" ecosystem as:

$$\frac{d_iS}{dt} = \frac{1}{T}[W + GPP(\mathbf{C}, \mathbf{W_f})] \tag{11.6}$$

where the scalar W is a convolution of the inflows $\mathbf{W_f}$ and $\mathbf{q}$ — that is, the total anthropogenic inflow. Note that the convolution may also depend on the vector $\mathbf{C}$, since these internal concentrations are maintained by the inflows $\mathbf{q}$.

In accordance with the "entropy pump" concept, a certain part of the produced entropy is released at this point by the "entropy pump" with the power $|d_eS/dt| = GPP_0/T$, so that the total entropy balance is:

$$\frac{dS}{dt} = \sigma = \frac{1}{T}[W + GPP(\mathbf{C}, \mathbf{W_f}) - GPP_0] \tag{11.7}$$

We assume that despite anthropogenic perturbation, the disturbed ecosystem is tending toward a steady state again. If we accept this, we must also assume that the transition from natural to anthropogenic ecosystem is performed sufficiently fast so that the "tuning" of the entropy pump does not change. In other words, the internal entropy production will correspond to the "anthropogenic" ecosystem while the entropy export remains as a reference natural ecosystem. This misbalance really exists, since the power of entropy pump is bounded above by the value of GPP for the reference ecosystem. In this situation the overproduction of entropy cannot be exported to the systems' environment, and the system has to start deteriorating. But since the considered ecosystem has to be in a dynamic equilibrium, then there is only single way to resolve this contradiction: to destroy the environment.

A system can sustain or improve its organization if, and only if, the (inevitably) produced entropy is exported into the environment. Therefore, from a thermodynamic point of view, environmental degradation is a necessary condition for the survival of the system. To avoid any misunderstanding we will not use the expression "environmental degradation." We regard humans to be a part of the system that we are studying and would like to protect. From this point of view, we can join the "anthropogenic" ecosystem (e.g., the crop field) and its neighboring environment (soils, water, etc.) into the whole system, keeping its former name. In this case we can talk about the system's degradation.

The quantity of the entropy overproduction, σ, could therefore be used as an indicator (index, criterion) of the degradation of ecosystems under anthropogenic pressure (Svirezhev and Svirejeva-Hopkins, 1997), or an "entropy fee" which has to be paid by society (actually suffering from the degradation of environment) for modern industrial technologies.

The degradation may be manifested in different ways: as a chemical pollution of soil and water, soil erosion, a fall of productivity, etc. Nevertheless, although the method allows evaluating the system's degradation in general, we cannot predict concrete ways of degradation. For instance, we cannot say principally that shares of the total entropy overproduction will be responsible for soil erosion or a decrease in pH, etc. Therefore it will be difficult to forecast which component of the agro-ecosystem, for instance, will be the most sensitive in reaction to anthropogenic pressure.

As concerns the agro-ecosystems, the main conclusion of this section is as follows. It is obvious that by increasing the input of artificial energy we, by the same token, in accordance with Pimentel' relations, can also increase agricultural production. Note that this increase does not have an upper boundary and can continue infinitely. However in reality, this is not the case and there are certain limits determined by the second law of thermodynamics. In other words, we pay the cost for increasing of agricultural productivity, which is a degradation of the physical environment, in particular, soil degradation.

Of course, there is another way to balance the entropy production within the system. We can introduce an artificial energy and soil reclamation, pollution control (or, generally, ecological technologies). Using the entropy calculation we can estimate the necessary investments (in energy units).

11.4 WHAT IS A "REFERENCE ECOSYSTEM"?

When we talk about a "reference ecosystem" we take into account a completely natural ecosystem, without any anthropogenic load impacts. To find such an ecosystem today in industrialized countries is almost impossible (except possibly on the territories of natural parks). All so-called natural ecosystems today are under anthropogenic pressure (stress, impact, pollution, etc.). All these stresses started to act relatively recently (in the last 100 to 150 years) in comparison with characteristic relaxation times of the biosphere, so that we can assume with rather high probability that the mechanisms responsible for the functioning of "entropy pump" have not yet adapted to the new situation. On the other hand, plants, which are the main components of natural ecosystem, react to anthropogenic stress very quickly, as a rule, by reducing their productivity.

It is intuitively clear that "natural" grassland, located at the same geographical point, could serve as a "reference" natural ecosystem for a crop field. One can see that their architectures are very similar: the similar radiation regimes, the similar patterns of turbulent flows, the similar processes of evapotranspiration, the similar types of soil and their chemical compositions. So, the GPP value of grassland, located at the close vicinity of considered agro-ecosystem could be used as some phenomenological value for GPP_0.

For the more correct definition of a reference ecosystem we could use the ergodicity paradigm: instead of considering two spatially close ecosystems, we may take two temporally close ecosystems that are connected by the "relation of succession." The latter means that these ecosystems are two sequential stages of one succession. Later on we shall call this pair as "successionally close" ecosystems. Then a natural ecosystem, which is successionally close to agriculture, should be considered to be a "reference" one (see chapter 2).

Let us imagine that energy and chemical fluxes into an anthropogenic ecosystem were interrupted. Then a succession from the latter towards a natural ecosystem (grassland, steppe, etc.), which is typical for this location, starts here. If the anthropogenic ecosystem is an agro-ecosystem then this succession is commonly called "old field succession" (Odum, 1983).

Formally, a final stage of the succession may be considered as a reference ecosystem. For instance, if an agro-ecosystem is surrounded by forest then a final stage will be forest, so that the reference ecosystem also will be forest. However, this reasoning is slightly flawed. The point is that, on one hand, if we want to stay in the frameworks of the concept of successional closeness, we have to assume that at any stage of a succession the system has to be in a dynamic equilibrium — that is, the successional transition has to be quasi-stationary. On the other hand, a succession is the transition process between two equilibriums. Therefore, we have encountered a contradiction. However, since the temporal scale of ecological succession is much greater than the same of anthropogenic processes, we can consider a succession as a quasi-stationary process. Since we can join onto a single "successionally close" pair only close steady states, then their vicinities must be significantly intersected, and the temporal scale of a quasi-stationary transition from a natural to anthropogenic ecosystem and vice versa must be small (in comparison to the temporal scale of succession).

Finally, we define the reference ecosystem as a natural ecosystem, which is the first stage of a succession of an anthropogenic ecosystem. The succession is caused by the interruption of anthropogenic pressure. Both ecosystems are successionally close. From this point of view a grass–shrubs ecosystem (i.e., not a forest) will be a reference ecosystem in relation to surrounding forest agro-ecosystem.

Thermodynamically, the succession is a typical reversible process. However, if the anthropogenic ecosystem is in a state of severe degradation, the succession moves along another way, towards another type of ecosystem, which differs from a "natural" one. This is quite natural however, since the environmental conditions have been strongly perturbed (for instance, as a result of soil degradation). This is an irreversible situation that we do not consider. So, the concept of "successional closeness" means that we remain in the framework of "reversible" thermodynamics.

11.5 AGRO-ECOSYSTEM: THE LIMITS OF AGRICULTURE INTENSIFICATION AND ITS ENTROPY COST

As we have already mentioned, the intensification of agriculture (the increase of crop production) correlates with an increase of artificial energy flow in the ecosystem. Indeed, the increase in fertilizer input, usage of complex infrastructure, pesticides, herbicides, etc. — that is, all that is called a "modern agriculture technology," results in greater crop production. This is a typical pattern of the developed agriculture in industrial country (industrial agro-ecosystem).

The total flow of additional artificial energy into the agro-ecosystem is the convolution W of all energy and matter inflows, which can be represented as a sum of two items: $W = W_f + W_{ch}$. If the first item W_f, which is a convolution of the vector $\mathbf{W_f}$, can be associated with the direct inflows of such type of artificial energy as electricity, fossil fuels, etc., and called an *energyload*); then the second item W_{ch} is associated with the inflows of chemical elements that maintain molar concentrations $\mathbf{C}$ within the system, and called a *chemicalload*. The flowchart of an agro-ecosystem is shown in Figure 11.1.

Since all direct inflows are measured in standard energy units then the convolution of $\mathbf{W_f}$ is defined simply as:

$$W_f = \sum_i W_i^f$$

The problem is how to calculate W_{ch}. The first way is a calculation of some energy equivalent of W_{ch} using the conversion coefficients by D. Pimentel et al. By applying the special algorithm, fertilizers, pesticides, etc. are represented in energy units ("gray energy"). As a result we can join the values W_f and W_{ch} into a single value $W = W_f + W_{ch}$. If W_{ch} is defined by the first way then we

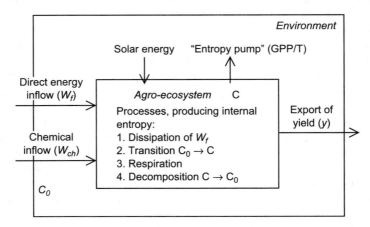

Figure 11.1 Energy flows and entropy production within agro-ecosystem, and entropy exchange between agro-ecosystem and its environment.

assume that finally this energy also transfers into heat (as a result of multiple chemical reactions), then the second item in the equation for entropy production is equal to W_{ch}/T, and the total entropy production will be equal to W/T.

In the second way we use a thermodynamic model of the chemical transition. We assume that at the beginning the residence ecosystem is in the state of chemical equilibrium, $\mathbf{C}^0\{C_1^0, \ldots, C_n^0\}$. At an initial point the concentrations are changed from $\mathbf{C}^0\{C_1^0, \ldots, C_n^0\}$ to $\mathbf{C}\{C_1, \ldots, C_n\}$ very quickly in order not to disturb other equilibriums. The residence ecosystem immediately becomes a nonequilibrium agro-ecosystem, which is capable of performing useful work, which is equal to the "chemical" exergy (Jørgensen, 1992; see also chapter 2):

$$\text{Ex}_{ch} = RT \sum_{i=1}^{n} \left[C_i \ln\left(\frac{C_i}{C_i^0}\right) - (C_i - C_i^0) \right] \tag{11.8}$$

We assume that the work is being performed during time τ, then $W_{ch} = \text{Ex}_{ch}/\tau$. Without loss of generality we can set τ at one year, so that $W_{ch} = \text{Ex}_{ch}$.

If we take the annual value of the GPP of the agro-ecosystem, then the NPP is equal to $(1-r)\text{GPP}$ (r is the respiration coefficient), and the term $rP1$ describes the respiration losses. The kth fraction of the Net Primary Production (NPP) is being extracted from the system as the crop yield, $y = k(1-r)P_1$, so that only the remaining fraction, $(1-k)(1-r)\text{GPP} = [(1-s)/s]y$ where $s = k(1-r)$, is transferred to litter and soil as dead organic matter. If the stationary hypothesis is accepted then we have to assume that the corresponding amounts of litter and soil organic matter are decomposed. Concerning the exported fraction of production we assume that namely this fraction does not take part in the local entropy production. Therefore, the total entropy production is equal to:

$$\sigma = \frac{dS}{dt} = \frac{1}{T}\left(W_f + W_{ch} + \frac{1-s}{s}y - \text{GPP}_0 \right) \tag{11.9}$$

If we compare the equation with above Equation 11.7 then we can see that the term GPP in Equation 11.7 is replaced by the term $[(1-s)/s]y = (1-k)(1-r)\text{GPP}$ — that is, the fraction of the GPP is exported out of the system and does not participate in the thermodynamic processes occurring in the system. This is a principal difference between the "anthropogenic ecosystem" considered in section 11.2, which is closed with respect to created biomass, and the agro-ecosystem, where the part of its biomass is exported as a crop production.

The terms in the right side of Equation 11.9 are not independent: there is also Pimentel's relationship:

$$y = \eta(W_f + W_{ch}) = \eta W$$

If we take it into account then the expression for σ may be represented in two different forms:

$$\sigma = \frac{1}{T}\left[W\left(1 - \eta + \frac{\eta}{s}\right) - \text{GPP}_0\right] \tag{11.10a}$$

or

$$\sigma = \frac{1}{T}\left[y\left(\frac{1}{\eta} + \frac{1}{s} - 1\right) - \text{GPP}_0\right] \tag{11.10b}$$

If $\sigma > 0$ then we have a situation of entropy overproduction. The product σT is usually called *dissipativefunction*.

What is difference between these expressions? If we assume that the coefficients η and s that define the energy efficiency of agro-ecosystem and the type of implemented agriculture, as well as productivity of some reference natural system, GPP_0, are fixed, then the entropy overproduction will only depend on either the energy inflow (input) or the yield (output) — see Equations 11.10.a and 11.10b.

11.6 CONCEPT OF SUSTAINABLE AGRICULTURE: THE THERMODYNAMIC CRITERION

The agro-ecosystem will be in a dynamic equilibrium and it will exist for an infinitely long time if the annual overproduction of entropy is equal to zero ($\sigma = 0$). If the system overproduces σ units of entropy per year, then in order for the system to exist for sufficiently long time, we have to assume the existence of special mechanisms discharging the excess of entropy. Since usually the growth of entropy is associated with amplification of degradation processes then it is natural to assume that these mechanisms are connected with agro-ecosystem's degradation. On the other hand, if $\sigma = 0$ then the agro-ecosystem is functioning without the accumulation of entropy — that is, it is not degrading. This is a typical situation of *sustainability*.

Therefore, Equation 11.10 under the condition $\sigma = 0$ gives us the value of such energy inflow:

$$W_{sust} = W_{sust} = \frac{\text{GPP}_0}{1 - \eta + (\eta/s)} \tag{11.11}$$

which provides sustainability of agro-ecosystem. The analogous criteria of sustainability under the common title "the limit energy load" were suggested by Bulatkin (1982), Simmons (cited by Bulatkin, 1982), and Novikov (1984). It is interesting that in spite of the fact that they were using different concepts (on the whole, economical), their evaluations were very close (15 GJ/ha by

Simmons, 14 GJ/ha by Novikov and Bulatkin). Note the "limit energy load" concept is a typical empirical one, and this load is the maximal value of the total anthropogenic impact (including tillage, fertilization, irrigation, pest control, harvesting, grain transportation, drying, etc.) on one hectare of agricultural land. It is assumed that when the anthropogenic impact exceeds this limit then the agro-ecosystem begins to deteriorate (soil acidification and erosion, chemical contamination, etc.).

Letting $\sigma = 0$ in Equation 11.10b, i.e., equating the right side of this equation to zero we obtain:

$$y_{sust} = \frac{\text{GPP}_0}{(1/s) + (1/\eta) - 1} \tag{11.12}$$

This is the evaluation of a "sustainable" yield — that is, the maximal crop production, which could be obtained without a degradation of agro-ecosystem (i.e., in a "sustainable" manner).

11.7 SOIL DEGRADATION: THERMODYNAMIC MODEL

Let us keep in mind that a system that accumulates entropy cannot exist for a long time and will be inevitably self-destroyed. In order for the system, which overproduces σ of entropy every year, to exist for a long time, we have to assume the existence of special mechanisms that discharge the excess of entropy. It is natural to assume that this is a degradation of the system and its environment, in particular the degradation of soil. The main degradation process, which includes all degradation processes, is soil erosion. If we have a thermodynamic model of soil erosion, we can estimate the annual erosion losses resulting from intensive agriculture.

How do we estimate the quantity of entropy contained in one ton of erosive soil? Consider the soil erosion as a sum of two processes. The first is a burning of all organic carbon that is contained in soil, and, as a consequence, the destruction of soil aggregates down to ground particles (e.g., to sandy ones). Note that availability or absence of organic component is the main characteristic that distinguishes soil from the ground. This is a serious argument why we can consider the burning of organic matter as the first stage of soil degradation.

The second process is a mechanical destruction of ground particles from z_s to z_w ($z_w \ll z_s$) in size where z_s is the size of standard ground particles (usually $z_s \sim 10^{-1}$ cm) and z_w is the size of dust particles, which can be taken away by wind and be washed out by water (usually $z_s \sim 10^{-4}$ cm). The erosive soil may be transported out off the system, thereby "sucking" the excess entropy out of the system. We assume that the transportation is realized by means of processes that are external in relation to the system. In other words, their work is performed at the expense of energy, which does not directly participate in the production of biomass and its decomposition. Since wind and water take away the erosive soil, this is a fully correct assumption.

If p is the relative content of carbon in 1 ton of soil, then the "burning" of carbon gives us the following first item of dissipative function: $\sigma_1^{er} T = h_c p \times 10^6$ where h_s is the specific enthalpy of organic carbon, $h_s = 4.2 \times 10^4 \, \text{J/gC}$, then:

$$\sigma_1^{er} T \approx 4.2 \times 10^{10} p \, \text{J} \tag{11.13}$$

This item corresponds to a chemical destruction.

In order to estimate the second item of the total dissipative function, corresponding to the mechanical destruction we use the "urn model": Every particle is considered as one urn, which contains r molecules of SiO_2 with molecular weight $\mu = 60$ (sand). We assume that all the urns contain the same number of molecules. If we also assume that all particles are the closest packed in one volume V, then the number of urns is equal to $m = 1/z^3$ where z is the particle size. Then Gibbs' entropy (entropy of mixture) (i.e., entropy that corresponds to the distribution of N molecules among m urns) is equal to $S^G = kN \ln m$ (Landau and Lifshitz, 1995). It is obvious that the particles breaking from size z_s to size z_w can be represented as an increase in number of urns from m_s to m_w. This corresponds to the increase in entropy by the value:

$$\Delta S^G = kN(\ln m_w - \ln m_s) = kN \ln\left(\frac{z_s}{z_w}\right)^3 = 3kN \ln\left(\frac{z_s}{z_w}\right) \tag{11.14}$$

By taking into account that if N is the number of molecules contained in one mole then $kN = R$, where $R \approx 8.31 \, \text{J/mol·K}$ is the gas constant. For one ton of ground we then get:

$$\sigma_2^{er} T = \frac{10^6}{60} \times 3 \times 8.31 \times 293 \ln\left(\frac{z_s}{z_w}\right) \approx 0.122 \times 10^9 \ln\left(\frac{z_s}{z_w}\right) \text{J} \tag{11.15}$$

Finally, the total dissipative function corresponding to the erosion of one ton of soil is equal to:

$$\sigma^{er} T = \sigma_1^{er} T + \sigma_2^{er} T = \left\{ 4.2 \times 10^{10} p + 0.122 \times 10^9 \ln\left(\frac{z_s}{z_w}\right) \right\} \text{J} \tag{11.16}$$

If we assume that the unique manifestation of the degradation is the loss of soil, then it is not a problem to estimate how many tons of soil are annually lost as a result of the intensive crop production. Since we know the annual value of entropy overproduction σ then the amount of lost soil (in tons) will be equal to $L_{er} = \sigma/\sigma^{er}$.

It is obvious that any intensive agriculture with high crop production is accompanied by losses of significant amount of soil. In order to avoid these irreversible losses, the various strong restrictions are implemented in many countries. For instance, according to U.S. standards, no more than 10 tons of soil resulted in agricultural activity may be lost from one hectare (E. Odum, 1983). However, with the loss of soil, there are at the same time other degradation processes such as a soil contamination by different chemical substances (e.g., herbicides, pesticides, etc.). In particular, one of the forms of contamination is the acidification as a result of fertilizer overuse. For instance, if the value of soil loss L_{er} is known then the remaining entropy, $R_\sigma = \sigma - L_{er}\sigma^{er}$, has to be compensated by these processes.

11.8 "ENTROPY FEE" FOR INTENSIVE AGRICULTURE

There are increasing talks about how chemical "no-till" agriculture actually allows topsoil to accrete. It is being touted as a "sustainable" form of agriculture. Let us consider this concept from a thermodynamic point of view. If we look at the formula for $\sigma^{er}T$ you see that the second item corresponds to mechanical destruction. The main reason for such destruction is tilling. "No-till" agriculture means that the value $\sigma^{er}T$ is reduced by approximately 30%. The first impression is that we achieve our goal and reduce soil erosion. However, it is impossible to get something free of charge, and if we want to maintain the same crop production we must increase energy input (W) up to the former value replacing the mechanical component by a chemical one. As a result, we obtain approximately the same value of the entropy overproduction (σ), but there would be another way of compensation. For instance, instead of soil erosion we would get an increase in chemical contamination.

Within frameworks of the thermodynamic approach we can calculate entropy of these processes as well. For example, the entropy contribution to the acidification of soil can be calculated in terms of appropriate chemical potentials. However, (and this is the principal constraint of thermodynamic approach), we cannot predict the way in which the degradation will be realized: by the strong mechanical degradation of soil and weak chemical pollution, the high acidification of soil, the strong chemical contamination by pesticides and fertilizers, or some intermediate variants. Every way and combination type is possible. For the solution of this problem some additional information is needed.

On the other hand, this approach gives the possibility of estimating an "entropy fee," which humans pay for high crop yield (i.e., for agriculture intensification). Overproduction of entropy can be compensated by the processes of system's degradation, in particular by soil degradation. It is known that the loss of ~40% of soil tends to speed up by five to seven times, the fall of crop yield (Dobrovolsky, 1974). This is a typical agricultural disaster. But it is a disaster from anthropocentric point of view; from the physics laws point of view, a fall of crop yield as a reason of soil degradation is a natural reaction of a physical system tending to decrease an internal production of entropy and to minimize its overproduction. It is the consequences of Prigogine's theorem.

We suspect that this statement summons very serious objections, especially among biologists. Their essential argument may be the following: Certainly, Prigogine's theorem is applicable to the physical world, but its application to the biological realm is very doubtful. It is the juxtaposition of the countervailing tendencies of the physical and biological realms that lends tragic overtones to the overwhelming of the biological trend by that of the physical.

In general we agree with above-mentioned sentence but for self-justification, we speak as strong followers of reductionism. We understand that this point of view is biological "heresy" but we consider an agricultural system solely as a physical one. In the framework of our approach we reduce a variety of all processes inside an agro-system to one process of heat production and its dissipation. In other words, we measure the entropy production by a total thermal effect of different physical and chemical processes taking place in the system. We understand that it is a very serious simplification of the biological realm, but as it seems to us, this approach gives some practical results. Namely in this sense we talk about applicability of Prigogine's theorem.

11.9 HUNGARIAN MAIZE AGRICULTURE

Here we would like to demonstrate how to apply this method for the analysis of concrete agro-ecosystem. In particular, we shall analyze the Hungarian maize agriculture in 1980s (Svirezhev, 1990, 1998, 2000, 2001).

The average yield of maize was 4.9 tons of dry matter (dm) per hectare. Since the energy content of one ton dm for maize is equal to 1.5×10^{10} J (Pimentel, 1980) then the energy content of yield will be equal to 0.735×10^{11} J/ha. On the other hand, the artificial energy inflow for maize production is equal to $W = 0.27 \times 10^{11}$ J/ha. Therefore, the energy efficiency coefficient $\eta = 2.7$. Since the ratio of main product (grain) to byproducts (straws, roots) is approximately 50:50 (%) then $k = 0.5$. For a maize crop in the temperate zone $r \cong 0.4$ (Lang, 1990) then $s = k(1 - r) = 0.3$. The Hungarian steppe community ("puszta") is a successionally close (reference) ecosystem for a cornfield after its cultivation is stopped (grassland of the temperate zone) and can be considered as a reference natural ecosystem in given case. Its GPP is equal to $P_0 = 2800 \text{ kcal/m}^2 = 1.18 \times 10^{11}$ J/ha (Lang, 1990). By substituting these values into (3.7) we get $\sigma T = 0.81 \times 10^{11}$ J/ha. Therefore, compensation for environmental degradation requires the 200% increment in energy input with all the additional energy spent only for soil reclamation, pollution control, etc., with no increase in the crop production. Note that here and later on we are dealing with annual values, so that all the values (entropy, energy production, etc.) are calculated for a one-year period. If an agro-ecosystem with crop rotation is considered then these values have to be averaged over the rotation period, and the entropy balance also has to be calculated for the total rotation.

The main condition of a sustainable agro-ecosystem is $\sigma = 0$. From Equations 11.1 and 11.11 we get $W_{sust} = 16 \text{ GJ/ha}$ and $y_{sust} = 2.9 \text{ tons dm/ha}$.

We see that the value of $W_{sust} = 16\,\text{GJ/ha}$ is very close to the estimations of "limit energy load," 14 to 15 GJ/ha (see section 11.6). Bearing in mind that the contemporary value of maize yield in U.S. is equal to 3 tons, and after the "black storms" of the 30s, modern agricultural technologies do not lead to the strong soil erosion, then we again get a new surprising coincidence. Note that these coincidences allow us to solve one more problem (see below).

Note the "limit energy load" concept is a typical empirical one and expresses the maximal value of the total anthropogenic impact (including tillage, fertilization, irrigation, pest control, harvesting, grain transportation, drying, etc.) per hectare of agricultural land. All these values are evaluated in energy units (in accordance with Pimentel's method). It is assumed that when the anthropogenic impact exceeds this limit then the agro-ecosystem begins to fail (soil acidification and erosion, chemical contamination, etc.).

There is an implicit contradiction in all these considerations. If we compare the energy content of one gram of dry matter (grain) given by Pimentel (15 kJ) and its enthalpy, which equals to 21.7 kJ (Odum, 1983) then we can see a significant difference. For instance, the energy content of yield in enthalpy units will be equal to $y' = 21.7 \times 4.9 \times 10^9\,\text{J/ha} \approx 1.06 \times 10^{11}\,\text{J/ha}$. This is very different to the value $0.735 \times 10^{11}\,\text{J/ha}$. Enthalpy units are normally used when the GPPs of natural ecosystems are estimated. For instance, we are aware that the value $P_0 = 2800\,\text{kcal/m}^2$ is expressed in enthalpy units. If we keep in mind Equation 11.9 then the expression for the balance of entropy production may be represented as:

$$\sigma T = W + ay - \text{GPP}_0, \quad a = (1 - s)/s \tag{11.17}$$

When we above calculated the value of σT, the terms W and y were expressed in Pimentel's units, and the term P_0 in enthalpy units. In fact, the transition in Equation 11.17 only to either Pimentel's or enthalpy units influences the value of σT. For instance, the substitution of y' by y in Equation 11.17 results in $\sigma'T \approx 1.56 \times 10^{11}\,\text{J/ha}$, instead of $\sigma T \approx 0.81 \times 10^{11}\,\text{J/ha}$. On the other hand, we saw above that even partially correct use of these different energy units gives us a few of interesting coincidences. We can conclude from this that our empirical approach reflects some characteristics of reality, and now we have the following choices: either to deal with the whole situation as it is or to attempt using the new estimation of yield and, at the same time, to save our previous results. We selected the second way.

By re-writing Equation 11.17 in the form $\sigma T + P_0 - W = ay$ and changing only the right side of the equality, y to y', we see that in order to keep the same left side we have to replace a with a' so that $ay = a'y'$ and $a' = a(y/y')$. Since $a = (1 - s)/s \approx 2.67$ and $y/y' \approx 0.69$ then $a' = 1.84$ and $s' = 0.35$. Bearing in mind that $s = k(1 - r)$ and keeping the previous value of respiration coefficient, $r = 0.4$, we get $s' = 0.35$ if only to set $k' = 0.58$. Note however that the previous value of $k = 0.5$ was obtained under the assumption that the biomasses of grain, straw, and roots (expressed in unit of dm) were comparable (i.e., their specific enthalpies are equal). Their enthalpies are different, however: 21.7 kJ for grain and 15.9 kJ for straw and roots. If we now calculate the coefficient k with respect to the energy content, we get $k = 21.7/(21.7 + 15.9) \approx 0.58$. We see that this value coincides with k'. Thus, this contradiction was resolved, and all our results were saved. It is easily to check, note only that in this case $\eta' = y'/W = 1.06/0.27 \approx 3.9$.

Let us estimate entropy that corresponds to the destruction of one ton of soil in the Hungarian case. By substituting the value of $p \approx 4\%$ (Lang, 1990) into Equation 11.16, we obtain:

$$\sigma^{er} T = \sigma_1^{er} T + \sigma_2^{er} T = \{4.2 \times 4 \times 10^{10} + 0.122 \times \ln(10^3) \times 10^9\}$$
$$\approx 0.168 \times 10^{10} + 0.084 \times 10^{10} = 0.252 \times 10^{10} \, \text{J} \qquad (11.18)$$

where the first and the second items correspond to chemical and mechanical destruction. Then the annual total loss of soil per one hectare due to erosion is equal to $\sigma/\sigma^{er} \approx 32$ tons. Therefore, the high maize production would cost us 32 tons of soil loss annually.

It is obvious that the value of 32 tons per hectare is an extreme value: the actual losses are less, (approximately 13 to 15 tons). This means that there are other degradation processes such as environmental pollution, soil acidification (the latter is very significant for Hungary), etc.

11.10 AGRICULTURE IN NORTHERN GERMANY (STEINBORN AND SVIREZHEV, 2000)

If in the previous section we applied the thermodynamics method to the analysis of monocultural (single-crop) agro-ecosystem on a scale of the entire country, then here we shall study a polycultural (multiple crops) agro-ecosystem located at a comparatively small typical rural territory (about 3.3 km^2). The area is the watershed of Lake Belau in North Germany (near Kiel) comprised of arable lands (grassland, crop fields, hedges). In the years 1988 and 1997, the data regarding all anthropogenic inflows (inputs) and outflows (outputs) were collected. By using Pimentel's method we could estimate the total energy inflow.

From 1988 until 1997, the size of the arable land in the study site did not change remarkably. Changes in land management practices were also small. In 1997 the grassland increased by 7% of the total area. *Fabaceae* and beets were no longer cultivated and wheat replaced barley as the dominant crop. For most of the crops the yield increased whereas the total energy input was reduced. This may be explained by the following: since the most part of energy input is in the form of organic and inorganic fertilizers, the reduction of total input in the year 1997 is mostly due to a mean decrease of organic fertilizer from 21.8 GJ/ha to 7.6 GJ/ha. All the yields were expressed in energy units by Pimentel's method. These data are shown in Table 11.1, in which the energy efficiency coefficients η are also shown.

Table 11.1 Land management, input and output of energy (area near the Lake Belau)

	Land use (in %)		Input W [GJ/ha]		Output y [GJ/ha]		η (output/input)	
	1988	1997	1988	1997	1988	1997	1988	1997
Fabaceae	4	—	36	—	160	—	4.5	—
Beet	4	—	123	—	196	—	3.2	—
Maize	9	8	97	32	168	169	1.7	5.3
Rape	3	1	48	37	118	131	2.4	3.6
Oats	2	4	55	22	141	147	2.6	6.7
Rye	6	10	27	19	181	194	6.8	10.0
Barley	24	3	47	29	170	192	3.6	6.7
Wheat	8	28	72	29	191	217	2.7	7.5
Annual grass	2	5	39	25	122	169	3.1	5.5
Permanent grass	37	41	35	32	90.7	89.2	2.6	2.6
Mean			50	29	139	147	3.0	5.4

Mean energy input was obtained by averaging all the cultures and was equal to 49.9 GJ/ha in 1988 and 28.9 GJ/ha in 1997. It is interesting that the latter value is very close to the Hungary case study.

To set an example, the scheme of further calculations will be demonstrated for wheat in the year 1988. In order to calculate the entropy balance of a wheat field with Equations 11.10a or 11.10b, we have to know either the values of W or y, and also η, $s = k(1-r)$, P_0 and T. The energy input for wheat in 1988 is equal to 71.6 GJ/ha, the output, which is the crop production y (grain plus the main part of straw), is equal to 191 GJ/ha (see Table 11.1). Using these values we calculate the ratio $\eta = $ output/input $= 191/71.6 \approx 2.7$. We get $k = 0.74$ (Hydro Agri Dülmen AG 1993; Weisheit, 1995; Vetter, 1952; Wachendorf, 1995); and since $r = 0.3$, then $s = k(1-r) \approx 0.52$. After Sach (1999) the productivity of a reference "natural" grassland in the area, P_0, is equal to 106 GJ/ha, and the average temperature of air during vegetation period is $T = 285$ K. Substituting all these values into (11.9) we get $\sigma \approx 0.51$ GJ/ha* K per year. The value of $\sigma T \approx 145$ GJ/ha, it is higher by 80% than in Hungary. It becomes clear if to take into account that the energy input in Germany (71.6 GJ/ha) is higher than in Hungary (27 GJ/ha) by 277%! It is also clear that the German agriculture is more "ecological," (i.e., more "sustainable") then the Hungarian one. The conclusion is especially visible if we consider that the first is producing two units of "dissipative" (or "degradative") energy per one unit of input energy ($145/71.6 \approx 2$), and the second is producing three units (81 GJ/27 GJ $= 3$).

Analogously we calculate the entropy overproduction for all crop production systems. The results are shown in Table 11.2. In order to visualize these data we represent them in the form of the following maps (see Figure 11.2).

Looking at Table 11.2, we can see that the general tendency is either that entropy excess decreased (for maize, oats, wheat and permanent grass) or it did not change (for rape, rye, barley, and annual grass). If we compare their mean values (in order to estimate a general tendency for the whole region) we can see that the total energy input, W, has decreased (from 50 to 29 GJ/ha) by

Table 11.2 Overproduction of entropy, σ, in 1988 and 1997 for different crops (area near the Lake Belau). Comparison of actual inputs and actual yields

	Overproduction of entropy σ [GJ/ha*K]		Energy input W [GJ/ha]		Energy output y [GJ/ha]	
	1988	1997	1988	1997	1988	1997
Fabaceae	0.23	—	36	—	160	—
Beet	0.43	—	123	—	196	—
Maize	0.55	0.32	97	32	168	169
Rape	0.28	0.29	48	37	118	131
Oats	0.34	0.26	55	22	141	147
Rye	0.36	0.38	27	19	181	194
Barley	0.36	0.37	47	29	170	192
Wheat	0.51	0.45	72	29	191	217
Annual grass	0.23	0.22	39	25	122	169
Permanent grass	0.09	0.05	35	32	90.7	89.2
Mean	0.28	0.25	50	29	139	147

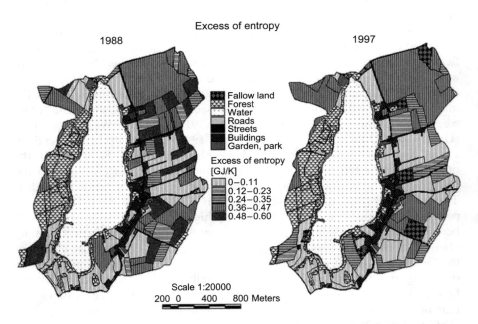

Excess of entropy

1988

1997

Fallow land
Forest
Water
Roads
Streets
Buildings
Garden, park

Excess of entropy [GJ/K]
0–0.11
0.12–0.23
0.24–0.35
0.36–0.47
0.48–0.60

Scale 1:20000
200 0   400   800 Meters

Figure 11.2 Overproduction (excess) of entropy in the rural area near the Lake Belau (northern Germany) in 1987 and 1997.

approximately 49% within nine years. Nevertheless, the mean crop yield was enlarged in the same period (from 139 to 147 GJ/ha), although insignificantly, by approximately 6%. As a result, overproduction of entropy, which measures the impact of agriculture on its environment, slightly decreased (approximately by 11%). On the whole, we can speak about slow evolution of the regional system towards a sustainable state, but a picture of this movement is heterogeneous.

On the one hand, maize, oats, and wheat show a decrease of entropy overproduction. In the case of rye, barley, and rape, there is even a very small increase of entropy. For these crops, yield increased significantly although energy input was lower. This is a consequence of technological improvements (an increase of coefficient η). In fact, the decrease of energy input under simultaneous increase in crop production allows for the maintenance of entropy balance.

Permanent grass has a peculiar position in this set, since its basic entropy excess is minimal (0.09 in comparison with a mean of 0.28 in 1988 and 0.05 in comparison with the a mean of 0.25 in 1997). This is not a surprise if we bear in mind that the permanent grass system is very close to the reference one.

Note that our approach does not include cattle breeding, which actually plays an important role on the study site. Therefore the presented evaluation of grasslands provides only a part of the story as the yields represent the inputs of cattle breeding. Moreover, animal production needs some other input, such as extra fodder and shed keeping.

From the thermodynamic point of view the study site has reached a higher level of sustainability, although the agro-ecosystems still produce more entropy than they are able to export into the environment.

If one wishes to react to an unbalanced entropy budget, there are the following possibilities. First, anthropogenic energy input could be reduced. The overproduction of entropy would then also be reduced proportionally. In addition, the crop production could remain at the same level. Note that if the actual production must be maintained, the system could be transited to a sustainable state by reducing the energy input down to a limit critical load. This would have meant an average reduction from 50 to 34 GJ/ha/year in 1988, and from 31 to 24 GJ/ha/year in 1997, would have been necessary.

Second, crop production could be reduced. However, this (of course) would counteract the general aims of agriculture. Moreover, this is not a suitable measure, as the high production in 1997 was more than compensated by the economical use of anthropogenic inputs.

Third, export of biomass could be increased (this corresponds to an increase of k, for instance, by removing more straw with the harvest). However, this could decrease the soil organic carbon in the long run, which has a negative effect on soil fertility and therefore cannot be the aim of sustainable agriculture. Moreover the process of extraction of residues (straw and roots) requires a lot of additional energy. Therefore when increasing k we must also increase W! As a consequence, we do not necessarily decrease the σ, but it may also increase.

As a conclusion, the reduction of anthropogenic energy input seems to be the most useful strategy. The study on hand proves the effectiveness of this tool. On the one side, the waste of fertilizer was reduced and on the other, the yield increased. This could, however, partly be a result of different weather conditions.

The local excess of entropy depends on several parameters. The value of GPP was calculated using data about crop yield and empirical ratios between roots, straw, and harvested parts of the plant. In addition, a general respiration coefficient was applied, which had been determined by Weisheit (1995) for several herbaceous species of the study site. As plants can adopt root/shoot ratio (Vetter, 1952) and respiration coefficient (Larcher, 1995) to weather and soil conditions, these values only allow a rough estimation of the real situation.

The resulting overproduction of entropy depends significantly on the selected reference system. To accord with the hypothesis, the reference system must be a natural system that is similar to the examined agro-ecosystem. Only in this case will the anthropogenic ecosystem have the same property of exporting entropy to its environment without degradation. Under the climate and soil conditions in the watershed of Lake Belau, a natural succession would lead to a forest ecosystem (like on all farmland in central Europe). Since forest is not successionally close to farmland, we selected long-term fallow grassland as a reference, which seems to be a good compromise.

11.11 AGRICULTURE IN SACHSEN-ANHALT (EASTERN GERMANY) AND THE DYNAMICS OF ENTROPY OVERPRODUCTION (LINDENSCHMIDT ET AL., 2001)

Up to now we considered static cases (the two time points for northern Germany case study do not allow us to construct a curve of dynamics of the entropy overproduction) when a system is in steady state and environmental conditions are not changing: there are not any environmental and social revolutions. But if there is such a revolution then it would be interesting to estimate how this event influences the slow evolution of an agro-ecosystem, a macroscopic state of which is described by the entropy overproduction. Fortunately, there are all the necessary data for the agro-ecosystem in eastern Germany, which were collected from 1981 till 1996. We know that this time interval includes such "revolution" as the German reunification in 1990 to 1991.

The area of studied region is about $30 \, km^2$ and is covered (mainly) by winter wheat and sugar beets fields. A balance of artificial energy inputs, both mechanical (e.g., fuel and machinery) and chemical (e.g., fertilizers and pesticides), and outputs (yield) has been observed for these cultures over the course of 16 years. The dynamics of yield for winter wheat and sugar beets are shown in Figure 11.3. The mean entropy overproduction for each year and the entire region is calculated by the method described in the previous section. Its dynamics are shown in Figure 11.3.

We can see that the dynamics of yield and entropy overproduction are similar. This means that the technology did not principally change in the course of the entire period of observation, and the reunification did not lead to a significant "ecologization" of farming practices in this region. The drastic drop of entropy overproduction in the period 1989 to 1991 is the result of

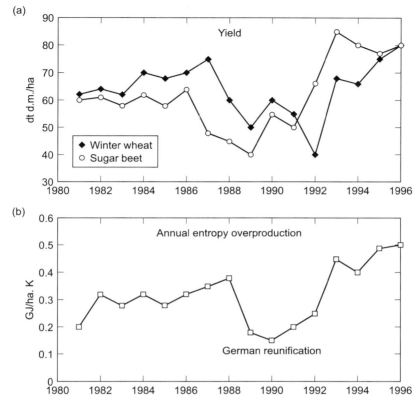

Figure 11.3   (a) Dynamics of beets and winter wheat yields; (b) dynamics of joint entropy overproduction on beets and winter wheat fields (eastern Germany 1981 to 1996).

decrease in the energy load on the system, and as a consequence the drop of yield. It is interesting that if we compare the entropy overproduction in 1995 to 1996, then its value is close to the same value for the northern Germany case study: 0.51 GJ/ha × K.

REFERENCES

Budyko, M. *Global Ecology*. Mysl', Moscow, 1977, 328 p.

Bulatkin, G.A. *Energy Efficiency of Fertilisers in Agroecosystems*. The USSR Academy of Sciences, Moscow — Puschino, 1982 (in Russian).

D'Agostini, L.R. and Schlindwein, S.L. Dialetica da avaliacao do uso e manejo das terras. Da classiicacao interpretativa a um indicador de sustentablidade. Technical report, Ed. da USFC, Florrianopolis, 1996, 100 p.

Dobrovolsky, G.V. *Soil Geography*. Moscow University Press, Moscow, 1974 (in Russian).

Eulenstein, F. "Ökologische und sozioökonomische Analyse der Konsequenzen von Nutzungsänderungen: Landschaftsindikator Energie," in *Agrarlandschaftswandel in Nordost-Deutschland unter veränderten Rahmenbedingungen*, Bork, H.-R., Dalchow, C., Kächele, H., Piorr, H.-P., and Wenkel, K.-O., Eds., Ernst & Sohn, Berlin, 1995, pp. 264–285.

Eulenstein, F., Schlindwein, S.L., and Olejnik, J. From the land use evaluation to a quantitative indicator of sustainability. *Archives of Agronomy and Soil Science* 49, 445–456, 2003a.

Eulenstein, F., Haberstock, W., Steinborn, W., Svirezhev, Y, Olejnik, J., Schlindwein, S.L., and Pomaz, V. Perspectives from energetic-thermodynamic analysis of land use systems. *Archives of Agronomy and Soil Science* 49, 663–677, 2003b.

Glansdorff, P. and Prigogine, I. *Thermodynamics of Structure, Stability and Fluctuations*. Wiley, New York, 1971.

Hydro Agri Dülmen AG. *Faustzahlen für Landwirtschaft und Gartenbau*, 12th ed. Landwirtschaftsverlag, Münster-Hiltrup, 1993.

Jørgensen, S.E. *Integration of Ecosystem Theories: A Pattern*. Kluwer, Dordrecht, 1992, 383 p.

Jørgensen, S.E. and Svirezhev, Y.M. *Towards a Thermodynamic Theory for Ecological Systems*. Elsevier, Amsterdam, 2004, 370 p.

Landau, L.D. and Lifshitz, E.M. *Statistical Physics. Vol. 5: Course of Theoretical Physics*. Nauka, Moscow, 1995 (in Russian).

Lang, I. (1990). Personal communication.

Larcher, W. *Ökophysiologie der Pflanzen*, 5. UTB, Auflage, 1995, 394 p.

Lindeman, R.L. The trophic dynamic aspect of ecology. *Ecology* 23, 399–418, 1942.

Lindenschmidt, K.E., Franko, U., and Rode, M. Entropy as an indicator for sustainability. NATO Research Workshop on Water Resources Planning, Wageningen, 2001.

Novikov, Y.F. Energy balance of agro-industrialised systems and bio-energetics of agroecosystems. *The USSR Academy of Agriculture Sciences Report* 5, 4–9, 1984 (in Russian).

Odum, H.T. *Systems Ecology*. Wiley, New York, 1983, 644 p.

Pedlosky, J. *Geophysical Fluid Dynamics*. Springer-Verlag, New York, 1979, 619 p.

Pimentel, D. *Handbook of Energy Utilisation in Agriculture*. CRC Press, Boca Raton, FL, 1980, 475 p.

Pimentel, D., Hurd, L.E., and Bellotti, A.C. Food production and energetic crisis. *Science* 182, 443–449, 1973.

Sach, W. Vegetation und Nährstoffdynamik unterschiedlich genutzten Grünlandes in Schleswig-Holstein. *Dissertationes Botanicae* 308, 1999, 311 p.

Sommerfeld, A. Themodynamik und Statistik. Vorlesungen über Theoretische Physik, Band V. Spinger-Verlag, Wiesbaden, 1952, 480 p.

Smil, V., Nachman, P., and Long, T.V., II, *Energy Analysis and Agriculture*. Westview Press, Boulder, CO, 1983.

Steinborn, W. and Svirezhev, Y. Entropy as an indicator of sustainability in agro-ecosystems: North Germany case study. *Ecological Modelling* 133, 247–257, 2000.

Svirezhev, Y. Entropy as a measure of environment degradation. *Proceedings of an International Conference on Contaminated Soils*, Karlsruhe, Germany, 1990, additional volume, 26–27.

Svirezhev, Y. "Thermodynamic orientors: how to use thermodynamics concept in ecology," in *Eco Targets, Goal Functions and Orientors*, Müller, F. and Leupelt, M., Eds. Springer-Verlag, Berlin, 1998, 102–122.

Svirezhev, Y. Thermodynamics and ecology. *Ecological Modelling* 132, 11–22, 2000.

Svirezhev, Y. and Svirejeva-Hopkins, A. Sustainable biosphere: critical overview of basic concept of sustainability. *Ecological Modelling* 106, 47–61, 1998.

Vetter, H. Menge und Zusammensetzung der Ernterückstände bei den wichtigsten Kulturpflanzen und verschiedenen Fruchtfolgen auf Lehmboden und rohhumushaltigem Heidesand. PhD Thesis, Kiel University, 1952, 217 p.

Wachendorf, C. Eigenschaften und Dynamik der organischen Bodensubstanz ausgewählter Böden unterschiedlicher Nutzung einer norddeutschen Moränen landschaft. *Ecosystems* 13, 1995, 130–144.

Weisheit, K. Kohlenstoffdynamik am Grünlandstandort: untersucht an 4 dominanten Grasarten. PhD Thesis, Kiel University, 1995, 141 p.

CHAPTER 12

Ecosystem Indicators for the Integrated Management of Landscape Health and Integrity

F. Müller

In the following chapter an attempt is described to represent the state of ecosystems and landscapes on a holistic, systems-oriented basis. The general guidelines for the derivation of the ecosystem indicators originate in thermodynamic ecosystem theory, in empirical ecosystem analysis, and in the concept of ecosystem health/integrity. The respective fundamental concepts are principles of self-organization, the ecological orientor approach, an integration of structural and functional items and the normative idea of ecological risk prevention. These principles are explained in a first part of the chapter which leads to a presentation of the indicator set, which aims at a depiction of the self-organizing capacity of ecological entities. In the second part, some applications of the indicator set are shown. They refer to the ecosystem scale (with a comparison of a forest ecosystem and an arable land ecosystem), to the landscape scale (characterizing different wetland ecosystem types in a northern German watershed) and to the development of sustainable landscape management regimes in northern Fennoskandia (consequences of different approaches for reindeer herding in an ecological, social and economic context).

1-56670-665-3/05/$0.00 + $1.50

12.1 INTRODUCTION

Throughout the past few decades, ecosystem approaches seem to have grown out of puberty: For a rising number of ecologists the high complexity of ecological systems has not only become an accepted fact, but also an interesting object of investigation. In parallel, a successful reductionistic methodology has been accomplished steadily by holistic concepts which stress systems approaches and syntheses, and which elucidate the linkages between the multiple compartments of ecological and human-environmental systems within structural, functional, and organizational entities. For instance, in Germany five ecosystem research centers have been installed and supported within the past few decades (e.g., Fränzle, 1998; Fritz, 1999; Gollan and Heindl, 1998; Hantschel et al., 1998; Widey, 1998; Wiggering, 2001) and additional research projects have been carried out in national parks (e.g., Kerner et al., 1991), biosphere reservations (e.g., Schönthaler et al., 2001), and coastal districts (e.g., Dittmann et al., 1998; Kellermann et al., 1998). With these initiatives, the comprehension and the acceptance of ecosystem approaches has made a big step forward in Germany (for an overview see Schönthaler et al., 2003).

Also in environmental practice, ecosystemic attitudes are becoming more and more favorable: While in the past, environmental activities were restricted to specific ecological resorts, today — in the age of the sustainability principle — we can find resort-spanning environmental politics. Instead of a concentration on environmental sectors, ecosystems are becoming focal objects, and interdisciplinary cooperation is increasing continuously. The same is true of environmental practice (see Schönthaler et al., 2003).

The major problem of these modern approaches is to cope with the enormous complexity of environmental systems, which arises from the various elements, subsystems, and interrelations that ecosystems provide. Hence scientific approaches to reduce this complexity with a valid and theory-based methodology have become basic requirements for a highly qualitative development of systemic approaches in science, technology, and practice (see Müller and Li, in press). One concept to reduce the complexity of ecological and human-environmental systems is a representation of the most significant parameters of an observer-defined system by indicators, which are quantified variables that provide information on a certain phenomenon with a synoptic distinctness (Radermacher et al., 1998). Often, indicators are used if the indicandum — the focal object of the demanded information — is too complex to be measured directly or if its features are not accessible with the available methodologies.

There are certain acknowledged requirements for indicators. For instance, they should be easily measurable, they should be able to be aggregated, and they should depict the investigated relationships in an understandable manner. The indicandum should be clearly and unambiguously represented by the indicators. These variables should comprise an optimal sensitivity, include normative loadings in a defined extent only, and they should provide a high utility for early warning purposes (Wiggering and Müller, 2004). As Table 12.1

Table 12.1 Some criteria and requirements for ecological indicators. The listed items should be realized to an optimum degree to produce an applicable indicator system according to Müller and Wiggering (2004)

Political relevance	High level of aggregation
Political independence	Target-based orientation
Spatial comparability	Usable measuring requirements
Temporal comparability	Usable requirements for quantification
Sensitivity concerning the indicandum	Unequivocal assignment of effects
Capability of being verified	Capability of being reproduced
Validity	Spatio-temporal representativeness
Capability of being aggregated	Methodological transparency
Transparency for users	Comprehensibility

shows, there are many further needs for the quality of indicator sets, which often can only barely be met if complex interrelations have to be represented.

Concerning these requirements, the existing holistic indicator sets comprise different potentials, advances, and limitations. For example, with respect to indicator complexity, on the one hand we can find very complex indicator sets with a very high number of proposed variables (e.g., Schönthaler et al., 2001; Statistisches Bundesamt et al., 2002), and on the other there are approaches that include a reduction up to one parameter, only (e.g., Müller, 1998; Jørgensen, 2000; Ulanowicz, 2000; Odum et al., 2000). Between these indicator systems there is a broad wingspan according to the necessary database, the demanded measuring efforts, the complexity of the aggregation methodology, and the comprehensibility of the results as well as the cognitive transparency for the users.

Within this polarization, we have tried to find a representative holistic indicator set on the basis of the concepts, results, and the theoretical background of a research and development project entitled "Ecosystem Research in the Bornhöved Lakes District" (Fränzle, 1998, 2000). Secondary investigations have been executed in the research and development project "Macro Indicators to Represent the State of the Environment for the National Environmental-Economic Accounting System of Germany" (Statistisches Bundesamt et al., 2002). The respective investigations led to a set of eight ecosystem variables, which are suitable for representing the focal element of the pressure-state response and the drivers pressure state impact-response indicator approaches — the state of ecosystems on an integrative level. The indicators are proposed to be used as representatives for the capacity of self-organization in ecological systems which is the selected indicandum to depict the degree of integrity or health in ecological entities.

This chapter tries to demonstrate the derivation and application of the aggregated ecosystem indicator set. The basic principles and the specific requirements for the indicator selection will first be described. These resulting conceptual forcing functions come from ecosystem analysis, ecosystem theory, and from the normative principles of ecosystem integrity. The respective framework for indicator selection will be clarified, and thereafter the indicators will be presented together with some information on the utilized methodologies for their quantifications on different scales. On this basis, some case studies will be presented, beginning with a comparison of different ecosystems and continued by a description of applications on the landscape scale. The potentials of the indicator set for monitoring schemes will also be discussed, and finally an application in sustainable landscape management will be described. The chapter will end with a discussion and a prospect to future developments.

12.2 BASIC PRINCIPLES FOR THE INDICATOR DERIVATION

Besides the requirements summarized in Table 12.1, three principle pillars have been considered as basic conceptual "points of departure" for indicator derivation.

The first guideline, which guarantees a high applicability and a general correctness, origins in fundamental ideas from ecosystem theory: ecosystems are comprehended as self-organizing entities, and the degree of self-organizing processes and their effects have been chosen as an aggregated measure to represent the systems' actual states. The basic theoretical principles of this approach stem from thermodynamic fundamentals of self-organization and from the orientor principle, which is also used by many other concepts published in this book.

A second pillar is built up by the methodologies of ecosystem analysis: to depict ecological entities in a holistic manner, structure as well as function has to be taken into account, the latter representing the performance of the ecosystems.

Finally, for a utilization in environmental management, the basic approaches which emerge from these principles have to be reflected on a normative level. As the factual evaluation of the concrete indicator values is a societal (not an ecological) task, a useful indicator set has to be based on political concepts and targets. In this case, the preconditions for environmental decision-making are formulated by a specific definition of ecological integrity (Barkmann et al., 2001) which includes several items that are valid for the ecosystem health approach as well.

12.2.1 Ecosystem Theory — The Conceptual Background

To reach an optimal applicability of scientific methodology, theoretical considerations seem to be a good starting point, even if applicable indicators for practical purposes have to be developed. In ecosystem theory there are many different approaches (see Jørgensen, 1996; Müller, 1997) which can easily be condensed and aggregated within the theory of self-organization. This approach does not only provide a unifying concept of ecosystem dynamics, it also depicts a high agreement with basic ideas from the ecosystem health concept (see Table 12.2) that stresses the creativity of nature, which is nothing else than the potential for self-organization.

Table 12.2 Axioms of ecosystem health. The listed parameters reflect the basic system related fundamentals of the health approach, which are also valid for the concept of ecological integrity according to Costanza et al. (1993)

Dynamism:	Nature is a set of processes, more than a composition of structures
Relatedness:	Nature is a network of interactions
Hierarchy:	Nature is built up by complex hierarchies of spatio-temporal scales
Creativity:	Nature consists of self-organizing systems
Different fragilities:	Nature includes various sets of different resiliences

In a generalized outline of the selected theoretical concept, the order of ecological systems emerges from spontaneous processes which operate without consciously regulating influences from the system's environment. Actually these processes are constrained by human activities (see Müller et al., 1997a, 1997b; Müller and Nielsen, 2000) but although such constraints can reduce the degrees of freedom for ecosystem development, the self-organized processes cannot be set aside. The consequences of these processes have been condensed within the orientor approach (Bossel, 1998; Müller and Leupelt, 1998), a systems-based theory about ecosystem development, which is founded on the general ideas of nonequilibrium thermodynamics (Jørgensen, 1996, 2000; Schneider and Kay, 1994; Kay, 2000) and network development (Fath and Pattten, 1998) on the one hand and succession theory on the other (e.g., Odum, 1969; Dierssen, 2000).

Self-organized systems are capable of creating structures and gradients if they receive a throughflow of exergy (usable energy, or the energy fraction of a system which can be transferred into mechanical work, see Jørgensen 2000). The typical exergy input path into ecosystems is solar radiation. This "high-quality" energy fraction is transformed within metabolic reactions (e.g., respiration, heat export), producing nonconvertible energy fractions (entropy) which are exported into the environment of the system. As a result of these energy conversion processes, under certain circumstances (Ebeling, 1989) gradients (structures) are built up and maintained. There are two extreme thermodynamic principles that take these conditions into account and which postulate an optimizing behavior of open, biological systems. Jørgensen (2000) states that self-organized ecological systems tend to move away from thermodynamic equilibrium, that is build up ordered structures and store the imported exergy within biomass, detritus, and information (e.g., genetic information) which can be indicated by structural diversities. In addition, Schneider and Kay (1994) state that the degradation of the applied gradients is an emerging function of self-organized systems.

As a consequence of these physical principles, throughout the undisturbed complexifying development of ecosystems — between Holling's exploitation and conservation stages (Holling, 1986; Gunderson and Holling, 2002) — there are certain characteristics which are increasing steadily and slowly. These features are developing towards an attractor state which is restricted by the specific site conditions and the prevailing ecological functions. As the development seems to be regularly oriented towards that attractor basin, the respective state variables are called orientors (Bossel, 2000).

Using these ecosystem features as indicators, the naturalness of an ecosystem's development can be depicted. Figure 12.1 shows some of these orientors. In general it can be postulated that throughout an undisturbed development, the complexity of the ecosystems will increase asymptotically up to the state of maturity (Odum, 1969). Within this development, exergy storage will be rising on a materialistic level as well as on a structural basis: more and more gradients are built up. With this increasing structural diversity, the diversity of flows and the system's ascendancy (Ulanowicz, 2000) will grow as well as certain network features (Fath and Patten, 2000), and therefore the energy necessary for the maintenance of the developing system will also increase. Therefore, exergy storage as well as exergy degradation are typical orientors, and their dynamics can be explained in a contemporary manner. These basic thermodynamic principles have many consequences on other ecosystem features. For instance, the food web will become more and more complex, heterogeneity, species richness, and connectedness will be rising, and many other attributes, as shown in Figure 12.1 will follow a similar long-term trajectory.

This orientation is a theoretical principle which can rarely be found in reality due to the continuous effects of disturbances. Particularly in the case of high external inputs, the orientor values might decrease rapidly, proceeding into a retrogressive direction. In the following sequence, an adaptive or resilient system will find the optimization trajectory again, while a heavily disturbed ecosystem might not be able to improve the values of the orientors. Therefore the robustness of ecosystems can be indicated by the orientors as well. Consequently, their values are also suitable for representing the ecological risk correlated to external inputs or changes to the prevailing boundary conditions. However, we have to be aware of the fact that high orientor values do not guarantee a high stability or a high buffer capacity. Following Holling's ideas on ecosystem resilience and development, at the mature stage complex ecosystems become "brittle," their adaptivity decreases because of the high internal connectedness and the respective interdependencies. Thus, the dynamics of external variables can force the mature system to break down and start with another developmental sequence.

An indication for ecosystem self-organization has been proposed in only a small number of case studies. Most of them refer to the concepts of ecosystem health (e.g., Rapport, 1989; Haskell et al., 1993; Rapport and Moll, 2000) or ecological integrity (e.g., Karr, 1981; Woodley et al., 1993). Besides multivariate approaches (e.g., Schneider and Kay, 1994; Kay, 1993, 2000) and aggregated approaches (e.g., Costanza, 1993) some authors propose to use highly integrated variables like exergy (Jørgensen, 2000), emergy (Odum et al., 2000; Ulgiati et al., 2003) or ascendancy (Ulanowicz, 2000). These bright concepts are very original, they are discussed very actively, and they can cope with the concept of emergent properties. However, there are tremendous

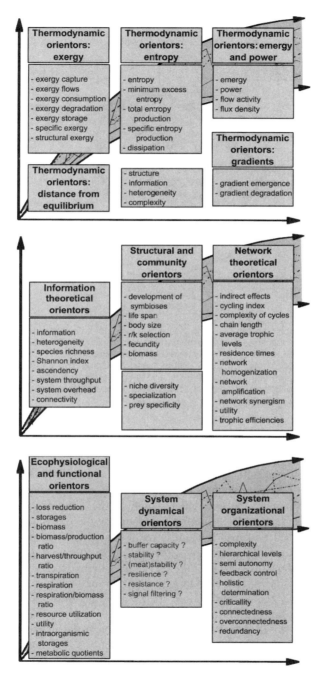

Figure 12.1 Ecological orientors from different theoretical origins. The listed ecosystem properties regularly show an optimizing behavior during the long-term development in undisturbed situations according to Müller and Jørgensen (2000).

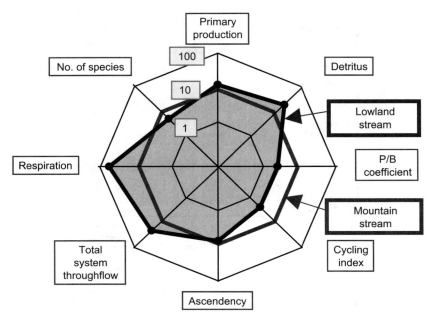

Figure 12.2   Amoeba diagram depicting the relative indicator values for a mountain stream and a lowland stream on the basis of a trophic ECOPATH model which has been applied to data sets from Meyer (1992) and Pöpperl (1996). The model has been calibrated and run by R. Pöpperl and S. Opitz. The mountain stream values represent 100% in the graphics, and the comparison depicts the consequences of eutrophication for some orientor values of the northern German lowland stream.

problems, data requirements and modeling demands when trying to apply them in practice.

One example of multivariate orientor applications is shown in Figure 12.2. Two different German stream ecosystems are compared on the basis of emergent ecosystem properties which can take the function of orientors. The depicted values are based on intensive measurements in a Black Forest stream from Meyer (1992) and in a lowland stream ecosystem within the Bornhöved Lakes district in northern Germany (Pöpperl, 1996). These data have been used to run the model software ECOPATH 3.0, which describes the food web structures, quantifying the standing stock, production and consumption of the elements and the whole system as well as the flow of matter between the ecosystem compartments (average annual rates per m^2). Additionally, the model can quantify a series of holistic ecosystem properties.

The diagram elucidates that there are enormous differences between the investigated ecosystems. Especially, concerning the primary production based parameters (primary production, respiration, total system throughflow) the lowland stream provides typical values for a strongly eutrophicated ecosystem. On the other hand, the more complex structure (number of species), the relative diversity of flows and related parameters (cycling index, p/b coefficient) show

Table 12.3 Some arguments stressing the methodological significance of ecosystem approaches in environmental management, as they can provide a better consideration of the following items

---

Indirect effects (e.g., webs of reactions concerning forest dieback)
Chronic effects (e.g., accumulation of toxic substances)
Delocalized effects (e.g., forest effects of ammonia from slurry)

Integration of ecological processes and relations into planning procedures
Representation of ecological complexity
Consideration of features of self-organization
Aggregation of structure and function

Integration of different ecological media (e.g., soil-vegetation-atmosphere)
Integration of different environmental sectors (e.g., immission and erosion)
Utilization of improved extents and resolutions
- in terms of time (multiple interacting temporal scales)
- in terms of space (multiple interacting spatial scales)
- in terms of content and disciplines (multiple scientific approaches)
- in terms of analytical depth (multiple levels-of-aggregation and reduction)

---

that the mountain stream represents a much higher degree of ecosystem integrity.

12.2.2   Ecosystem Analysis — The Empirical Background

Besides the theoretical considerations, there are other good reasons to use an ecosystem approach for environmental assessments. In Table 12.3, some of these motivations are listed. Various case studies from forest dieback research, ecotoxicology, and eutrophication research have documented that indirect effects, chronic effects, and delocalized effects are much more significant than direct interactions (see Patten, 1992). Furthermore, many disturbances do not affect just one environmental sector, but the whole ensemble of ecological compartments via webs of interactions and consequences. Last but not least, the ecosystem approach makes it possible to include phenomena like self-organization, emergent properties and ecological complexity (Fränzle, 2000). Therefore, the conceptual combination of structural and functional approaches into an organizational concept is a fine starting point to fulfill the empirical requirements for health or integrity indication (Costanza et al., 2000; Golley, 2000; Müller and Windhorst, 2000).

The respective scientific approaches focus on "models of networks consisting of biotic and abiotic interactions in a certain area" (Jørgensen and Müller, 2000; Müller and Breckling, 1997). Schönthaler et al. (2003) have defined ecosystem research as a "media spanning research of element and energy cycling, of structures and dynamics, of control mechanisms and of criteria for ecosystem resilience with the aim to learn how to understand the steering and feedback processes in ecological entities." Kaiser et al. (2002) have accomplished this description in the following way: "Ecosystem research analyses the interactions of biological ecosystem components with each other, with their inanimate environment and with man. It delivers basic knowledge on structure, dynamics, element and energy flows, ecosystem stability, and resilience."

Apart from structural aspects (e.g., items of abiotic and biotic heterogeneity and their dynamics), ecosystem research investigates the imports, exports, storages, and the internal flows of energy, water, and nutrients (e.g., carbon, nitrogen, potassium, calcium, sodium, magnesium) through the compartments of ecological entities (e.g., soil horizons, the unsaturated zone, the groundwater layer, plants on different structural levels and in different layers, but also with different internal functional subunits, animals with different positions in the food webs, or micro-organisms which can be found in different spatial compartments, and the atmospheric compartment) including the derivation of efficiency and cycling attributes (e.g., different ratios of biomass, respiration, production, water movement, cycling index).

As there are various variables that can be taken into account to measure these items, and as they are linked within very complex webs of interactions, it is hard to select a small number of indicators which are capable of representing the whole variety of aspects describing the state of ecological systems. To proceed with this task, a combination has to be made which reflects the theoretical items, the empirical requirements and the normative targets of the indicator set.

12.2.3 Ecosystem Health and Ecological Integrity — The Normative Background

As the aspired indicators have to be used as information sources in environmental decision-making, societal and normative arguments are also important prerequisites of their selection. The indicators have to refer to the leading concept of environmental management, which is actually the global political principle of sustainable development. It has been discussed in various papers and political statements (e.g., Hauff, 1987; WCED, 1987; Daily, 1997; Costanza, 2000), and in essence we are asked to utilize natural resources in a way that enables future generations to access these resources in at least similar mode as applied today. The main conceptual innovations of the sustainability principle are the interdisciplinary linkage of social and natural items and the large spatio-temporal scales which have to be taken into account. Thus some specific requirements arise from this principle (summarized in Table 12.4).

An important outcome of the described self-organized processes in the ecosphere is the potential of man utilizing the outputs of ecosystems' performances. Ecosystem structure and function provide certain environmental services, which are the benefits people obtain from ecosystem organization (being the basic requirements for human life) (see Costanza et al., 2000; Millennium Assessment Board, 2003). One potential classification of these services is based on the works of De Groot (1992): from his point of view the performance of ecosystems can be distinguished into the following classes:

- *General provisions (carrier services)*. Ecosystem structures are providing space and suitable substrates for human activities.

Table 12.4 Basic features and requirements of sustainable landscape management strategies according to Müller and Li (in press)

Long-term strategies...	... think in generations
Multi-scale strategies...	... compare human vs. ecological time scales
Interdisciplinary strategies...	... realize that ecology is only one part
Holistic strategies...	... consider structure *and* function
Realistic strategies...	... include uncertainties
Nature-oriented strategies...	... take nature as a model
Theory-based strategies...	... make sure correctness
Hierarchical strategies...	... realize constraints and scales
Goal-oriented strategies...	... joint definition of the targets

- *Products.* Ecosystem development provides natural resources for human use.
- *Information.* Ecosystems are providing cultural attributes.
- *Regulations.* Ecosystem functions are regulating the availability of basic demands for human life. All ecological processes can be assigned to this category as they buffer external influences in a way that enables man to continue life in an environment with suitable climatic, chemical and physical conditions.

Taking into account the terms and concepts mentioned in the last chapter, it is possible to use an alternative formulation for the ecological components of sustainable development: "Meet the needs of future generations," in this context means, "Keep available the ecosystem services on a long-term, intergenerational and a broad scale, intragenerational level."

From a synoptic viewpoint at these four categories, one fact becomes obvious: All ecosystem services are strongly dependent on the performance of the regulation function. The correlated processes do not only influence production rates, but in the long run they also determine the potentials of ecosystems to provide carrier and information services. And if we finally link all argumentations of this chapter, it becomes clear that the respective benefits are strictly dependent on the degrees and the potentials of the fundamental self-organizing processes. To maintain these services, the ability for future self-organizing processes within the respective system has to be preserved (Kay, 1993). This demand is considered as a focal point of modern environmental management models, such as ecosystem health or ecological integrity. In a recent paper, Barkmann et al. (2001) have defined ecological integrity as a political target for the preservation against nonspecific ecological risks, which are general disturbances of the self-organizing capacity of ecological systems. Thus the goal should be a support and preservation of those processes and structures that are essential prerequisites of the ecological ability for self-organization.

12.3 THE SELECTED INDICATOR SET

The three basic pillars for the presented indicator selection result in a set of variables that are able to depict the state of ecosystems on the basis of their features concerning the degree of self-organization and the potential to proceed in this way. Referring to the orientors presented in Figure 12.1, it becomes obvious that many of them cannot be easily measured or even modeled under normal circumstances. Some orientors can only be calculated on the basis of very comprehensive data sets which are measured on a very small number of sites. Other orientors can only be quantified by model applications. Therefore the selected orientors have to be represented by variables that are accessible by traditional methods of ecosystem quantification. Consequently, the next step of indicator derivation is a "translation" of the thermodynamic, organizational network, and information theoretical items into ecosystem analytical variables. Within this step, it has to be reflected that the number of indicators should be reduced as far as possible (see Table 12.1). Thus many of the ecosystem variables depicted in Figure 12.1 cannot be taken into account. Instead, a small set consisting of the most important items which can be calculated or measured in many local instances is what we have to look for. This set should furthermore be based on the focal variables that are usually investigated in ecosystem research and that can be made accessible in comprehensive monitoring networks (Müller et al., 2000). The general subsystems that should be taken into account to represent ecosystem organization are listed below as elements of ecosystem orientation:

1. *Ecosystem structure.* While ecosystems are evolving, the number of integrated species is regularly increasing steadily and the abiotic features are becoming more and more complex. This development is accompanied by a rising degree of information, heterogeneity, and complexity. Also, specific life forms (e.g., symbiosis) and specific types of organisms (r/k strategists, organisms with increasing life spans and body masses) become predominant throughout the orienting development.

2. *Ecosystem function.* Due to the increasing number of structural elements, the translocation processes of energy, water, and matter are becoming more and more complex, the significance of biological storages is growing as well as the degree of storage in general, and consequently the residence times of the input fractions are increasing. These processes influence the budgets of the respective fractions, which can be measured by input–output analysis. Due to the high degree of mutual adaptation throughout the long developmental time the efficiencies of the single transfer reactions are rising, cycling is optimized, and losses of matter are therefore reduced. The respective ecosystem functions are usually investigated within three classes of processes which are interrelated to a very high degree:

   a. *Ecosystem energy balance.* Exergy capture (uptake of utilizable energy) is rising during the undisturbed development, the total system throughput is growing (the "maximum power principle," see Odum et al., 2000) as well as the articulation of the flows (ascendancy, see Ulanowicz, 2000). Due to the high number of processors and the growing amount of biomass, the energetic demand for maintenance processes and respiration is growing as well (entropy production, see Svirezhev and Steinborn, 2000; Steinborn, 2001).

    b. *Ecosystem water balance.* Throughout the undisturbed development of ecosystems and landscapes, more and more elements have to be provided with water. This means that water flows through the vegetation compartments show a typical orientor behavior (Kutsch et al., 1998). These fluxes provide another high significance because they demonstrate an important prerequisite for all cycling activities in terrestrial ecosystems: the water uptake by plants, which is regulated by the degree of transpiration.

    c. *Ecosystem matter balance.* Imported nutrients are transferred within the biotic community with a growing partition throughout undisturbed ecosystem development. Therefore the biological nutrient fractions are rising as well as the abiotic carbon and nutrient storages, the cycling rate is growing and the efficiencies are being improved. As a result, the loss of nutrients is reduced.

On the basis of these features, a general indicator set to describe the ecosystem or landscape state in terrestrial environments, has been derived (see Table 12.5). The basic hypothesis concerning this set is that a holistic representation of the degree and the capacity for complicating ecological processes on the basis of an accessible number of indicators can be fulfilled by these variables. They also represent the basic trends of ecosystem development, thus they show the developmental stage of an ecosystem or a landscape. As a whole this variable set represents the degree of self-organization in the investigated system. Hence it can be postulated that (with the exception of mature stages which are in fact very seldom in our cultural landscape) the potential for future self-organization can be depicted with this indicator set.

Of course this parameter set cannot provide a complete indication of sustainability, because the social and economic subsystems are not taken into account (e.g., driving force or response indicators). Also external inputs and other pressures are not represented. But the focal ecological branch of

Table 12.5 Proposed indicators to represent the organizational state of ecosystems and landscapes. The nominated key variables can be regarded as an optimal indicator set. If these parameters are not available other variables may be chosen to reflect the respective indicandum. Doing this, the observer must realize that the quality of the indicator–indicandum relations may be sinking

Orientor group	Indicator	Potential key variable(s)
Biotic structure	Biodiversity	Number of species
Abiotic structure	Biotope heterogeneity	Index of heterogeneity
Energy balance	Exergy capture	Gross or net primary production
	Entropy production	Entropy production after Aoki
		Entropy production after Svirezhev and Steinborn (2001)
		Output by evapotranspiration and respiration
	Metabolic efficiency	Respiration per biomass
Water balance	Biotic water flows	Transpiration per evapotranspiration
Matter balance	Nutrient loss	Nitrate leaching
	Storage capacity	Intrabiotic nitrogen
		Soil organic carbon

sustainability can be described on the basis of the orientor state indication. In spite of this strategic restriction, the integrity indication provides potential linkages to the human-based indicators of the DPSIR scheme. A comparison with the basic ecosystems services after de Groot (1992) shows that regulation services are well represented in this indicator set and that there are high interrelations with the production services while carrier and information services are not represented in a satisfactory manner.

12.4 CASE STUDIES AND APPLICATIONS

12.4.1 Indicating Health and Integrity on the Ecosystem Scale

This indicator set has been applied within several case studies on different scales, whereby the linkages between data sources, model outputs and indicator demands have been an object of methodological optimization throughout the past few years. In the following paragraphs one example will be discussed from the ecosystem research project on Bornhöved Lakes, which was conducted between 1988 and 2001 in northern Germany. Within the main research area, Altekoppel, comparative empirical ecosystem studies were carried out in agro-ecosystems and forests (Hörmann et al., 1992). A precise description of the methodologies used for the indicator quantification can be found in Schimming and von Stamm (1993), Baumann (2001), and Barkmann (2001). The respective measurements have been conducted by numerous colleagues from the Bornhöved Lakes project (see also http://www.ecology.uni-kiel.de) whose investigations are summarized in Hörmann et al. (1992), Breckling and Asshoff (1996) and Fränzle et al. (in press), to name a few examples.

In the following case study some results from a 100-year-old beach forest and a directly neighboring arable land ecosystem will be demonstrated. Both ecosystems had a similar agricultural use before the forest was planted. Thus the question is which ecosystem features and which ranges of the self-organization capacity have been modified by the different land-use schemes (see also Kutsch et al., 2001; Kutsch et al., 1998; Windhorst et al., 2004).

Figure 12.3 shows the differences between the two ecosystems with respect to their biocenotic structure. This variable represents the biotic complexity of ecosystems, and it reflects the amount of exergy stored in information. Nearly all investigated organism groups show higher numbers of species in the forest ecosystem. One exception is the group of small mammals, who can find very good food conditions in the arable land and who are well adapted to this ecosystem type. The second structural indicator is the abiotic heterogeneity which was calculated with a GIS-based neighborhood method after Reiche (Baumann, 2001). While the index of the forest ecosystem is 0.56 (referring to the soil organic matter), the maize field has a value of only 0.08. Also, corresponding to the soil chemical constituents H^+, Ca^{2+}, Mg^{2+}, K^+, and phosphate, the forest soil heterogeneity is higher than the respective value on

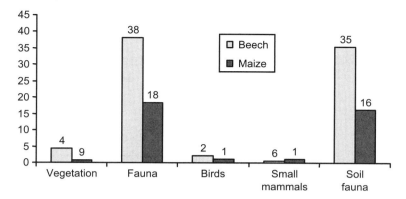

Figure 12.3 Comparison of the species numbers in some community groups of the investigated ecosystems. Data have been compiled from Hörmann et al. (1992).

the arable land (Reiche et al., 2001). Therefore we can constitute very high differences concerning the structural patterns of these ecosystems.

Investigating the storage capacities of the two ecosystems, the biomass and the intrabiotic nutrients were used as indicators. They are capable of representing the ecosystem pools as another compartment of exergy storage, the chemical buffer capacities, and the availability of nutrients for the further development of the system. The depicted data are based on direct measurements, yield analyses and modeling results (see Baumann, 2001). The living biomass varied from $131\,t\,C/ha$ in the beech forest to $6.5\,t\,C/ha$ in the arable land, and the relations for the soil organic carbon is $80\,t\,C/ha$ vs. $56\,t\,C/ha$, respectively. The correlated ecosystem comparison concerning the intrabiotic nutrients is sketched in Figure 12.4. It shows that the higher values can be found in the forest ecosystem for both nitrogen and phosphorus compounds.

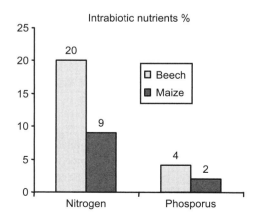

Figure 12.4 Comparison of the intrabiotic nutrient contents of the investigated ecosystems. Data from Kutsch et al. (1998).

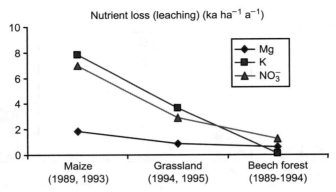

Figure 12.5 Comparison of the nutrient loss from the investigated ecosystems. Data from H. Wetzel and C.G. Schimming (unpublished). The figure demonstrates that there are significant differences between the ecosystems concerning the leaching quantities of the three chemical compounds.

Another important functional parameter used is the loss of nutrients. It shows the irreversible export of chemical compounds as well as the efficiency of the recycling regime in the ecosystem. Data from Figure 12.5 are based on chemical analyses of the soil solution and model applications concerning the balances and the output path into the atmosphere. The figure shows that there are enormous differences between the two systems. Of course, this is a consequence of the different import and export regimes. But besides these extreme examples, the loss of nutrients seems to be a very general effect resulting from ecological disturbances. This may be caused by opening the food webs and cycles, which usually become more and more closed in undisturbed developmental phases. Nutrient loss is therefore a suitable candidate for a key indicator of ecosystem health.

Similar results were obtained concerning the biotic water flows which represent a biological efficiency measure and which symbolizes the basic prerequisites for all cycling processes. The data are based on hydrological and microclimatological measurements and transpiration modeling with a two-layer Soil-Vegetation-Atmosphere model (Herbst et al., 1999). The percentage of transpiration from the total evapotranspiration loss was 63% in the case of the forest ecosystem and 34% in the case of the field. This signals the distinct significance of biological flows in the site budgets of water. This item could also be comprehended as an ecosystemic water-use efficiency because it is strongly correlated with the capacity of nutrient cycling, and because transpiration is a very important factor of the temperature regulation of ecosystems.

The metabolic efficiency (respiration/biomass) of the forest was much higher than the efficiency of the arable land ecosystem. This elucidates the different degrees of flow organization and the energetic demand to maintain the existing structure. The entropy production was calculated in a methodology after Aoki (1998) and on the basis of the exergy radiation balance (Steinborn, 2001). While the first method does not produce a satisfying sensitivity, the

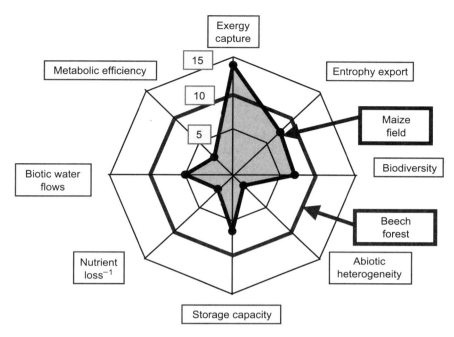

Figure 12.6 Synopsis of the indicator values for the two compared ecosystems. The beech values have been taken as reference values (100%).

radiation balance approach can discriminate both ecosystems very well (see Baumann, 2001).

A synopsis of the indicator values is presented in Figure 12.6. Looking at the whole figure, it is obvious that all values of the forest ecosystem are higher than the respective numbers of the arable land system with one exception: exergy capture. This indicandum has been represented by the gross primary production. The high value of the arable-land ecosystem demonstrates that the farmer has been successful in optimizing the production of his site. The consequences of this economic orientation can be seen in all other variables, summarizing they show that the degree of self-organization of the forest — and with this the ecological integrity — is much higher than it is in the field. In the case of new external disturbances this system bears a much higher risk of retrogressive changes than the forest which represents a higher state of self-organizing capacity.

12.4.2 Indicating Landscape Health

While the case study stated before is totally based on small-scaled measured data, additional approaches have been developed on the landscape scale where the demands for empirical measurements are much smaller. To extend the indicator system to the landscape level, it was linked with the GIS-coupled modeling system DILAMO (digital landscape analysis and modeling, see

Reiche, 1996). Using this instrument, many of the explained indicators can be calculated on the landscape scale. The integrated models "Wasmod" and "Stomod" have been stepwise enabled to calculate the parameters described in Table 12.5 in a validated and reliable manner. This methodology has been applied in different areas: M. Meyer (2000) used the modeling procedure to foresee the outcome of three land-use scenarios for the whole Bornhöved Lakes district. He could show that the nutrient budget indicators in particular show large differences due to distinct land-use strategies (see also Reiche and Windhorst, in press). The municipality of Plön in northern Germany was analyzed by Barkmann (2001) to show the dynamics of the integrity indicators in different years, using the same methodology. He could also underline that the loss of nutrients in particular seems to be very sensitive to changes in ecosystem structure and function. U. Meyer (2001) has conducted a similar study for two catchments in the biosphere reservation Rhön in central Germany (Schönthaler et al., 2001).

Taking a similar approach, Schrautzer et al. (in press) have derived landscape balances for the different ecosystem types of the Bornhöved Lakes district, including water, matter, and energy budgets for the whole watershed. With this contribution, the methodological linkage between the modeling systems, the GIS and the proposed indicator set has been transferred into a highly applicable form. This case study is based on an ecosystem classification which has been conducted for all terrestrial ecosystems in the catchment of Lake Belau (447 ha). This watershed includes a high proportion of wetland ecosystems. The ecosystem classification (conducted by U. Heinrich, J. Schrautzer and H.P. Blume, see Fränzle et al. (in press)) takes into account the vegetation types, soil criteria and the dominant land-use structures. The resulting ecosystem types have been calibrated with data on the groundwater table, the carbon/nitrogen ratios, the pH values and the soil constant values of the soil compartments. The result was a map of ecosystem type, which has been elaborated with a GIS.

Basing on that classification the resulting ecosystem types have been analyzed with the computer-based "digital landscape analysis system" (Reiche, 1996). The four information layers of soil, topography, linear landscape elements, and land use have been utilized to produce more detailed digital maps which were joint with the classification maps. In the next step, the modeling system Wasmod–Stomod (Reiche, 1996) was used to simulate the dynamics of water budgets, nutrient, and carbon fluxes based on a 30-year series of daily data about meteorological and hydrological forcing functions. The model outputs were validated by measured data in some of the systems (Schrautzer, 2002). Furthermore, the model outputs were extended to include data sets concerning the ecosystem indicators by the following variables:

- Exergy capture: net primary production (NPP)
- Entropy production: microbial soil respiration
- Storage capacity: nitrogen balance, carbon balance

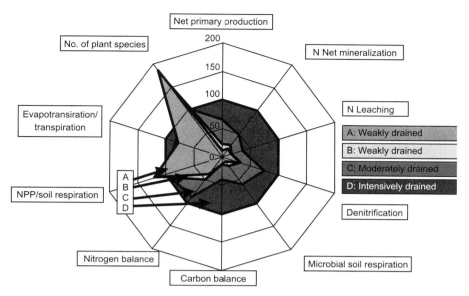

Figure 12.7   Synoptic illustration of the development of ecosystem indicators in a retrogressive succession of wet grasslands in the watershed of Lake Belau, according to Schrautzer et al. (in press). The most disturbed system type (the eutrophicated and strongly drained grasslands) is used as a reference state (100%).

- Ecosystem efficiency: evapotranspiration/transpiration, NPP/soil respiration
- Nutrient loss: nitrogen net mineralization, nitrogen leaching, denitrification
- Ecosystem structure: no. of plant species (measured values)

    In the following example, these indicators have been used to investigate different stages of a retrogressive wetland succession. The wet grasslands of the Bornhöved Lakes district are managed in a way that includes the following measures: drainage, fertilization, grazing, and mowing in a steep gradient of ecosystem disturbances. The systems have been classified due to these external input regimes, and in Figure 12.7 the consequences can be seen in a synoptic manner: With the farmer's target — improving the production and the yield of the systems — the NPP is growing by a factor of 10, the structural indicator is decreasing enormously throughout the retrogression. Also the efficiency measures (NPP/soil respiration) are going down, and the biotic water flow gets smaller. On the other hand, the development of the nitrogen and carbon balances demonstrates that the system is turning from a sink function into a source, the storage capacity is being reduced, and the loss of carbon and nitrogen compounds (all indicators on the right side of the figure) is rising enormously. With these figures we can state an enormous decrease in ecosystem health, and as many of the processes are irreversible, the capacity for future self-organization is reduced to a very small degree.

12.4.3 Application in Sustainable Landscape Management

The final case study demonstrated in this paper is taken from a European project about strategies for a sustainable reindeer herding in northern Fennoscandia (RENAMN, see http://www.urova.fi/home/renman/). In this case, the indication of ecosystem health has been accomplished by social and economic data to build a science-based fundament for the outstanding land-use decision-making processes in the region. Besides big ethnic problems between the Sami native inhabitants and the Scandinavian population, the key problems of reindeer herding can be put down to the fact that in the past few decade, there has been an immense loss of grazing land for the big reindeer herds. Causes are growing demands for electricity (hydroelectric power plants with huge artificial lakes), an increasing demand for tourist areas, a nonsustainable, clear-cut-based intensive forestry, and fence systems that reduce the original mobility of the reindeer herds to an extreme degree (see Burkhard et al., 2003, Burkhard and Müller, in press). As a result, there is now a relatively high number of animals within a rather small area. Additionally, the traditional differentiation of summer and winter pastures (which were situated at very distant areas) is no longer realized. Consequently, the reindeer herds today are in danger of destroying their traditional winter fodder during the summer grazing periods: Lichens are a specific fodder which becomes the focal food during the end of the winter season. During the summer seasons, the lichens can dry very rapidly, getting brittle and easily disturbed. If the herds use the winter grazing grounds during a dry summer period, the ground lichens can be easily destroyed. Besides these problems, the abundances of the tree lichens, which are an alternative winter food, has decreased enormously due to the intensive forestry practices.

In this conflict field between different land-use strategies, we have carried out a systems analysis concerning land-use structure, ecological items, and social and economic problems. The methodology was based on landscape mapping, measurements of ecological variables, and modeling with the Wasmod–Stomod system (Reiche, 1996) which was introduced previously. Three scenarios were carried out: (1) a business as usual strategy; (2) an intensification; and (3) a reduction of reindeer herding. While the ecological data still are in calculation, the other items have been investigated on the basis of expert interviews. The experts were asked to foresee the consequences of the scenario conditions within a time span of 25 years by estimating the development on a scale from -5 (high decrease in indicator values) to $+5$ (high increase), using the indicators depicted in Figure 12.8. This is one example for the scenario outcomes, referring to the scenario C, "Reduction of reindeer herding."

The land-use amoeba demonstrates an impressive decrease in reindeer herding, while before all forestry will be intensified, new lakes are suspected to be built, and tourism as well as mining will have a higher significance in the land-use structure. Concerning the consequences for the Sami population, a very high economic risk has been postulated: All values of the economic

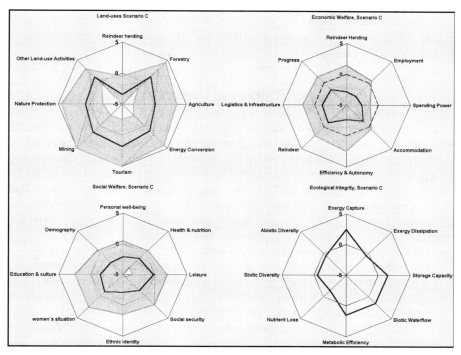

Figure 12.8 An overview of the indicator values concerning the scenario C (reduction of reindeer herding). The values for the indicator groups land use, economic and social welfare have been taken from the expert interviews in Lapland. Here the gray background shows the range of the estimations (between +5 and −5) while the fat line demonstrates the medium values of all consultations. 0 represents the actual state. The ecological values are recently being calculated in detail. Here the initial expert hypotheses are demonstrated.

amoeba are decreasing, the employment situation might become fatal, and the autonomy of the Lapland region will rapidly go down. Furthermore, the social amoeba demonstrates that the experts are afraid of a high demographic loss of population, that the Sami ethnic identity will be diminished, and social security and health will be confronted with huge problems. The last part of the figure concerns the ecological outcome. Here the expert opinions, which match the first trend that became visible throughout the modeling exercises, are shown in qualitative diagram. Due to the expected dominance of lichens, which will follow a reduction of the herd sizes, the structural features supposable will decrease as will the loss of nutrients (because they are not transformed by the reindeer populations in the prevailing level). Production and storage will increase and the consequence also the storage compartments will have a higher capacity. While the net primary production (the total yield) will take advantages, the respiration of the system will decrease due to the low potential of the lichens. But all these changes are minor transformations.

In summary, the ecological consequences of the different herding regimes are not very significant. Thus the focal argumentation for the finding of a sustainable management of Lapland's landscapes is an economic and social question. To solve this problem, our indicators can hopefully be a helpful tool.

12.5 DISCUSSION AND CONCLUSIONS

In this chapter an approach for a holistic indication of ecosystem health features with a medium number indicator set has been described. It is based on ecosystem theory, empirical ecosystem research, and on a self-organization-based aspect of ecological integrity. Some applications have been briefly demonstrated on different scale levels demonstrating a methodological shift from comprehensive ecosystem measurements to model applications on the landscape level. Finally, a case study has been used to show potential applications in environmental management as a part of the search for sustainable landscape management strategies.

The indicator system was used to show the consequences of different land-use systems throughout a developmental duration of 100 years, where the optimization of agricultural production has lead to a loss of many other important functional and structural features. It could also be shown that the economically oriented management of wet grasslands modifies the selected ecological attributes enormously — that is, it provokes the development of a source function of these ecosystems which subsequently produce a high fraction of entropic flows (carbon loss, nitrogen loss) into their environment. Finally, the reindeer management example demonstrates that structural changes must not always be accomplished by big functional modifications and that in some cases the human dimensions of sustainability are much more significant than the ecological ones.

Recently, we have tried tried to develop another application of the indicator set in environmental monitoring. There are many advantages in changing the sectoral monitoring approaches to an ecosystem-based concept. Attempts are made to monitor networks of the German counties (Bundesländer), in biosphere reservation and national parks, and the indicator set has also been implemented into a new concept of environmental-economic accounting in Germany. Regrettably, these applications are still far from being realized, because the change to a holistic attitude in official networks is hampered by many psychological and administrative restrictions.

In parallel with these applications, the variable set has to remain in development. If we look back at Table 12.1, it becomes obvious that some of the mentioned requirements have not been fulfilled up to now. For instance: the comprehensibility and the methodological transparency have to be improved; the indicators provide different sensitivities for specific environmental developments; and due to the theoretical and complex background, the high political relevance of the indicators has not become obvious for politicians up until now. Consequently, aside from the scientific tasks of improving the indication, the conviction of the potential users to change their concepts towards a higher consideration of ecosystem attributes and towards a fruitful application of the health or integrity concepts, will be a main task of future activities.

REFERENCES

Aoki, I. Entropy and exergy in the development of living systems: a case study of lake ecosystems. *J. Phys. Soc. Japan* 67, 2132–2139, 1998.

Barkmann, J. Modellierung und Indikation nachhaltiger Landschaftsentwicklung — Beiträge zu den Grundlagen angewandter Ökosystemforschung. Dissertation, University of Kiel, 2001.

Barkmann, J., Baumann, R., Meyer, U., Müller, F., and Windhorst, W. Ökologische Integrität: Risikovorsorge im Nachhaltigen Landschaftsmanagement. *Gaia* 10, 97–108, 2001.

Baumann, R. Konzept zur Indikation der Selbstorganisationsfähigkeit terrestrischer Ökosysteme anhand von Daten des Ökosystemforschungsprojekts Bornhöveder Seenkette. Dissertation, University of Kiel, 2001.

Bossel, H. "Ecological orientors: emergence of basic orientors in evolutionary self-organization," in *Eco Targets, Goal Functions and Orientors*, Müller, F. and Leupelt, M., Eds. Springer, Berlin, 1998, pp. 19–33.

Bossel, H. "Sustainability: application of systems theoretical aspects to societal development," in *Handbook of Ecosystem Theories and Management*, Jørgensen, S.E. and Müller, F., Eds. CRC Press, Boca Raton, FL, pp. 519–536.

Breckling, B. and Asshoff, M. Modellbildung und Simulation im Projektzentrum Ökosystemforschung. *Ecosys Bd.* 5, 1996, 342 p.

Burkhard, B., Kumpula, T., and Müller, F. Renman — an integrative study in Northern Scandinavia. *Ecosys.* 10, 116–124, 2003.

Burkhard, B. and F. Müller Systems analysis and modelling of reindeer husbandry in northern Finland. *Rangifer* (in press).

Costanza, R. "Towards an operational definition of ecosystem health," in *Ecosystem Health* Costanza, R., Norton, B.G., and Haskell, B.D., Eds. Island Press, Washington, 1993, pp. 239–256.

Costanza, R. Societal goals and the valuation of ecosystem services. *Ecosystems* 3, 4–10, 2000.

Costanza, R., Cleveland, C., and Perrings, C. "Ecosystem and economic theories in ecological economics," in *Handbook of Ecosystem Theories and Management*, Jørgensen, S.E. and Müller, F., Eds. CRC Press, Boca Raton, FL, 2000, pp. 547–560.

Costanza, R., Norton, B.G., and Haskell, B.D. *Ecosystem Health*. Island Press, Washington, 1993.

Daily, G.C. *Nature's Services: Societal Dependence on Natural Systems*. Island Press, Washington, D.C., 1997.

De Groot, R.S. *Functions of Nature*. Wolters-Noorhoff, Dordrecht, 1992.

Dierssen, K. "Ecosystems as states of ecological successions," in *Handbook of Ecosystem Theories and Management*, Jørgensen, S.E. and Müller, F., Eds. CRC Press, Boca Raton, 2000, 427 p.

Dittmann, S., Schleier, U., Günther, S.P., Villbrandt, M., Niesel, V., Hild, A., Grimm, V., Bietz H., and Dohn, C. *Elastizität des Ökosystems Wattenmeer*. Projektsynthese. Forschungsbericht für den BMBF, Bonn, 1998.

Ebeling, W. *Chaos–Ordnung–Information*. Verlag Hari Deutsch, Frankfurt/M, 1989.

Fath, B. and Patten, B.C. "Network orientors: a utility goal function based on network synergism," in *Eco Targets, Goal Functions and Orientors*, Müller, F. and Leupelt, M. Springer, Berlin, 1998, pp. 161–176.

Fath, B. and Patten, B.C. "Ecosystem theory: network environ analysis," in *Handbook of Ecosystem Theories and Management*, Jørgensen, S.E. and Müller, F., Eds. CRC Press, Boca Raton, 2000, pp. 345–360.

Fränzle, O. "Ökosystemforschung im Bereich der Bornhöveder Seenkette," in *Handbuch der ökosystemforschung*, Fränzle, O., Müller, F., and Schröder, W., Eds., Chapter V-4.3. ecomed-Verlag, Landsberg, 1998.

Fränzle, O. "Ecosystem research," in *Handbook of Ecosystem Theories and Management*, Jørgensen, S.E. and Müller, F., Eds. Boca Raton, FL, 2000, pp. 89–102.

Fränzle, O., Kappen, L., Blume, H.P., and Dierssen, K., Eds. Ecosystem organisation in a complex landscape. *Ecol. Stud.* (in press).

Fritz, P. "UFZ-Forschungszentrum Leipzig-Halle," in *Handbuch der Ökosystemforschung*, Fränzle, O., Müller, F., and Schröder, W., Eds., Chapter V-4.5. ecomed-Verlag, Landsberg, 1999.

Gollan, T. and Heindl, B. "Bayreuther Institut für Terrestrische ökosystemforschung" in *Handbuch der Ökosystemforschung*, Fränzle, O., Müller, F., and Schröder, W., Eds., Chapter V-4.6. Landsberg, 1998.

Golley, F. "Ecosystem structure," in *Handbook of Ecosystem Theories and Management*, Jørgensen, S.E. and Müller, F., Eds. CRC Press, Boca Raton, 2000, pp. 21–32.

Gundersson, L.H. and Holling, C.S. *Panarchy*. Island Press, Washington, D.C., 2002.

Hantschel, R., Kainz, M., and Filser, J. "Forschungsverbund Agrarökosysteme München," in *Handbuch der Ökosystemforschung*, Fränzle, O., Müller, F., and Schröder, W., Eds., Chapter V-4.7. Landsberg, 1998.

Haskell, B.D., Norton, B.G., and Costanza, R. "Introduction: what is ecosystem health and why should we worry about it?" in *Ecosystem Health*, Costanza, R., Norton, B.G., and Haskell, B.D., Eds. Island Press, Washington, 1993, pp. 3–22.

Hauff, V. Unsere Gemeinsame Zukunft. Der Brundlandt-Bericht der Weltkommission für Umwelt und Entwicklung. Greven, 1987.

Herbst, M., Eschenbach, C., and Kappen, L. Water use in neighbouring stands of beech and black alder. *Ann. For. Sci.* 56, 107–120, 1999.

Holling, C.S. "The resilience of terrestrial ecosystems: local surprise and global change." in *Sustainable Development of the Biosphere*, Clark, W.M. and Munn, R.E., Eds. IBP Reports, Oxford, 1986, pp. 292–320.

Hörmann, G., Irmler, U., Müller, F., Piotrowski, J., Pöpperl, R., Reiche, E.W., Schernewski, G., Schimming, C.G., Schrautzer, J., and Windhorst, W. Ökosystemforschung im Bereich der Bornhöveder Seenkette. Arbeitsbericht 1988–1991. *Ecosys* 1, 1992, 338 p.

Jørgensen, S.E. *Integration of Ecosystem Theories: A Pattern*. Kluwer, Dordrecht, 1996.

Jørgensen, S.E. "The tentative fourth law of thermodynamics," in *Handbook of Ecosystem Theories and Management*, Jørgensen, S.E. and Müller, F., Eds. CRC Press, Boca Raton, 2000, pp. 161–176.

Jørgensen, S.E. and Müller, F. "Ecosystems as complex systems," in *Handbook of Ecosystem Theories and Management*, Jørgensen, S.E. and Müller, F., Eds. CRC Press, Boca Raton, 2000, pp. 5–20.

Jørgensen, S.E. and Müller, F. "Ecological orientors — a path to environmental applications of ecosystem theories," in *Handbook of Ecosystem Theories and Management*, Jørgensen, S.E. and Müller, F., Eds. CRC Press, Boca Raton, 2000, pp. 561–576.

Kaiser, M., Mages-Delle, T., and Oeschger, R. Gesamtsynthese Ökosystemforschung Wattenmeer. UBA-Texte 45/02, Berlin, 2002.

Karr, J.R. Assessment of biotic integrity using fish communities. *Fisheries* 6, 21–27, 1981.

Kay, J.J. "On the nature of ecological integrity: some closing comments," in *Ecological Integrity and the Management of Ecosystems*, Woodley, S., Kay, J., and Francis, G., Eds. University of Waterloo and Canadian Park Service, Ottawa, 1993.

Kay, J.J. "Ecosystems as self-organised holarchic open systems: narratives and the second law of thermodynamics," in *Handbook of Ecosystem Theories and Management*, Jørgensen, S.E. and Müller, F., Eds. CRC Press, Boca Raton, 2000, pp. 135–160.

Kellermann, A. Gätje, C., and Schrey, E. "Ökosystemforschung im Schleswig-Holsteinischen Wattenmeer," in *Handbuch der Ökosystemforschung*, Fränzle, O., Müller, F., and Schröder, W., Eds., Chapter V-4.1.1. ecomed-Verlag, Landsberg, 1998.

Kerner, H.-F., Spandau, L., and Köppel, J. Methoden zur Angewandten Ökosystemforschung entwickelt im MAB 6 — Projekt Ökosystemforschung Berchtesgaden. MAB Mitteilungen 35.1 und 35.2, Bonn, 1991.

Kutsch, W., Dilly, O., Steinborn, W., and Müller, F. "Quantifying ecosystem maturity: a case study," in *Eco Targets, Goal Functions and Orientors*, Müller, F. and Leupelt, M., Eds. Springer, Berlin, 1998, pp. 209–231.

Kutsch, W.L., Steinborn, W., Herbst, M., Baumann, R., Barkmann, J., and Kappen, L. Environmental indication: a field test of an ecosystem approach to quantify biological self-organization. *Ecosystems* 4, 49–66, 2001.

Meyer, E. Die benthischen Invertebraten in einem kleinen Fließgewässer am Beispiel eines Schwarzwaldbaches: Biozönotische Dtruktur, Populationsdynamik, Produktion und Stellung im trophischen Gefüge. Habil Thesis, Konstanz, 1992.

Meyer, M. Entwicklung und Formulierung von Planungsszenarien für die Landnutzung im Bereich der Bornhöveder Seenkette. Dissertation, University of Kiel, 2000.

Meyer, U. Landschaftsökologische Modellierung als Auswertungsinstrument in der ökosystemaren Umweltbeobachtung — Beispielsfall Biosphärenreservat Rhön. *Ecosys* suppl. 36, 2001, 156 p.

Millennium Assessment Board. *Ecosystems and Human Well-Being*. Island Press, Washington, 2003.

Müller, F. State of the art in ecosystem theory. *Ecol. Model.* 100, 135–161, 1997.

Müller, F. "Ableitung von integrativen Indikatoren der Funktionalität von Ökosystemen und Ökosystemkomplexen für die Beschreibung des Umweltzustandes im Rahmen der Umweltökonomischen Gesamtrechnungen (UGR)," in *Statistisches Bundesamt* (Hrsg.), Beiträge zu den Umweltökonomischen Gesamtrechnungen, Bd. 2, Wiesbaden, 1998.

Müller, F. and Breckling, B. "Der Ökosystembegriff aus heutiger Sicht," in *Handbuch der Umweltwissenschaften*, Fränzle, O., Müller, F., and Schröder, W., Eds. Chapter II-2.2. ecomed-Verlag, Landsberg, 1997.

Müller, F., Breckling, B., Bredemeier, M., Grimm, V., Malchow, H., Nielsen, S.N., and Reiche, E.W. "Ökosystemare Selbstorganisation," in *Handbuch der Ökosystemforschung*, Fränzle, O., Müller, F., and Schröder, W., Hrsg, Chapter III-2.4. Landsberg, 1997a.

Müller, F., Breckling, B., Bredemeier, M., Grimm, V., Malchow, H., Nielsen, S.N., and Reiche, E.W. "Emergente Ökosystemeigenschaften," in *Handbuch der Ökosystemforschung*, Fränzle, O., Müller, F., and Schröder, W., Hrsg, Chapter III-2.5. Landsberg, 1997b.

Müller, F. and Leupelt, M. *Eco Targets, Goal Functions and Orientors.* Berlin, Heidelberg, New York, 1998.

Müller, F. and Li, B.L. Complex systems approaches to study human–environmental interactions — issues and problems. *Proceedings of the Intecol World Conference, Seoul, 2002,* (in press).

Müller, F., Hoffmann-Kroll, R., and Wiggering, H. Indicating ecosystem integrity — from ecosystem theories to eco targets, models, indicators and variables. *Ecol. Model.* 130, 13–23, 2000.

Müller, F. and Nielsen, S.N. "Ecosystems as subjects of self-organising processes," in *Handbook of Ecosystem Theories and Management*, Jørgensen, S.E. and Müller, F., Eds. CRC Press, Boca Raton, 2000, pp. 177–194.

Müller, F. and Wiggering, H. "Umweltindikatoren als Maßstäbe zur Bewertung von Umweltzuständen und –entwicklungen," in *Umweltziele und Indikatoren*, Wiggering, H. and Müller, F., Eds. Springer, Berlin, 2004, pp. 121–129.

Müller, F. and W. Windhorst "Ecosystems as functional entities," in *Handbook of Ecosystem Theories and Management*, Jørgensen, S.E. and Müller, F., Eds. CRC Press, Boca Raton, 2000, pp. 33–50.

Odum, E.P. The strategy of ecosystem development. *Science* 104, 262–270, 1969.

Odum, H.T., Brown, M.T. and Ulgiati, S. "Ecosystems as energetic systems," in *Handbook of Ecosystem Theories and Management*, Jørgensen, S.E. and Müller, F., Eds. CRC Press, Boca Raton, 2000, pp. 283–302.

Patten, B.C. Energy, emergy and environs. *Ecol. Model.* 62, 29–69, 1992.

Pöpperl, R. Functional feeding groups of a macroinvertebrate community in a Northern German lake outlet. (Lake Belau, Schleswig-Holstein). *Int. Revue ges. Hydrobiol.* 81, 183–198, 1996.

Radermacher, W., Zieschank, R., Hoffmann-Müller, R., Noyhus, J.v., Schäfer, D., and Seibel, S. Entwicklung eines Indikatorensystems für den zustand der Umwelt in der Bundesrepublik Deutschland. Beiträge zu den umweltökonomischen Gesamtrechnungen, Bd. 5, Statistisches Bundesamt, Wiesbaden, 1998.

Rapport, D.J. What constitutes ecosystem health? *Perspect. Biol. Med.* 33, 120–132, 1989.

Rapport, D.J. and Moll, R. "Applications of ecosystem theory and modelling to assess ecosystem health," in *Handbook of Ecosystem Theories and Management*, Jørgensen, S.E. and Müller, F., Eds. CRC Press, Boca Raton, FL, 2000, pp. 487–496.

Reiche, E.W. Wasmod. Ein Modellsystem zur gebietsbezogenen Simulation von Wasser- und Stoffflüssen. *Ecosys* 4, 143–163, 1996.

Reiche, E.W., Müller, F., Dibbern, I., and Kerrinnes, A. "Spatial heterogeneity in forest soils and understory communities of the Bornhöved Lakes District," in *Ecosystem Approaches to Landscape Management in Central Europe*, Tenhunen, J., Lenz, R., and Hantschel, R., Eds. *Ecological Studies* 147, 2001.

Reiche, E.W. and Windhorst, W. "Ecosystem based planning scenarios," in *Ecosystem Organisation in a Heterogeneous Landscape*, Fränzle, O., Blume, H.P., Kappen, L., and Dierssen, K., Eds. *Ecological studies* (in press).

Schimming, C.G. and von Stamm, S. Arbeitsbericht des Projektzentrums Ökosystem-forschung, Anhang I: Untersuchungsmethoden. Interne Mitteilungen aus dem FE-Vorhaben Ökosystemforschung im Bereich der Bornhöveder Seenkette, Report, Kiel Ecology Centre, 1993.

Schneider, E.D. and Kay, J.J. Life as a manifestation of the second law of thermodynamics. *Math. Comp. Model.* 19, 25–48, 1994.

Schönthaler, K., Meyer, U., Windhorst, W., Reichenbach, M., Pokorny, D., and Schuller, D. *Modellhafte Umsetzung und Konkretisierung der Konzeption für eine Ökosystemare Umweltbeobachtung am Beispiel des länderübergreifenden Biosphärenreservats Rhön.* Umweltbundesamt, Berlin, 2001.

Schönthaler, K., Müller, F., and Barkmann, J. Synopsis of systems approaches to environmental research — German contribution to ecosystem management. UBA-Texte 85/03, 2003.

Schrautzer, J. Niedermoore Schleswig-Holsteins: Charakterisierung und Beurteilung ihrer Funktionen im Landschaftshaushalt. Habil Thesis, University of Kiel, 2002.

Schrautzer, J., Müller, F., and Reiche, E.W. Testing orientor-based indicators in retrogressive successions in an agricultural landscape. *Ecol. Stud.* (in press).

Statistisches Bundesamt, Forschungsstelle Für Umweltpolitik Der Fu Berlin and Ökologiezentrum Der Universität Kiel. Makroindikatoren des Umweltzu-stands. Beiträge zu den Umweltökonomischen Gesamtrechnungen, Bd. 10, 2002.

Steinborn, W. Quantifizierung von Ökosystemeigenschaften als Grundlage für die Umweltbewertung. Dissertation, University of Kiel, 2001.

Svirezhev, Y.M. and Steinborn, W. Exergy of solar radiation: thermodynamic approach. *Ecol. Model.* 145, 101–110, 2001.

Ulanowicz, R.E. "Ascendency: a measure of ecosystem performance," in *Handbook of Ecosystem Theories and Management*, Jørgensen, S.E. and Müller, F., Eds. Boca Raton, FL, 2000, pp. 303–316.

Ulgiati, S., Brown, M.T., Giampietro, M., Herendeen, R.A., and Mayumi, K. *Advances in Energy Studies: Reconsidering the Importance of Energy.* SGIE Detoriali, Padova, 2003.

Widey, G.-A. "Forschungszentrum Waldökosysteme der Universität Göttingen," in *Handbuch der ökosystemforschung*, Fränzle, O., Müller, F., and Schröder, W., Hrsg, Chapter V-4.4. economed-Verlag, Landsberg, 1998.

Wiggering, H. "Zentrum für Agrarlandschafts- und Landnutzungsforschung," in *Jahresbericht 2000/2001.* ALF, Müncheberg, 2001.

Wiggering, H. and Müller, F., *Umweltziele und Indikatoren.* Springer, Berlin, 2004.

Windhorst, W., Müller, F., and Wiggering, H. "Umweltziele und Indikatoren für den Ökosystemschutz." in *Umweltziele und Indikatoren*, Wiggering, H. and Müller, F., Eds. Springer, Berlin, 2004, pp. 345–373.

Woodley, S., Kay, J., and Francis, G. *Ecological Integrity and the Management of Ecosystems.* St. Lucie Press, Ottawa, 1993.

World Commission on Environment and Development. *Our Common Future.* Oxford University Press, Oxford, 1987.

CHAPTER 13

Multi-Scale Resilience Estimates for Health Assessment of Real Habitats in a Landscape

G. Zurlini, N. Zaccarelli, and I. Petrosillo

Vegetation or habitat types are ecological phases that can assume multiple states. Transformations from one type of phase to another are called ecological phase transitions. If an ecological phase maintains its condition of normality in the linked processes and functions that constitute ecosystems then it is believed to be healthy. An adaptive cycle, such as that given in Holling's model, has been proposed as a fundamental unit for understanding complex systems. Such model alternates between long periods of aggregation and transformation of resources and shorter periods that create opportunities for innovation. The likelihood of shifts among different domains largely depends on domain resilience, measurable by the size of scale domains, but these do not provide any indication on resistance — the external pressure to displace a system by a given amount. We argue that the type, magnitude, length, and timing of external pressure, its predictability, the exposure of the habitat, and the habitat's inherent resistance, have important interactive relationships which determine resilience, and in turn, ecosystem health. Different resilience levels are expected to be intertwined with different scale domains of real habitats in relation to the type and intensity of natural and human disturbances from management activities and land manipulation. In this paper, we provide an operational framework to derive operational indices of short-term retrospective resilience of real grasslands in a northern Italy watershed, from multiscale analysis of landscape patterns, to find scale domains for habitat edges where change is most likely — that is, where resilience is lowest and fragility highest. This is achieved through cross-scale algorithms such as fractal analysis coupled with change detection of ecological response indices. The framework implements the integration of habitat-edge fractal geometry, the fitting of empirical power functions by piecewise regressions, and change-detection procedures as a method to find scale domains for grassland habitat edges where change is most likely and consequently resilience is lowest. Changes due to external pressure significantly related to habitat scale domains, according to their scaling properties resulting from the interaction between ecological, physical, and social controls shaping the systems. Grassland scale domains provided evidence and support for identifying and explaining scale-invariant ecological processes at various scales, from which much insight could be gained for characterizing grassland adaptive cycles and capabilities to resist disturbances to facilitate ecosystem health assessment.

13.1 INTRODUCTION

The rapid progress made in the conceptual, technical, and organizational requirements for generating synoptic multi-scale views and explanations of the Earth's surface, and for linking remote sensing at multi-resolution levels from satellite and airborne imageries, geographical information systems, spatial analysis of landscape patterns, and habitat classification methods, provides an outstanding potential support to:

1. Identify real landscape patches as habitats and land use types
2. Detect ecological processes by remotely sensed response variables
3. Relate response variables to habitats, by observing at different times ecological changes in habitat pattern as well as in the scales of habitat pattern (Simmons et al., 1992).

Multi-scale studies are increasingly conducted (Wu and Qi, 2000), which give emphasis to the identification of scale domains (Li, 2000; Brown et al., 2002), that are self-similarity regions of the scale spectrum over which, for a particular phenomenon, patterns do not change or change monotonically with scale. Thresholds are, in general, difficult to delineate across scales, because it remains difficult to detect multiple scales of variability in ecological data and to relate these scales to the processes generating the patterns (Levin, 1992; Ward and Salz, 1994). The likelihood of sharp shifts is linked to an ecosystem's resilience, which is the capacity of a system to undergo disturbance and maintain its functions and controls (Gunderson and Holling, 2002). The importance of a clear and measurable definition of resilience has become paramount (Carpenter et al., 2001) for evaluating the health of an ecosystem, defined as being stable and sustainable, maintaining its organization and autonomy over time, and its resilience to stress (Costanza, 1992; Mageau et al., 1995). Scaling domains of habitats can be identified, for instance, by shifts in the fractal dimension of patch edges, and can indicate a substantial change in processes generating and maintaining landscape patches at different scales (Krummel et al., 1987; Sugihara and May, 1990; Milne, 1991), so that different processes dominate at different scales (Peterson, 2000). One way to appreciate the interaction between pattern and processes is to look at temporal changes detected by remote sensing, and whether they are significantly associated with different scale domains. If such processes change in type and intensity across scales, the ability of ecosystems to resist lasting change caused by disturbances — their resilience (Gunderson et al., 1997; Gunderson and Holling, 2002) — will change accordingly, so that habitat resilience and scaling are expected to be intertwined (Peterson, 2000).

This study was designed to address some specific questions:

1. Can we objectively and accurately identify scale breaks delimiting ecologically equivalent scales in real habitat patches in a landscape?
2. Are temporal changes detected significantly related to habitat scale domains, providing evidence on the types of biophysical and social controls shaping the systems?
3. If so, can we derive an operational index of short-term retrospective resilience, through cross-scale algorithms like fractal analysis coupled with remotely sensed change detection, to find scale domains where change is most likely — that is, where resilience is lowest?

To make this approach practical, we need:

1. A really effective classification procedure for habitat recognition from general and vague categories of habitats to more specific categories
2. A statistically objective procedure for identifying shifts in scale domains
3. Suitable ecological response variables for change detection.

We describe an operational framework for the accurate identification of self-similar domains in few specific grassland habitats, and for estimating their displayed short-term resilience. First, we looked for scale domains in real grassland patches of a stream watershed in northern Italy (Zurlini et al., 2001), resulting from long-term natural and man-induced interactive disturbance regimes. We then quantified short-term intensity changes of habitat scale domains, based on remotely-sensed ecological response indices. This framework implements the integration of edge fractal analysis, the fitting of power laws by piecewise regressions and hypothesis testing of scale shifts (Grossi et al., 1999; 2001), together with procedures of change detection, as a method to find scale domains for grassland edges where change is most likely. Together they represent a framework for spatially defining critical landscape thresholds and scale domains, habitat adaptive cycles (Gunderson and Holling, 2002), and habitat resilience by which scale-dependent ecological models could be developed and applied. By introducing this approach, we address some basic concepts of ecological phases and multiple states, with a general discussion on self-similarity regions, fractal analysis, and on the statistical procedures for the objective identification of shifts among scale domains, as well as on resilience and its practical measure. The detailed and often complex composition of real landscape habitat mosaics in terms of habitat types and land use has been rarely considered in the understanding of the relationships between landscape pattern and process response variables. Much of the insight obtained is related to the coupling of change-detection procedures with the availability of detailed habitat type distribution in a stream watershed. The potential of such approach for ecosystem health assessment, planning, and management of habitats mosaics is also discussed.

13.2 RATIONALE

13.2.1 Ecological Phases, States, and Scale Domains

From several long-term observations, experimentations, and comparative studies of many sites, it is now evident that alternate and alternative states arise in a wide variety of ecosystems, such as lakes, marine fisheries, benthic systems, wetlands, forests, savannas, and rangelands (Gunderson and Holling, 2002). A phase state of a system at a particular instant in time is the collection of values of the state variables at that time (Grimm et al., 1992; Walker et al., 2002), different from other states the system can visit over and over again (alternate), or from those typical of other systems (alternative). Vegetation or habitat types are considered ecological phases which can assume multiple states, and transformations from one type to another (alternative) correspond to ecological phase transitions, which change the integral structures of the systems (Li, 2002). Multiple states (alternate) can be assumed by ecological phases without losing their basic identity. For example, a forest stand or grassland may remain a forest patch or grassland over and over again, each with its own dynamic states of rapid growth, conservation, collapse, and reorganization as proposed by Holling's adaptive cycle model (Holling et al., 1995). For grasslands, such model proposes that as young grasses grow without grazing or cutting, they gradually become denser and accumulate fuel, and thus become increasingly susceptible to fire. After a fire, the system is reorganized as vegetation resprouts from roots or seeds, producing new grassland. All these states are deemed as multiple configuration states (attractors) of the same ecological phase. Another example is provided by a simple meta-model describing different common ecological states of coral reefs, and the factors that may cause or maintain these states (McClanahan et al., 2002). Individual coral reefs that are exposed to a combination of human and natural influences may be a mosaic of several states. Which state the ecosystem currently assumes is function of its history and of the driving forces operating.

Multi-scale analysis corresponds to the detection of self-similar scale domains of alternate states, a central point for the development of a scalar theory in ecology (Levin, 1992; Holling, 1992; Wiens, 1995). Such self-similarity or fractality implies a particular kind of structural composition or dynamic behavior — that is, the fundamental features of the system exhibit an invariant, hierarchical organization that holds over a wide range of spatial scales (Gell-Mann, 1994; Li, 2000). A spatial ecological phase transition, or ecotone, is a "zone of transition between adjacent ecological systems, having a set of characteristics uniquely defined by space and time scales and by the strength of the interactions between adjacent ecological systems," (di Castri et al., 1988). Therefore the nature of a habitat's edge is not just a property of a specific habitat, but is the outcome of interactions at the landscape level. Ecological phases like vegetation or habitat types are dynamic in space and time, each trying to expand and invade adjacent ones whenever environmental and management conditions are beneficial to one of the adjacent ecosystems (Risser, 1995). They can have different regions of the scale spectrum over which there are several possible ecological states, equivalent or self-similar for a particular phenomenon, and which do not change or change monotonically with changes in scale. This would allow drawing the same ecological conclusions statistically from any scale (Sugihara and May, 1990; Milne, 1991; Li, 2000).

13.2.2 Resilience and Resistance

Abrupt shifts among several very different (alternative) stable domains are plausible in local and regional ecosystems more susceptible to changes; the likelihood of such shifts depends on resilience and resistance (see chapter 2), whereas the costs of such shifts depend on the degree of and duration for reversibility from one domain to another (Gunderson and Holling, 2002). Two systems, or two states of the same system, may have the same resilience but differ in their resistance. We can surmise that if the same external pressure is applied to two systems with different intrinsic resistances, they will show a different ability, or resilience, to resist lasting change caused by disturbances. Resilience estimates differ from ecological indicators in that they refer to socio-ecological systems and ecosystem services (Costanza et al., 1997) they provide (Carpenter et al., 2001).

Most studies in the literature refer to theoretical approaches, using resilience as a metaphor or a theoretical construct (Carpenter et al., 2001). Where resilience has been defined operationally, this has occurred in a few cases within a mathematical model of a particular system (Carpenter and Cottingham, 1997; Peterson et al., 1998; Janssen et al., 2000; Casagrandi and Rinaldi, 2002). In this context, bifurcation analysis of simple dynamic models has been often suggested or adopted, together with the size of stability domains, or the magnitude of disturbance the system can tolerate and still persist before the system changes its structure by changing the variables and processes that control behavior (Peterson et al., 1998; Gunderson and Holling, 2002).

However, not all those definitions, even though measurable in models, are operationally measurable in the field. In an operational sense, resilience needs to be considered in a specific context. As discussed by Carpenter et al. (2001), it requires defining the resilience of what to what. One important distinction, along with those on space–time scales advanced by Carpenter et al. (2001), is whether resilience has to be measured prospectively — to predict the ability of ecosystems to resist lasting change caused by disturbances, or retrospectively — to evaluate such ability as observed by past exposure to external pressures.

13.3 STUDY AREA AND METHODS

13.3.1 The Baganza Stream Watershed

The Baganza watershed was selected as pilot study area of the Map of Italian Nature (MIN) program (Zurlini et al., 1999; 2001), since it is a good representative of the typical series of watersheds located along the same side of the northern Apennines ridge. The watershed is approximately 174.63 km^2, and is located on the Emilian side of the northern Apennines (Figure 13.1), with a main stream 57 km long and a progressive elevation gradient in the southwest direction which varies from 57 m in the flat to the piedmont, up to 1943 m at the highest peak in the Apennines mountains. Mean monthly temperature varies from -0.6 to $-17.1°C$ in the mountain range, and from $+1.5$ to $+24.7°C$ in the lowland. Mean rainfall varies from 40 to 95 mm per year with most of the rain occurring during the fall and the spring seasons, with no deficit of evapo-transpiration during summer. Snow is usually present for four months above 1,400 m.

In the past few centuries, due to human influence on Mediterranean ecosystems and the slow abandonment of agricultural and pastoral practices, plant communities have been shaped into a mosaic-like pattern composed of different man-induced degradation and regeneration stages (Naveh and Liebermann, 1994). In the past, this watershed was almost fully covered by ancient forests, still present during the ducat of Parma at the end of the eighteenth century. Around the end of the nineteenth century, much of the forests in the piedmont and hills were cleared for building the many miles of the national railway network. Many cleared areas were maintained as grasslands with pastoral practices with sheep and cattle breeding on natural or cultivated pastures. In the last century, cattle breeding on pastures prevailed due to the increasing market success of diary products.

Intensive agricultural land use is currently prevalent in the lowlands and the nearby Baganza stream, whereas abandonment of agricultural and pastoral practices in the hills and mountains is still in progress. Conservation and endangered species legislation at the national and regional level reduce the possibility of clearing the land, whereas they are allowed to maintain pastures in the high-hill and mountain range.

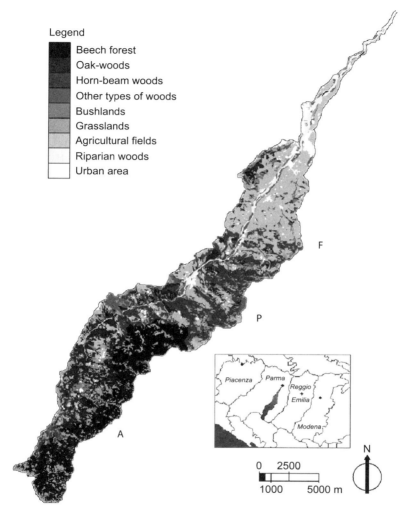

Figure 13.1 Location of the Val Baganza watershed and distribution of large habitat classes (modified from Zurlini et al., 2001). F is the flat, with urban/agricultural matrix; P is the piedmont, with agricultural/grassland/woods matrix; and A is the Apennines mountain range, with grassland/forest matrix. The list of main habitats corine habitat is given in Table 13.1.

13.3.2 Corine Habitats

Using synoptic multi-scale views and classifications of the Earth's surface now available, researchers, land managers, and land-use planners can quantitatively place landscape units, from general and vague categories such as "forests" to more specific categories such as "Illyrian Holm-oak woodland, *Orno-Quercetum Hilicis* dominated formations," in their large-area contexts. Remote sensing technologies represent the primary data source for habitat identification and landscape analysis, but often suffer from the Modifiable Areal Unit Problem (MAUP, Openshaw, 1984), that is a potential source of error that can affect spatial studies which utilise aggregate data sources. It states that a number of different and often arbitrary ways exist by which an area can be divided or aggregated into nonoverlapping areal units.

We used the CORINE habitat classification (EU/DG XI, 1991) to identify ecosystems as patches (Tansley, 1935) for generating digital thematic maps as Geographic Information System GIS coverages of mosaics of contiguous patches. To avoid MAUP effects, the final delineation of habitat mosaics was performed by an iterative process based on integrated evidence from processed satellite imagery, aerial photos, hyperspectral imagery, existing vegetation and geological soil maps, digital elevation models (DEM), and field reconnaissance (Zurlini et al., 1999). The detailed CORINE habitat distribution for the Baganza watershed was available at a scale of 1:25,000 (Zurlini et al., 2001), in a revised and more detailed form with respect to the original habitat classification used in Grossi et al. (1999; 2001), with 2,327 irregular patches belonging to 69 different CORINE habitat and habitat mosaic types (Table 13.1).

The flat and piedmont sections are dominated by agricultural fields, with few relatively natural habitats, represented by typical wet woodlands (Figure 13.1). Hop-horn beam (CORINE code 41.812) mixed to *Quercus pubescens* (41.7314) woods, are dominant in the hills, while neutrophile beech forests (41.1744) are most frequent in the mountain range above 900 to 1000 m. Three of the most frequent grassland habitats in the watershed were considered for subsequent analyses (EU/DG XI 1991; Sburlino et al., 1993):

1. Lowland hay meadows (CORINE code 38.2) present with 378 patches
2. Northern Apennine *Mesobromion* (CORINE code 34.3266) with 77 patches
3. *Brachypodium* grassland (CORINE code 36.334) with 131 patches, corresponding roughly to increasing elevation gradients and to decreasing human influence and control (Figure 13.2).

So-called lowland hay meadows are rich mesophile grasslands in the lowland, hills and submountain ranges, regularly manured, and when necessary irrigated, well-drained under direct human control, with species such as *Arrhenaterum elatius*, *Trisetum flavescens*, and *Anthriscus sylvestris*. They often begin from seeding of leguminous grasses or mixed fodder, and after are regularly cut in time for cattle breeding in farms. Northern Apennine *Mesobromion* are poor closed mesophile grasslands, sparse and rich in *Bromus erectus* and orchids, in local semiarid environments naturally exposed to drought and limited by the amount of organic matter in soil; they are not under direct human disturbances, apart from infrequent cutting, and grazing and manuring by cattle (which is an important source of organic matter). When lowland hay meadows are abandoned, they become *Mesobromion* grasslands. *Brachypodium* grasslands are subalpine thermophile siliceous habitats, often found on skeleton soils, and are not under direct human influence apart from

Table 13.1   List of the main CORINE habitat type identified in the Baganza watershed (modified from Zurlini et al., 2001)

CORINE code	CORINE habitat type
42.1B1	*Abies alba* reforestations
41.812	Supra-mediterranean hop-hornbeam woods
41.813	Montane hop-hornbeam woods
41.74	Quercus cerris woods
41.1744	Beech forests
42.67	Black pine reforestation
44.614	Italian poplar galleries
83.324	Locust tree plantations
41.731	Semi-xerophile *Quercus pubescens* woods
41.7312	Xerophile *Quercus pubescens* woods
44.122	Mediterranean purple willow scrub
31.431	*Juniperion nanae* scrub
31.81	Medio-European rich-soils thickets
31.811	Blackthorn-bramble scrub
31.88	Common Juniper scrub
32.A	Spanish-broom fields
34.3266	Northern Apennine *Mesobromion*
34.3267	Sub-Mediterranean *Mesobromion*
36.334	Sub-alpine thermophile siliceous grasslands with *Brachipodium genuense*
38.1	Mesophile pastures
38.13	Overgrown pastures
38.2	Lowland high meadows
61.311	Rough-grass screes
61.3124	Submontane calcareous screes with *Calamagrostis varia*
61.3125	Sedo-Scleranthetea *Submontane* calcareous screes
61.3126	*Brometalia erecti* submontane calcareous screes
62.213	Hercynian serpentine cliffs
87.24	Ruderal communities with *Tussilago farfara*
87.23	Ruderal communities with *Melilotus albus*
87.29	Ruderal communities with *Agropyron repens*
82.11	Field crops
62.4	Bare inland cliffs
82.11	Plough field crops
86.2	Villages
86.3	Active industrial sites
86.41	Quarries

sporadic grazing by cows and sheep at lower altitudes, with carpet communities hardly browsed by cattle, and almost pure in *Brachypodium genuense*, typical of higher elevations and of the summits. Fire is not currently used as a practice for controlling scrub formation and seldom occurs in the watershed.

13.3.3   Empirical Patterns of Self-Similarity

Domains are delimited by relatively sharp transitions or critical points along the spatial scale continuum where a shift in the relative importance of variables influencing a process occurs (Meentemeyer, 1989; Wiens, 1989). To identify scales or hierarchical levels of landscape structures, some general statistical and spatial analysis methods, inherently multi-scaled, are available such as semi-variance analysis (Burrough, 1995; Meisel and Turner, 1998;

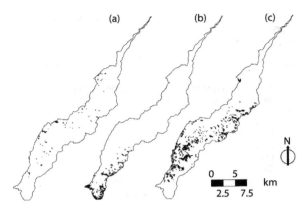

Figure 13.2 Distributions of: (A) *Mesobromiom* grasslands (CORINE code 34.3266); (B) *Brachypodium* grasslands (CORINE code 36.334); and (C) *lowland hay meadows* (CORINE code 38.2) in the Baganza watershed.

Bellehumeur and Legendre, 1998), multi-variate analysis of spatial autocorrelations (Burrough, 1983; Ver Hoef and Glen-Lewis, 1989), spectral analysis (Platt and Denman, 1975), wavelet analysis (Bradshaw and Spies, 1992), lacunarity analysis (Plotnick et al., 1993), scale variance (Wu et al., 2000), fractal analysis (Krummel et al., 1987), and fractal dimension combined with variograms (He et al., 1994).

Fractal analysis is a very useful tool for identifying hierarchical size scales of patches in nature, such as how to define boundaries between hierarchical levels and how to determine scaling rules for extrapolating within each level domain (Sugihara and May, 1990; Milne, 1991; Li, 2000). When natural "objects" like vegetation are not constrained by human activities and land manipulation, or by natural obstacles, they result in highly irregular shapes determined by iterative and diffusive growth, which can reproduce at different scales independently of size. In theory, a perfect fractal is self-similar at all scales, and it could be scaled up and down to infinity. Because of these limits to self-similarity, it is preferable to refer to these systems as fractal-like (Brown et al., 2002). Shifts in fractal dimension of irregular patch edges have been used to find substantial changes of spatial patterns at different scales (Krummel et al., 1987; Grossi et al., 1999; 2001). Krummel et al. (1987) were the first to develop a method for detecting different scaling regions in a landscape for a population of forest patches, based on perimeter-area relationships. Grossi et al. (1999) conceived a general statistical procedure to detect objectively the change points between different scaling domains in real patch populations, based on the selection of the best piecewise regression model using a set of statistical tests.

Given its significance within the framework of this paper, it seems worth providing a few details. Two distinct basic models were hypothesized to fit the data: continuous piecewise linear models and discontinuous piecewise linear models. To estimate the fractal dimension D of each scale domain, we used perimeter-area relationships as suggested by Lovejoy (1982). Given areas and perimeters of n patches, we can write the relationship as follows:

$$P_i = cA_i^{D/2}$$

where P_i and A_i are the perimeter and the area of the ith patch, respectively, and c is a constant. Taking the logarithm transform we get:

$$y_i = c + \frac{D}{2} x_i, \quad i = 1, 2, \ldots, n \tag{13.1}$$

where $y_i = \ln(P_i)$, and $x_i = \ln(A_i)$, so that D is twice the slope of a linear regression model by assuming self-affinity (Milne, 1991) — that is, all patches are similarly shaped independently of scale. Different hierarchical size scales of patches in nature can be identified by breakpoints, where parameters in Equation 13.1 change, which can be detected comparing this model to more complex models.

We considered five alternative models. If θ is the breakpoint of models with one breakpoint, and θ_1, θ_2, and ($\theta_1 < \theta_2$) are the first and second in models with two breakpoints:

$$y = \beta_0 I_{x \leq \theta} + \beta_0' I_{x > \theta} + \beta_1 x + \varepsilon \tag{13.2}$$

$$y = \beta_0 + \beta_1'(x I_{x \leq \theta} + \theta I_{x > \theta}) + \beta_1''(\theta I_{x \leq \theta} + x I_{x > \theta}) + \varepsilon \tag{13.3}$$

$$y = (\beta_0' + \beta_1' x) I_{x \leq \theta} + (\beta_0'' + \beta_1'' x) I_{x > \theta} + \varepsilon \tag{13.4}$$

$$y = \beta_0 + \beta_1'(x I_{x \leq \theta_{11}} + \theta_1 I_{x > \theta_1}) + \beta_1''(\theta_1 I_{x \leq \theta_1} + x I_{\theta_1 < x \leq \theta_2} + \theta_2 I_{x > \theta_2})$$
$$+ \beta_1'''(\theta_2 I_{x \leq \theta_2} + x I_{x > \theta_2}) + \varepsilon \tag{13.5}$$

$$y = (\beta_0' + \beta_1' x) I_{x \leq \theta_1} + (\beta_0'' + \beta_1'' x) I_{\theta_1 < x \leq \theta_2} + (\beta_0''' + \beta_1''' x) I_{x > \theta_2} + \varepsilon \tag{13.6}$$

where I is an indicator variable equal to one when the subscripted condition is true and equal to 0 otherwise; β_1', β_1'' and β_1''' are slopes — that is, half the fractal dimensions of the first, second, and third domain, respectively; β_0 is an intercept and ε the error term.

The simple continuous model given by 13.1 is called C_0, whereas the discontinuous model (13.2), called D_0, is a piecewise regression (Draper and Smith, 1998) with two parallel discontinuous segments and no change of slope, thereby with one single fractal domain. Models 13.3 and 13.5, called C_1 and C_2,

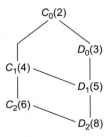

Figure 13.3   Nested collection of continuous and discontinuous piecewise linear models for hypothesis testing. The number of regression parameters to be estimate is between brackets (modified from Grossi et al., 1999).

have two and three continuous segments, and one and two breakpoints, respectively. Models 13.4 and 13.6, called D_1 and D_2, are models with two and three discontinuous segments, respectively, and so on.

More generally, let C_r, with $r = 0, 1, \ldots$, be the continuous piecewise linear model with r breakpoints and $(r+1)$ fractal domains, in C_r the number of parameters to be estimated is $2(r+1)$ with one intercept, r breakpoints, and $(r+1)$ slopes. Let D_r, with $r = 0, 1, 2, \ldots$, be the discontinuous piecewise linear model with r breakpoints, in D_r the number of parameters to be estimated is three (two intercepts and one slope) when $r = 0$, and $3r + 2$ — one slope and one intercept for each of $r + 1$ domains and r breakpoints — when $r \geq 1$. Therefore we can depict a nested collection of models (Figure 13.3).

Which of the nested models is the best is a typical problem of variable selection that, in multiple linear regressions, is usually based on the F test to measure the statistical significance of adding variables. If ω and Ω are two nested regression models having the same σ^2, with p and $p + q$ regression parameters, respectively, the null hypothesis $H_0 : \omega$ vs. the alternative hypothesis $H_A : \Omega$ can be tested using the following LR test statistic:

$$\lambda = n \ln \left(\frac{SSE_{\hat{\omega}}}{SSE_{\hat{\Omega}}} \right)$$

where $SSE_{\hat{\omega}}$ and $SSE_{\hat{\Omega}}$ are the residual sum of squares of ω and Ω, respectively. So, the rejection region can be expressed equivalently as:

$$\lambda > c_1 \tag{13.7}$$

or

$$F = \frac{n - p - q}{q} \left(\frac{SSE_{\hat{\omega}}}{SSE_{\hat{\Omega}}} - 1 \right) > c_2 \tag{13.8}$$

We wanted to test different null models $H_0 : \omega$ vs. alternative models $H_A : \Omega$, without knowing the exact distribution of the LR λ (in 13.7) or of the F statistic (in 13.8). Breakpoints in 13.2 to 13.6 were unknown parameters to be estimated like other regression parameters, and corresponding regression models results were not linear, so that in this case, the F distribution did not necessarily apply to variable selection procedures. The problem was studied using maximum likelihood and likelihood ratio (LR) tests, and simulations were conducted in order to check whether χ^2 and/or $F(q, n-p-q)$ were good approximations to the sampling distributions of λ and F. For this purpose, let $Y_i \sim N(\mu_i, \sigma^2)$, $i = 1, 2, \ldots, n$, be the dependent variable of a linear regression model where errors are Gaussian with $\mu_i = \mu(X_i, \Phi)$, where $\Phi = (\phi_1, \phi_2, \ldots, \phi_p)'$ is a vector of unknown parameters that can vary independently of the variance σ^2. The maximum likelihood estimate of Φ minimizes the residual sum of squares. We generated data from ω using habitat area as a regressor, and through opportune transformations of the dependent variable Y_i not affecting the null distributions (Grossi et al., 1999). Then the test statistics λ and F could be computed using $SSE_{\hat{\omega}}$ and $SSE_{\hat{\Omega}}$ from generated data, with 6000 replications for each alternative model Ω, when the null model is C_0, and 5000 otherwise. To select the best piecewise model for each habitat type, we compared hierarchically nested models (Figure 13.3) by computing the corresponding LR statistic. Some null models are possible:

1. With null model $\omega = C_0$, the alternative model Ω might be any of the more complex models C_1, C_2, D_0, D_1 and D_2
2. With $\omega = C_1$, the alternative model Ω might be any of C_2, D_0, D_1 and D_2
3. With $\omega = D_1$, the alternative model Ω could be only D_2
4. With $\omega = C_2$, the alternative model Ω could be only D_2. Both F and λ had empirical distributions which could be approximated by the nominal F and χ^2 distributions, respectively; therefore, we limited the analysis to the LR statistic.

13.3.4 Change Intensity Detection

Change detection is the comparison of the measurements computed from two co-registered remote sensing images of the same scene, by determining a quantity corresponding to the difference (or similarity) between two different times at the same location. A general equation for this metric may appear as follows (Skifstad and Jain, 1989):

$$D(x, y) = \varphi[f_{\tau 1}(x, y), f_{\tau 2}(x, y)] \tag{13.9}$$

where $D(x, y)$ is the difference metric, and $f_{\tau 1}$ is the metric computed at location (x, y) in image τ_i, where τ_i is a time index, and φ denotes a linear or nonlinear operation, which is often the absolute difference value.

In this paper we refer to $D(x, y)$ as the standardized change intensity image if φ denotes the standardized difference:

$$D(x, y) = |f_{\tau 1}(x, y) - f_{\tau 2}(x, y)| - |m| \bigg/ \sqrt{s_{\tau 1}^2 + s_{\tau 2}^2 - 2\mathrm{cov}_{\tau 1 \tau 2}} \tag{13.9}$$

where m is the mean of the differences, $s_{\tau 1}^2$ is the variance of the metric $f_{\tau 1}$ which is the Normalized Difference Vegetation Index (NDVI), (Rouse et al., 1974; Kerr and Osrtovsky, 2003), calculated as (band 4 − bond 3)/(band 4 + band 3) for both images and $cov_{\tau 1 \tau 2}$ is the covariance.

In order to capture mainly man-induced ecological changes, we used two five-year different dates of Landsat Thematic Mapper (TM) images of the study area: August 11 1990, and July 24 1995. Reflectances were used, since are the most correlated with ground data (Goward et al., 1991). NDVI exploits spectral responses in the red band and in the near infrared band channels and it is derived as the ratio of infrared minus red over infrared plus red values. It is strongly related to variables of most ecological interest such as the fraction of photosynthetically active radiation intercepted by vegetation (Fipar), leaf total nitrogen content, leaf area index (LAI), and, in general, to vegetative processes at ecosystem level (Law and Waring, 1994; Matson et al., 1994); it is also species specific, and reveals health and stress conditions of vegetation cover (Guyot, 1989).

Satellite images of 1995 were almost contemporary with reconnaissance activities in the field. Standardization was done to account for differences between dates due to climatic changes and other sources of added noise, not accounted for by prior geometric and atmospheric corrections. $D(x, y)$ in Model 13.9 was the spatially explicit response variable used for detecting either positive (gains) or negative (losses) in habitat scale domains; $D(x, y)$ was co-registered with the raster map of CORINE habitats to assign response variable values to each pixel of a particular habitat patch. One standard deviation is typically used as threshold for change detection (Fung and LeDrew, 1988), however, empirical distributions of $D(x, y)$ are not normal, but rather leptokurtic and skewed. We identified absolute change intensities $\overline{\Delta}$ as the medians of empirical 0.20, 0.10, and 0.05 percentiles, at both tails of the $D(x, y)$ distribution, to be sure that real changes occurring in the watershed were dealt with, avoiding background noise.

Percentiles less than 5% were not considered in order to avoid a few local extreme values. Changes detected through remote sensing techniques are real effects of ecological significance observed in five-year intervals due to extrinsic pressure, mostly given by human activities, which could have varied spectral responses affecting the NDVI metric.

13.3.5   Retrospective Resilience

When external pressures affect habitat intrinsic factors of resistance sensitivity, they might determine detectable habitat changes which are related to retrospective resilience or displayed fragility (Nilsson and Grelsson, 1995). Change intensities detected are just real effects observed in a specific time lag, but they do not allow distinguishing between external pressure and resistance factors, which could have determined change. The closer the time window, the better the possibility of recognizing driving forces causing pressure without confounding and overlapping.

Within an operational framework, we could think of resilience simply in terms of habitat intrinsic resistance coupled with extrinsic pressure. Intrinsic resistance factors might be identified, for instance, for communities and habitats to include size of community ranges, functional diversity, community and habitat rarity, habitat size, distribution and connectivity, edge complexity, source-sink habitat relationships, habitat fragmentation, and connectedness. The ecological memory of the system itself is also an important factor of resistance, as it allows persistence (Peterson, 2002). The type, magnitude, length, and timing of disturbance, its predictability, the exposure of the habitat, and the habitat's inherent resistance, have important interactive relationships which determine the propensity of the habitat to displace from its stability pit to another pit. Such propensity can be named vulnerability or fragility (Nilsson and Grelsson, 1995; Zurlini et al., 2003), which appears inversely related to resilience (Gunderson and Holling, 2002). In particular, as to retrospective resilience, we could simply think that the amount of extrinsic pressure (v) coupled with habitat intrinsic resistance (ρ) determines resilience; in other words, resilience can be deemed as proportional to the resistance per unit of external pressure, i.e., $\propto \rho/v$, as well as $\propto 1/\text{fragility}$.

So, for the $(r+1)$ fractal domain of a specific habitat, with r breakpoints, it follows that the absolute change intensity detected $\overline{\Delta}_{r+1} \propto v/\rho$, where v/ρ is the amount of external pressure per unit of resistance. Therefore, we used $1/\overline{\Delta}_{r+1}$ as an approximate estimate of resilience.

13.4 RESULTS

13.4.1 Best Regression Models and Scale Breaks

One important line of our investigation was a better characterization of the empirical patterns of self-similarity. Objective identification of scale breaks depended on selecting the best piecewise perimeter-area regression models. Models C_0, C_1, C_2, D_0, D_1 and D_2 were fitted for each grassland habitat and residual sum of squares (SSE) estimated. The LR statistics, the simulated critical values, as 0.95 percentiles of the empirical simulated LR statistic distribution, along with corresponding empirical p values, were computed for each null model and grassland habitat (Table 13.2). When C_0 is the null model, the p value is always less than 0.01 for *Brachypodium* and lowland hay meadows, while it is always over 0.50 for *Mesobromion*. Therefore, the hypothesis of a simple straight line (model C_0) was always rejected for *Brachypodium* and lowland hay meadows, but could not be rejected for *Mesobromion*, for which all patches apparently belonged to a single scale domain. This last habitat presented a narrower range of patch sizes missing larger patches as found for other grasslands. It did not apparently show any statistically significant shift in edge fractal dimension indicating no substantial change in scale with regard to processes generating and maintaining patches.

Table 13.2 Computed LR statistics, simulated critical values, and probability values for: (a) H_0: C_0; (b) H_0: D_0; (c) H_0: C_1; (d) H_0: D_1; (e) H_0: C_2 against the alternative hypothesis H_A. The sample size (number of patches) of generated data set was equal to the original size of the three grassland habitats considered. Critical values corresponded to the 0.95 percentiles of the empirical distribution. (A) *Mesobromiom* grasslands (CORINE code 34.3266); (B) *Brachypodium* grasslands (CORINE code 36.334); and (C) lowland hay meadows (CORINE code 38.2)

			H_A				
			C_1	C_2	D_0	D_1	D_2
(A)							
H_0	C_0	LR	1.7674	3.5625	4.3919	4.4199	11.6583
		LR$_{SIMUL}$ 95%	7.3021	12.4827	10.4206	12.6441	22.6046
		p value	0.6098	0.8588	0.5645	0.8242	0.8457
(B)							
H_0	C_0	LR	33.7691	38.3121	37.0409	38.5816	49.9436
		LR$_{SIMUL}$ 95%	7.3538	12.8895	10.6259	13.0548	23.4938
		p value	<0.001	<0.001	<0.001	<0.001	<0.001
	D_0	LR				1.5410	12.9030
		LR$_{SIMUL}$ 95%				5.2296	16.2808
		p value				0.2806	0.1526
	C_1	LR		4.5430		4.8130	16.1750
		LR$_{SIMUL}$ 95%		8.2295		7.0631	18.1489
		p value		0.2878		0.1586	0.0966
	D_1	LR					11.3620
		LR$_{SIMUL}$ 95%					13.5358
		p value					0.1296
	D_2	LR					11.6310
		LR$_{SIMUL}$ 95%					12.9376
		p value					0.0934
(C)							
H_0	C_0	LR	24.8380	33.3837	24.7804	27.5452	44.3722
		LR$_{SIMUL}$ 95%	7.4259	13.6475	11.0674	13.5946	25.4789
		p value	<0.001	<0.001	<0.001	<0.001	<0.001
	D_0	LR				2.7650	19.5900
		LR$_{SIMUL}$ 95%				4.6320	17.2541
		p value				0.1282	0.0182
	C_1	LR				2.7070	19.6340
		LR$_{SIMUL}$ 95%				8.5545	21.1631
		p value				0.6224	0.0946
	D_1	LR					16.827
		LR$_{SIMUL}$ 95%					15.4899
		p value					0.0274
	D_2	LR					10.9890
		LR$_{SIMUL}$ 95%					15.1130
		p value					0.2484

For the remaining grassland habitats further tests were necessary in order to select the best descriptor of the data, at least among the models we looked at. Different hypotheses were possible. The null hypothesis of D_0 could not be rejected at first glance for *Brachypodium* grassland (Table 13.2), so that discontinuous models might be plausible. The null hypothesis of D_0 must be instead rejected for lowland hay meadows, because D_2 was clearly better. However, model C_1 could not be rejected for both habitats at the 0.05 probability level. In the case of lowland hay meadows, the p value was very

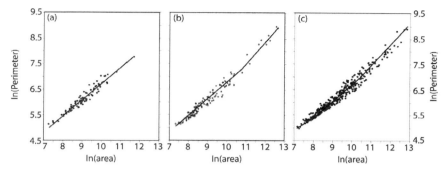

Figure 13.4 Scatterplots and corresponding piecewise regression model fitting for: (A) *Mesobromiom* grasslands (CORINE code 34.3266); (B) *Brachypodium* grasslands (CORINE code 36.334); and (C) lowland hay meadows (CORINE code 38.2) in the Baganza watershed.

near to 0.05 for model C_2, indicating that something more complicated than C_1 was needed for best fitting these patches, since D_2 was better than D_1, but not better than C_2. For *Brachypodium*, the conclusion was that the best model was C_1, with a clear change point at about 5.5 ha, whereas the best descriptor of lowland hay meadows was C_2, where the lowest fractal dimension was less than one but not statistically different from one at the 0.05 level. For *Brachypodium* we could say that, at the higher fractal domain, patches assumed a more complex shape, related to lower human disturbance associated with altitude. The single lower domain of this habitat had about the same scale range of the entire set of *Mesobromion* patches (Figure 13.4). We can therefore conclude that distinct scale domains could be objectively and accurately recognized by shifts in the boundary fractal dimension of real patches present in the watershed. Fractal dimensions, standard errors and fractal domain ranges of single grasslands are given in Table 13.3.

13.4.2 Change Intensity Detection

The relative higher temporal persistence of forested areas in the watershed was the main cause for the leptokurtosis of $D(x, y)$, while its slender right asymmetry was due to changes in agricultural land use. In general, on going from higher to lower percentiles, the probability of a real modification in land use increases respect to the hypothesis of simple phenologic changes.

At the empirical percentile of 10%, the major change observed was in the large CORINE habitat category of field crops (82.11), due to local agricultural practices and crop rotation. Beech forest (41.1744) pixels did not occur at 10% or lower percentiles, showing a higher temporal persistence, while large parts of pixels in the grasslands category belonged to mountain, sub-mountain, and lowland hay meadows. Changes observed are likely linked to the different timing of cutting, considering that a recently cut meadow might have a spectral response similar to bare soil.

Table 13.3 Fractal dimensions, standard errors, and fractal domain ranges for: (a) *Brachypodium* grasslands (CORINE code 36.334); and (b) lowland hay meadows (CORINE code 38.2). *Mesobromion* patches apparently belonged to a single scale domain with $D = 1.2$ and standard error $= 0.8$

Domain ($r+1$)	Value	St. error	Domain (m^2)
(a)			
1	1.292	0.0277	0 $\dashv$ 55143
2	1.879	0.079	>55143
(b)			
1	0.8786	0.111	0 $\dashv$ 3277
2	1.2622	0.021	3277 $\dashv$ 31611
3	1.5432	0.04	>31611

At the percentile of 5% watershed appeared clearly divided in two distinct sections (Figure 13.5). Mountain and sub-mountain elevation range showed few changes linked to agricultural practices of crop rotation, whereas in the hills and the flat there were several important changes linked to field and urban area dynamics, and to modifications of riparian habitats. As regards grasslands, lowland hay meadows (38.2) presented the highest dynamics of change, whereas other grasslands were considerably persistent with relatively few losses (Table 13.4). A substantial change in processes generating and maintaining landscape patches at different scales was revealed by a change in intensity detection, with the exclusion of *Mesobromion*. For this habitat there was no apparent substantial change in scale as regards generating and maintaining processes.

For *Brachypodium* grassland patches, with two scale domains, the highest change intensities were significantly related to smaller boundary fractal dimensions of scale domains and less complex patch geometry (Table 13.4). For lowland hay meadows, with three scale domains under direct human control (Figure 13.4), a different pattern appeared (Table 13.4): at 20% and 10% percentiles change intensities were higher at larger patches, whereas the highest change intensity (5%) was not related to specific scale domains. For this grassland, patches were generally close to roads and small villages, to reduce costs of management, so they resulted in more regularly shapes at all scales (Figure 13.4). *Brachypodium* grasslands, not under direct human influence, showed higher change intensity for scale domains with relatively smaller and less complex patch geometry.

13.4.3   Resilience of Habitat Scale Domains

Habitat resilience, as operationally defined here, is expected to be lowest for scale fractal domains where change is most likely. Different resilience or fragility levels were found to be associated with different scale domains of real habitats, according to human management activities and land manipulation (Table 13.5). *Brachypodium* grasslands showed a higher short-term retrospective resilience persistence at the upper than at the lower scale domain,

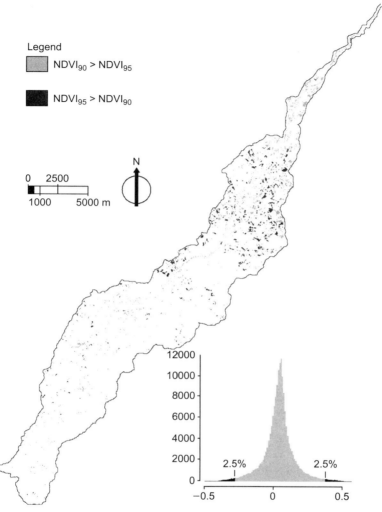

Figure 13.5 Standardized change intensity image (1995 to 1990) of the Baganza watershed at 0.05 percentiles, with black pixels (gains) and gray pixels (losses), along with standardized change intensity distribution (modified from Zurlini et al., 2001).

which had a smaller fractal dimension and a less complex patch geometry. *Mesobromion* habitats showed a single scale domain, with a range and resilience comparable to the lower *Brachypodium* domain. *Mesobromion* patches in the watershed are expected to resist change similarly independently of scale. Lowland hay meadows, despite its three scale domains, presented much lower short-term retrospective resilience levels across scales with respect to other grasslands. This is a managed ecosystem under direct human control and change is most likely due to management practices, thereby resilience is expected to be lowest and fragility highest.

Table 13.4 Medians of absolute change intensities $\overline{\Delta}_{r+1}$ of grasslands at empirical 0.20, 0.10, and 0.05 percentiles for each $(r+1)$ habitat scale domain, where $r = 0, 1, \ldots$, is the number of change points

Percentiles	20%			10%			5%		
(a) *Mesobromiom* grasslands (CORINE code 34.3266)									
Scale domain $(r+1)$	1			1			1		
Change median	0.94			1.15			1.26		
Pixel n.	363			184			97		
(b) *Brachypodium* grasslands (CORINE 36.334)									
Scale domain $(r+1)$	1	2		1	2		1	2	
Change median	0.95	0.90		1.15	1.05		1.27	1.21	
Pixel n.	119	764		74	368		45	176	
	z	p value		z	p value		z	p value	
Test Wilcox-Mann-Whitney	2.99	0.0028		−19.98	<0.001		−10.06	<0.001	
(c) *Low land hay meadows* (CORINE code 38.2)									
Scale domain $(r+1)$	1	2	3	1	2	3	1	2	3
Change median	1.04	1.07	1.17	1.47	1.54	1.64	1.75	1.80	1.84
Pixel n.	137	1367	1141	58	635	630	22	286	354
Test Kruskal Wallis	χ^2	Df	p value	χ^2	Df	p value	χ^2	Df	p value
	31.22	2	<0.001	20.68	2	<0.001	5.91	2	0.052

Table 13.5 Retrospective resilience estimates of grasslands as inverse of medians of absolute change intensities $(1/\overline{\Delta}_{r+1})$ at empirical 0.20, 0.10, and 0.05 percentiles for each $(r+1)$ habitat scale domain, where $r = 0, 1, \ldots$, is the number of change points

Resilience estimates $1/\overline{\Delta}_{r+1}$	Percentiles		
	20%	10%	5%
Scale domains $(r+1)$	1, 2, 3	1, 2, 3	1, 2, 3
Lowland hay meadows	0.95, 0.94, 0.86	0.68, 0.65, 0.61	0.57, 0.56, 0.54
Brachypodium	1.05, 1.11, -	0.87, 0.95, -	0.79, 0.83, -
Mesobromiom	1.063, -, -	0.87, -, -	0.79, -, -

13.5 GENERAL DISCUSSION AND CONCLUSION

13.5.1 Grassland Phase States

Alternate states were shown to arise in some real grassland habitats, and were objectively and accurately identified by scale breaks delimiting equivalent scales of states (Figure 13.4). Grassland habitats either under direct or indirect human influence presented different regions in the scale spectrum of ecological phases over which state patterns were self-similar as to edge fractal dimension. All habitat patches pertaining to each scale domain can be deemed as multiple configurations of the same ecological phase state, according to dominating processes which generate and maintain habitats. Individual grassland patches, exposed to a combination of human and natural influences, appeared as a mosaic of several states in the watershed (Figure 13.1).

Hierarchy theory (Allen and Starr, 1982; O'Neill et al., 1986) postulates that distinct levels in the ecological system should be reflected in corresponding distinct scales of patterns in space, as revealed for instance by multiple scales of vegetative pattern in plant communities (O'Neill et al., 1991; Simmons et al., 1992). Different processes dominate at different scales, and the study of scaling, through a better characterization of empirical power-law patterns, is believed to be one powerful way of simplifying ecological complexity and of understanding the physical and biological principles that regulate biodiversity (Brown et al., 2002). Certain synchronization is expected among patches in a real spatial mosaic of grassland patches since, apparently, only small amounts of local migration are required to induce broad-scale phase synchronization, with all populations phase-locking to the same collective rhythm (Blasius et al., 1999).

Of the two distinct statistical models used to fit perimeter-area grassland data, only the continuous piecewise linear model appeared successful. The procedure of Grossi et al. (1999) proved to be effective in detecting landscape patterns when applied to patch mosaics of the Baganza watershed, however, empirical distributions of the test statistics were obtained through simulation procedures; thereby results obtained were strictly dependent on the data used. Edge fractal dimension of habitat patches appeared to be a useful scaling indicator of scale domains and habitat state transitions; however, in Model 13.1 we assumed self-affinity (Milne, 1991) — that is, all patches would have the same shape independently of scale, while patches might have dissimilar shapes. That could be a significant source of deviations from perimeter-area relationships, along with the fact that large and small-scale patterns could readily exhibit different degrees of complexity, so that fine-scale variability can be obscured by broad-scale variability (Meisel and Turner, 1998; Wu et al., 2000); thereby edge patterns of a single patch can be differently scaled and shapes need not be strictly fractal or fractal-like. Many habitat patches were found to be close to change points between state domains, for which shifts into another scale domain were most likely. Those patches could be identified as most susceptible to "flip" into another phase state, and would require priority for intervention and monitoring.

13.5.2 Scale Domains and Processes

In Mediterranean regions, ecosystems have been shaped by the millennial historic and evolved interactions between man and nature, so that many forms of human disturbance are recognized to be important factors sustaining natural systems (Pickett and White, 1985), since they have been gradually embodied into the systems' memory by adaptive processes (Ulanovicz, 1997).

The effects of external pressure were significantly related to habitat scale domains, according to their scaling properties resulting from the interaction among ecological, physical, and social controls shaping the systems. Scaling of domains provided evidence and support for identifying and explaining scale invariant ecological interactive processes at various scales. So broad-scale processes appeared to impose a broad-scale pattern, observable on the whole plant community at higher scales, essentially provided by geomorphological and climatic factors (Delcourt and Delcourt, 1988; O'Neill et al., 1991) ruling the watershed in the mountain range rather independently of scarce human presences. Such broad-scale processes appeared to maintain broad-scale patterns of the more natural grassland habitats like *Brachypodium* grasslands at the upper scale domain, whereas at the lower scale and altitudes, highest change intensities were significantly related to less complex patch geometry, likely due to the proximity of managed patches. Thus patches in the lower scale domain appeared more fragile, and more susceptible to changes.

Such broad-scale processes, apparently, were not influencing lowland hay meadows, mainly ruled by human management, while we could not exclude certain influences on *Mesobromion* habitats. At intermediate scales, patterns in the watershed were more dependent on shape and location of land forms, and the distribution patterns of vegetation and livestock populations (Swanson et al., 1988). Human disturbance here was more evident. At those scales, *Mesobromion*, with a single scale domain, was poor and sparse in naturally stressed environments because of drought, and adapted to relatively extreme environmental conditions. For this reason, it could be insensitive to direct and indirect human disturbances even though occasionally grazed by cattle, and cut for its proximity to lowland hay meadows (Sburlino et al., 1993).

In the watershed, *Mesobromion* patches did not reach the size of the other grasslands, and if completely abandoned, they could change slowly into *Brachypodium* grasslands. At finer scales, constraints could be provided by either local disturbances or biotic interactions at the community, population, and individual levels (Danielson, 1991; Hansen and Urban, 1992). *Mesobromion* and smaller *Brachipodium* patches appeared to be more confined to specific elevation ranges (Figure 13.2), and so they were more influenced by shape and location of land forms as well as by local sources of disturbance.

In contrast, lowland hay meadows were widely spread from the flat up to the sub-mountain range of the watershed (Figure 13.2), and conditioned by intermediate and local processes; patches were near to roads and small villages (Sburlino et al., 1993), to reduce costs of management, so they were more intensively managed for cattle nourishment and resulted more regularly shaped than others at all scales. In this case, scaling was the result of human selection of suitable surface dimensions for hay production in the watershed.

13.5.3 Adaptive Cycle and Resilience

The second kind of resilience (Holling, 1973; Gunderson and Holling, 2002) appears to be more appropriate as a tool for thinking about systems with the premise that disturbance and change are normal rather than seeking to predict or find optimal or final stable states. As a metaphor to guide the case studies, we employed the adaptive cycle (Holling et al., 1995) as an example of self-organization within an ecosystem with alternate states and possible "flips" into alternative phases (Kay, 2000). This conceptual model incorporates both linear succession (Clements, 1916) and independent, species-level disordered behavior (Gleason, 1926), integrated into a complexity based framework with insights from catastrophe theory, chaos theory, and self-organization theory.

Fire, storm, or pest outbreaks can be seen as "natural" bifurcation points between attractors within a cycle such as the exploitation and conservation phases. In this model, resilience decreases on going towards the conservation phase, where the system becomes more brittle; it expands when the cycle goes rapidly into a "back-loop" to reorganize accumulated capital or to initiate a new cycle. Coherently with what is implied by the metaphor of a system's adaptive cycle, fragility is expected to be inversely related to resilience (Gunderson and Holling, 2002; Zurlini et al., 2003). However, not all adaptive cycles are the same and there are some exceptions (Gunderson and Holling, 2002).

Grasslands in the watershed appeared to deviate from an adaptive cycle and represented distinct departures or variants from that cycle. *Mesobromion* semi-arid grasslands, with one phase state, appeared to be ecosystems that were strongly influenced by episodic external inputs, mainly manuring by cattle which provide essential organic matter for vegetation growth. They appeared relatively resilient with little internal regulation and highly adaptive responses to opportunity, oscillating in the reorganization and exploitation phases (Gunderson and Holling, 2002). *Brachypodium* grasslands were eco-systems with two main phase states: the former corresponding to the higher scale domain of higher elevations and of the summits (Figure 13.2), mainly influenced by broad-scale climatic processes, with little internal regulation, highly adaptive responses to opportunity, and with the highest retrospective resilience; the second related to the lower scale domain at lower altitudes, influenced by episodic inputs such as occasional grazing and fertilization by manure, with certain internal regulation. It showed a very high persistence or very low cycling of phase states, and was characterized by the highest retrospective resilience in the watershed (Table 13.5).

Lowland hay meadows were productive ecosystems with predictable inputs and some internal regulation of external variability over certain scale ranges; they showed the full cycle of boom-and-bust dynamics (Gunderson and Holling, 2002), even twice a year. In this case, constraints were provided by cutting and manuring practices forcing the system through the same trajectory; natural variability of structuring variables such as grazing has been reduced to stabilize hay production so that they tended to become more spatially uniform and less functionally diverse, and in that way more sensitive to disturbances that otherwise could have been absorbed (Holling, 1986). Their resilience was low and fragility high.

Our approach provided resilience estimates giving evidence and support to this general picture. In the past, critical structuring variables such as grazing pressure by cattle and sheep along with cutting, helped maintain grassland habitats in time at different elevation ranges in the watershed. The slow abandonment of traditional agricultural practices and pastures might lead to pathology of disease management in crops and people (Holling, 1986). Abandoned lowland hay meadows in the hills appeared to be slowly undergoing a phase change into *Mesobromion* grasslands, because of reduced supply of organic matter to soils by cattle. In turn, *Mesobromion* grasslands could slowly turn into scrub-dominated formations and thickets (Table 13.1), with communities characteristic of *Carpinion* (hop-horn beam) forest edges, whereas at higher elevations they could become *Brachypodium* grasslands, for the competitive predominance of *Brachypodium genuense*.

There is an increasing need to identify and quantify nature and man-induced ecological processes, at various scales, and their corresponding fingerprint patterns in space, in order to help planning and management of landscape mosaics with a predictable effect on ecological processes (Tischendorf, 2001). In this respect, linking multi-scale spatial pattern analysis to change intensity detection seems a promising approach for assessing retrospective resilience, critical structuring variables of habitats, to address ecosystem health.

Recently, Walker et al. (2002) captured the current state of understanding on how to measure and manage for resilience in socio-ecological systems with a stakeholder-driven description of the system along with a set of scenarios and simple models to guide in the identification and manipulation of the system's resilience on an ongoing basis and during times of crisis. However, to develop an operational and measurable concept of resilience, it is still necessary to gain much more insight from empirical analyses (Carpenter et al., 2001). Today, the fundamental condition of ecological knowledge, as provided by the CORINE habitat classification, can join together with the availability of new multi-spectral remote sensing tools at high spatial and temporal resolution providing outstanding potential for high-frequency remote monitoring in ecosystem features related to ecosystem health.

ACKNOWLEDGMENTS

This work was partly conducted under a contract of the national project of Map of Italian Nature; in this respect O. Rossi is gratefully acknowledged. We are thankful to S. Marchiori, for discussion on an earlier version of the paper, and Marco Dadamo for figures and tables.

REFERENCES

Allen, T.F.H. and Starr, T.B. *Hierarchy: Perspectives for Ecological Complexity.* University of Chicago Press, Chicago, 1982.
Bellehumeur, C. and Legendre, P. Multiscale sources of variation in ecological variables: modelling spatial dispersion, elaborating sampling designs. *Landscape Ecol.* 13, 15–25, 1998.

Blasius, B., Huppert, A., and Stone, L. Complex dynamics and phase synchronization in spatially extended ecological systems. *Nature* 399, 354–359, 1999.

Bradshaw, G.A. and Spies, T.A. Characterizing canopy gap structure in forests using wavelet analysis. *J. Ecol.* 80, 205–215, 1992.

Brown J.H., Gupta, V.K., Li, B.-L., Milne, B.T., Restrepo, C., and West, G.B. The fractal nature of nature: power laws, ecological complexity and biodiversity. *Phil. Trans. R. Soc. Lond. B* 357, 619–626, 2002.

Burrough, P.A. Multiscale sources of spatial variation in soil. *J. Soil Sci.* 34, 577–597, 1983.

Burrough, P.A. "Spatial aspects of ecological data," in *Data Analysis in Community and Landscape Ecology*, Jongman, R.H.G., Ter Braak, C.J.F., and Van Tongeren, O.F.R., Eds. Cambridge University Press, Cambridge, 1995, pp. 213–265.

Carpenter, S.R. and Cottingham, K.L. Resilience and restoration of lakes. *Conserv. Ecol.* 1, 2, 1997. See http://www.consecol.org/vol1/iss/art2.

Carpenter, S.R., Walker, B., Anderies, J. M., and Abel, N. >From metaphor to measurement: resilience of what to what? *Ecosystems* 4, 765–781, 2001.

Casagrandi, R. and Rinaldi, S. A theoretical approach to tourism sustainability. *Conserv. Ecol.* 6, 13, 2002. See http://www.consecol.org/vol6/iss/art13.

Castri di, F., Hansen, A.J., and Holland, M.M. A new look at ecotones: emerging international projects on landscape boundaries. *Biol. Int.* 17, 47–106, 1988.

Clements, F.E. Plant Succession: An Analysis of the Development of Vegetation. Carnegie Institute of Washington Publ. 242. Facsimile reprint by Haffner, New York, 1916.

Costanza, R. "Toward an operational definition of ecosystem health," in *Ecosystem Health. New Goals for Environmental Management*, Costanza, R., Norton, B.G., and Haskell, B.D., Eds. Island Press, Washington, DC, 1992, pp. 239–256.

Costanza, R., d'Arge, R., de Groot, R., Farber, S., Grasso, M., Hannon, B., Limburg, K., Naeem, S., O'Neill, R.V., Paruelo, J., Raskin, R.G., Sutton, P., and van den Belt, M. The value of the world's ecosystem services and natural capital. *Nature* 387, 253–260, 1997.

Danielson, B.J. Communities in a landscape: the influence of habitat heterogeneity on the interactions between species. *Am. Naturalist* 138, 1105–1120, 1991.

Delcourt, H.R. and Delcourt, P.A. Quaternary landscape ecology: relevant scales in space and time. *Landscape Ecol.* 2, 23–44, 1988.

Draper, N.R. and Smith, H. *Applied Regression Analysis*, 3rd ed. Wiley, New York, 1998.

EU/DG XI. CORINE Biotopes Manual, habitats of the European Community. A method to identify and describe consistently sites of major importance for nature conservation. EUR 12587/3, Bruxelles, 1991.

Fung, T. and LeDrew, E. The determination of optimal threshold levels for change detection using various accuracy indices. *Photogr. Eng. Remote Sensing* 54, 1449–1454, 1998.

Gell-Mann, M. *The Quark and the Jaguar. Adventures in the Simple and the Complex.* Freeman, New York, 1994.

Gleason, H.A. The individualistic concept of the plant association. *Bull. Torrey Bot. Club* 53, 7–26, 1926.

Goward, S.N., Markham, B., Dye, D., Dulaney, W., and Yang, J. Normalized difference vegetation index measurements from the advanced very high resolution radiometer. *Rem. Sens. Env.* 35, 257–277, 1991.

Grimm, V., Schmidt, E., and Wissel, C. On the application of stability concepts in ecology. *Ecol. Model.* 63, 143–161, 1992.

Grossi, L., Patil, G.P., Rossi, O., Taillie, C., and Zurlini, G. Statistical Selection of Landcover Patch Perimeter Area Models for Multiscale Landscape Analysis. Technical Report 99-1030, pp. 18. Center for Statistical Ecology and Environmental Statistics, Penn State University, PA, 1999.

Grossi, L., Zurlini, G., and Rossi, O. Statistical detection of multiscale landscape patterns. *Environmental and Ecological Statistics* 8, 253–267, 2001.

Gunderson, L.H., Holling, C.S., Pritchard, L., and Peterson, G.D. Resilience in ecosystems, institutions, and societies. Beijer Discussion Paper Series, Number 95. Beijer International Institute of Ecological Economics, Royal Swedish Academy of Sciences, Stockholm, Sweden, 1997.

Gunderson, L.H. and Holling, C.S. Panarchy: understanding transformations in human and natural systems. Island Press, Washington, D.C., 2002.

Guyot, G. Signatures spectrales des surfaces naturelles. *Collection Teledetection Satellitaire, Paradigme* 5, 178, 1989.

Hansen, A.J. and Urban, D.L. Avian response to landscape pattern: the role of species' life histories. *Landscape Ecol.* 7, 163–181, 1992.

He, F., Legendre, P., and Bellehumeur, C. Diversity pattern and spatial scale: a study of a tropical rain forest of Malaysia. *Env. Ecol. Stat.* 1, 265–286, 1994.

Holling, C.S. Resilience and stability of ecological systems. *Ann. Rev. Ecol. Syst.* 4, 1–23, 1973.

Holling, C.S. "The resilience of terrestrial ecosystems: Local surprise and global change," in *Sustainable Development of the Biosphere*, Clark, W.C. and Munn, R.E., Eds. Cambridge University Press, Cambridge, 1986.

Holling, C.S. Cross-scale morphology, geometry, and dynamics of ecosystems. *Ecol. Monogr.* 62, 447–502, 1992.

Holling, C.S., Schindler, D.W, Walker, B.H., and Roughgarden, J. "Biodiversity in the functioning of ecosystems: an ecological primer and synthesis," in *Biodiversity Loss: Ecological and Economic Issues*, Perrings, C.A., Maler, K.G., Folke, C., Holling, C.S. and Jansson, B.O., Eds. Cambridge University Press, Cambridge, 1995, pp. 44–83.

Janssen, M.A., Walker, B.H., Langridge, J., and Abel, N. An adaptive agent model for analyzing co-evolution of management and policies in a complex rangeland system. *Ecol. Model.* 131, 249–268, 2000.

Kay, J. "Ecosystems as self-organizing holarchic open systems: narratives and the second law of thermodynamics," in *Handbook of Ecosystem Theories and Management*, Jørgensen, S.E. and Müller, F., Eds. CRC Press–Lewis Publishers, Boca Raton, 2000, pp. 135–160.

Kerr, J.T. and Ostrovsky, M. From space to species: ecological applications for remote sensing. *Trends in Ecology and Evolution* 18, 299–305, 2003.

Krummel, J.R., Gardner, H., Sugihara, G., O'Neill, R.V., and Coleman, P.R. Landscape patterns in a disturbed environment. *Oikos* 48, 321–324, 1987.

Law, B.E. and Waring, R.H. Remote sensing of leaf area index and radiation intercepted by understory vegetation. *Ecological Applications* 4, 272–279, 1994.

Levin, S.A. The problem of pattern and scale in ecology. *Ecology* 73, 1943–1976, 1992.

Li, B.-L. Fractal geometry applications in description and analysis of patch patterns and patch dynamics. *Ecol. Model.* 132, 33–50, 2000.

Li, B.-L. A theoretical framework of ecological phase transitions for characterizing tree-grass dynamics. *Acta Biotheoretica* 50, 141–154, 2002.

Lovejoy, S. Area-perimeter relation for rain and cloud areas. *Science* 216, 185–187, 1982.

Mageau, M.T., Costanza, R., and Ulanowicz, R.E. The development and initial testing of a quantitative assessment of ecosystem health. *Ecosyst. Health* 1, 201–213, 1995.

Matson, P., Johnson, L., Billow, C., Miller J., and Pu, R. Seasonal patterns and remote spectral estimation of canopy chemistry across the Oregon transect. *Ecological Applications* 4, 280–298, 1994.

McClanahan, T., Polunin, N., and Done, T. Ecological states and the resilience of coral reefs. *Conserv. Ecol.* 6, 18, 2002. See http://www.consecol.org/vol6/iss2/art18.

Meentemeyer, V. Geographical perspectives of space, time, and scale. *Landscape Ecol.* 3, 163–173, 1989.

Meisel, J.E. and Turner, M.G. Scale detection in real and artificial landscapes using semivariance analysis. *Landscape Ecol.* 13, 347–362, 1998.

Milne, B.T. "Lessons from applying fractal models to landscape patterns," in *Quantitative Methods in Landscape Ecology*, Turner, M.G. and Gardner, R.H., Eds. Springer-Verlag, Berlin, 1991, pp. 199–235.

Naveh, Z. and Liebermann, A. *Landscape Ecology: Theory and Application.* Springer-Verlag, New York, 1994.

Nilsson, C.N. and Grelsson, G. The fragility of ecosystems: a review. *J. App. Ecol.* 32, 677–692, 1995.

O'Neill, R.V., DeAngelis, D.L., Waide, J.B., and Allen, T.F.H. *A Hierarchical Concept of Ecosystems.* Princeton University Press, Princeton, 1986.

O'Neill, R.V., Turner, S.J., Cullinam, V.I., Coffin, D.P., Cook, T., Conley, W., Brunt, J., Thomas, J.M., Conley, M.R., and Goszz, J. Multiple landscape scales: an intersite comparison. *Landscape Ecol.* 5, 137–144, 1991.

Openshaw, S. *The Modifiable Areal Unit Problem.* Geo Books, Norwich, 1984.

Peterson, G.D., Allen, C.I.R., and Holling, C.S. Ecological resilience, biodiversity, and scale. *Ecosystems* 1, 6–18, 1998.

Peterson, G.D. Scaling ecological dynamics: self-organization, hierarchical structure, and ecological resilience. *Clim. Change* 44, 291–309, 2000.

Peterson, G.D. Contagious disturbance, ecological memory, and the emergence of landscape pattern. *Ecosystems* 5, 329–338, 2002.

Pickett, S.T.A. and White, P.S. *The Ecology of Natural Disturbance and Patch Dynamics.* Academic Press, Orlando, 1985.

Platt, T. and Denman, K.L. Spectral analysis in ecology. *Ann. Rev. Ecol. Systematics* 6, 189–210, 1984.

Plotnik, R.E., Gardner, R.H., and O'Neill, R.V. Lacunarity indices as measures of landscape texture. *Landscape Ecol.* 8, 201–211, 1993.

Risser, P.G. The status of the science examining ecotones. *BioScience* 45, 318–325, 1995.

Rouse, J.W., Haas, R.H., Shell, J.A., Deering, D.W., and Harlan, J.C. Monitoring the vernal advancement of retrogradation of natural vegetation. Final Report, NASA/GSFC, Greenbelt, MD, 1974.

Sburlino, G., Tornadore, N., Marchiori, S., and Zuin, M.C. La flora delle alte valli del fiume Taro e del torrente Cevo (appennino parmense) con osservazioni sulla vegetazione. *Atti Soc. Tosc. Sci. Nat. Mem., Serie B* 100, 49–170, 1993.

Skifstad, K.D. and Jain, R.C. Illumination independent change detection for real world image sequences. *Comput. Vis., Graph., Image Understand.* 46, 387–399, 1989.

Simmons, M.A., Cullinan, V.I., and Thomas, J.M. Satellite imagery as a tool to evaluate ecological scale. *Landscape Ecol.* 7, 77–85, 1992.

Sugihara, G. and May, R.M. Applications of fractals in ecology. *T.R.E.E.* 5, 79–86, 1990.

Swanson, F.J., Kratz, T.K., Caine, N., and Woodmansee, R.G. Landform effects on ecosystem patterns and processes. *BioScience* 38, 92–98, 1998.

Tansley, A.G. The use and abuse of vegetational concepts and terms. *Ecology* 16, 284–307, 1935.

Tischendorf, L. Can landscape indices predict ecological processes consistently? *Landscape Ecol.* 16, 235–254, 2001.

Ulanowicz, R.E. *Ecology. The Ascendent Perspective.* Columbia University Press, New York, 1997.

Ver Hoef, J.M. and Glenn-Lewis, D.C. Multiscale ordination: a method for detecting pattern at several scales. *Vegetatio* 82, 59–67, 1989.

Walker, B., Carpenter, S., Anderies, J., Abel, N., Cumming, G.S., Janssen, M., Lebel, L., Norberg, J., Peterson, G.D., and Pritchard, R. Resilience management in social-ecological systems: a working hypothesis for a participatory approach. *Conserv. Ecol.* 6, 14, 2002. See http://www.consecol.org/vol6/iss1/art14.

Ward, D. and Saltz, D. Foraging at different spatial scales: dorcas gazelles foraging for lilies in the Negev desert. *Ecology* 75, 45–58, 1994.

Wiens, J.A. Spatial scaling in ecology. *Funct. Ecol.* 3, 385–397, 1989.

Wiens, J.A. "Landscape mosaics and ecological theory." in *Mosaic Landscape and Ecological Processes*, Hansson, L., Fahrig, L., and Merriam, G., Eds. Chapman and Hall, London, 1995, pp. 1–26.

Wu, J. and Qi, Y. Dealing with scale in landscape analysis: an overview. *Geogr. Info. Sci.* 6, 1–5, 2000.

Wu, J., Jelinsky, D.E., Luck, M., and Tueller, P.T. Multiscale analysis of landscape heterogeneity: Scale variances and pattern metrics. *Geogr. Info. Sci.* 6, 6–19, 2000.

Zurlini, G., Amadio, V., and Rossi, O. A landscape approach to biodiversity and biological integrity planning: the map of the italian nature. *Ecosyst. Health* 5, 294–311, 1999.

Zurlini, G., Rossi, O., Grossi, L., and Amadio, V. "Carta della natura: analisi multiscalare di mosaici CORINE e loro valutazione in termini di valore e fragilità ecologica," in *Cartografia Multiscalare della Natura*, Rossi, O., Ed. Italian Society of Ecology, Tipografia Supergrafica, Parma, 2001, pp. 21–43.

Zurlini G., Rossi, O., and Amadio, V. "Landscape biodiversity and biological health risk assessment: the Map of Italian Nature," in *Managing for Healthy Ecosystems*, vol. II – *Issues and Methods*. Rapport, D., Lasley, B., Rolston, D., Nielsen, O., and Qualset, C., Eds. CRC Press–Lewis Publications, Boca Raton, FL, 2003, pp. 633–653.

CHAPTER 14

Emergy, Transformity, and Ecosystem Health

M.T. Brown and S. Ulgiati

14.1 INTRODUCTION

In this chapter, ecosystems are summarized as energetic systems and ecosystem health is discussed in relation to changes in structure, organization, and functional capacity as explained by changes in emergy, empower, and transformity. The living and nonliving parts and processes of the environment as they operate together are commonly called ecosystems. Examples are forests, wetlands, lakes, prairies, and coral reefs. Ecosystems circulate materials, transform energy, support populations, join components in network interactions, organize hierarchies and spatial centers, evolve and replicate information, and maintain structure in pulsing oscillations. Energy drives all these processes and energetic principles explain much of what is observed.

The living parts of ecosystems are interconnected, each receiving energy and materials from the other, interacting through feedback mechanisms to self-organize in space, time, and connectivity. Processes of energy transformation throughout the ecosystem build order, cycle materials, and sustain

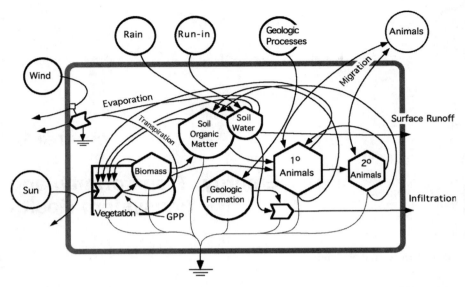

Figure 14.1   Generic ecosystem diagram showing driving energies, production, cycling, and the hierarchy of ecological components.

information, degrading energy in the process. The parts are organized in an energy hierarchy as shown in aggregated form in Figure 14.1. As energy flows from driving energy sources on the right to higher and higher order ecosystem components, it is transformed from sunlight to plant biomass, to first-level consumers, to second level and so forth. At each transformation second law losses decrease the available energy but the "quality" of energy remaining is increased.

14.2 A SYSTEMS VIEW OF ECOSYSTEM HEALTH

Conceptually, ecosystem health is related to integrity and sustainability. A healthy ecosystem is one that maintains both system structure and function in the presence of stress. Vigor, resilience and organization have been suggested as appropriate criteria for judging ecosystem health. Leopold (1949) referred to health of the "land organism" as "the capacity for internal self-renewal." Ehrenfeld (1993) suggested that "health is an idea that transcends scientific definition. It contains values, which are not amenable to scientific methods of exploration but are no less important or necessary because of that." Ecosystem health may be related to the totality of ecosystem structure and function and may only be understood within that framework.

The condition of landscapes and the ecosystems within them is strongly related to levels of human activity. Human-dominated activities and especially the intensity of land use can affect ecosystems through direct, secondary, and

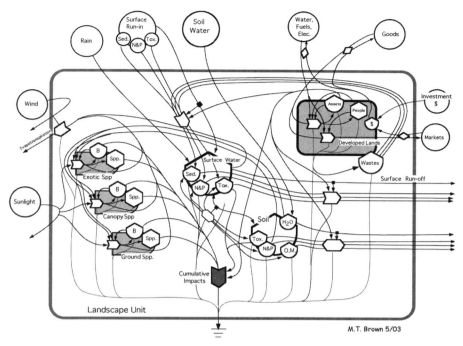

Figure 14.2 Landscape unit showing the effects of human activities on ecosystem structure and functions. The more intense the development, the larger the effects. B = biomass, Spp = species, Sed = sediments, N & P = nitrogen and phosphorus, Tox. = toxins, O.M. = organic matter.

cumulative impacts. Most landscapes are composed of patches of developed land and patches of wild ecosystems. While not directly converted, wild ecosystems very often experience cumulative secondary impacts that originate in developed areas and that spread outward into surrounding and adjacent undeveloped lands. The more developed a landscape, the greater the intensity of impacts.

The systems diagram in Figure 14.2 illustrates some of the impacts originating in developed lands that are experienced by surrounding and adjacent wild ecosystems. They come in the form of air- and water-born pollutants, physical damage, changes in the suite of environmental conditions (like changes in groundwater levels or increased flooding), or combinations of all of them. Pathways from the developed lands module on the right carry nutrients and toxins that affect surface and ground water which in turn negatively affect terrestrial and marine and aquatic systems. Other pathways interact directly with the biomass and species of wild ecosystems decreasing viability and quantity of each. Pathways that affect the inflow and outflow of surface and groundwater may alter hydrologic conditions, which in turn may negatively affect ecological systems. All these pathways of interaction affect ecosystem health.

14.3 EMERGY, TRANSFORMITY, AND HIERARCHY

Given next are definitions and a brief conceptual framework of the emergy synthesis theory (Odum, 1996) and systems ecology (Odum, 1983) that form the basis for understanding ecological systems within the context of ecosystem health.

14.3.1 Emergy and Transformity: Concepts and Definitions

That different forms of energy have different "qualities" is evident from their abilities to do work. While it is true that all energy can be converted to heat, it is not true that one form of energy is substitutable for another in all situations. For instance, plants cannot substitute fossil fuel for sunlight in photosynthetic production, nor can humans substitute sunlight energy for food or water. It should be obvious that the quality that makes an energy flow usable by one set of transformation processes makes it unusable for another set. Thus quality is related to the form of energy and its concentration; where higher quality is somewhat synonymous with higher concentration of energy and results in greater flexibility. So wood is more concentrated than detritus, coal more concentrated than wood, and electricity more concentrated than coal.

The concept of emergy accounts for the environmental services supporting process as well as for their convergence through a chain of energy and matter transformations in both space and time. By definition, emergy is the amount of energy of one type (usually solar) that is directly or indirectly required to provide a given flow or storage of energy or matter. The units of emergy are emjoules (abbreviated eJ) to distinguish them from energy joules (abbreviated J). Solar emergy is expressed in solar emergy joules (seJ, or solar emjoules). The flow of emergy is empower, in units of emjoules per time. Solar empower is solar emjoules per time (e.g., seJ/sec).

When the emergy required to make something is expressed as a ratio to the available energy of the product, the resulting ratio is called a transformity[1]. The solar emergy required to produce a unit flow or storage of available energy is called solar transformity and is expressed in solar emergy joules per joule of output flow (seJ/J). The transformity of solar radiation is assumed to be equal to one ($1.0\,\mathrm{seJ/J}$). Transformities of the main natural flows in the biosphere (wind, rain, ocean currents, geological cycles, etc) are calculated as the ratio of total emergy driving the biosphere, as a whole, to the actual energy of the flow under consideration (Odum, 1996). The total emergy driving the biosphere is

---

[1]The transformity was originally proposed as a measure of energy quality (Odum, 1976) and referred to as the energy quality ratio and the energy transformation ratio, but it was renamed transformity in 1983 (Odum et al., 1983). The ratio of emergy to matter produced by a process (i.e., seJ/g) is termed specific emergy. The general term for transformities and specific emergy is emergy intensity.

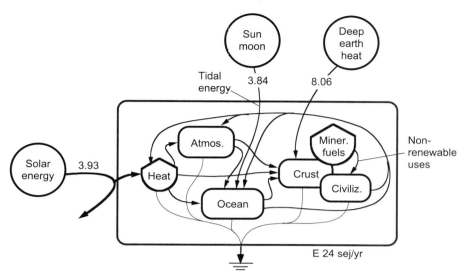

Figure 14.3 The main components of the bio-geosphere showing the driving energies and the interconnected cycling of energy and matter. The total emergy driving the bio-geosphere is the sum of solar, tidal and deep heat sources totaling 15.83 E24 seJ/year.

the sum of solar radiation, deep heat, and tidal momentum and is about 15.83 E24 seJ/year, based on a re-evaluation and subsequent recalculation of energy contributions done in the year 2000 (Odum et al., 2000)[2]. This total emergy is used as a driving force for all main biosphere scale processes (winds, rains, ocean currents, and geologic cycles), because these processes and the products they produce are coupled and cannot be generated one without the other (Figure 14.3).

Table 14.1 lists transformities (seJ/J) and specific emergy (seJ/g) of some of the main flows of emergy-driving ecological processes. Transformities and specific emergy given in the last column are ratios of the biosphere driving emergy in the second column to the annual production in the third column. Figure 14.3 shows in an aggregated way the emergy of the main biosphere flows that are, in turn, used to account for input flows to processes on smaller space-time scales, like processes in ecosystems as well as in human dominated systems (Ulgiati and Brown, 1999; Brown and Bardi, 2001; Brandt-Williams, 2002;

[2]Prior to 2000, the total emergy contribution to the geobiosphere that was used in calculating emergy intensities was 9.44 E24 seJ/yr. The increase in global emergy reference base to 15.83 E24 seJ/yr changes all the emergy intensities which directly and indirectly were derived from the value of global annual empower. Thus, to be consistent and to allow comparison with older values, emergy intensities calculated prior to the year 2000 are multiplied by 1.68 (the ratio of 15.83/9.44).

Table 14.1 Emergy of products of the global energy system (after Odum et al., 2000)

Product and units	Emergy* E24 seJ/year	Production units/year	Emergy/unit
Global latent heat: J	15.83	1.26 E24	1.3 E1 seJ/J
Global wind circulation: J	15.83	6.45 E21	2.5 E3 seJ/J
Hurricane: J	15.83	6.10 E20	2.6 E4 seJ/J
Global rain on land: g	15.83	1.09 E20	1.5 E5 seJ/g
Global rain on land (chem. pot.): J	15.83	5.19 E20	3.1 E4 seJ/J
Average river flow: g	15.83	3.96 E19	4.0 E5 seJ/g
Average river geopotential: J	15.83	3.40 E20	4.7 E4 seJ/J
Average river chem. potential: J	15.83	1.96 E20	8.1 E4 seJ/J
Average waves at the shore: J	15.83	3.10 E20	5.1 E4 seJ/J
Average ocean current: J	15.83	8.60 E17	1.8 E7 seJ/J

*Main empower of inputs to the geobiospheric system from Figure 14.1 not including nonrenewable consumption (fossil fuel and mineral use).

Kangas, 2002). The total emergy driving a process becomes a measure of the self-organization activity of the surrounding environment, converging to make that process possible. It is a measure of the environmental work necessary to provide a given resource. For example, the organic matter in forest soil represents the convergence of solar energy, rain, and winds driving the work processes of the forest over many years that has resulted in layer upon layer of detritus that ever so slowly decomposes into a storage of soil organic matter. It represents part of the past and present ecosystem's work that was necessary to make it available.

Example transformities of main ecosystem components are given in Tables 14.2 and 14.3. Table 14.2 lists components and processes of terrestrial ecosystems giving several transformities for each. Within each category transformities vary almost one order of magnitude reflecting the differences in total driving energy of each ecosystem type. The table is arranged in increasing quality of products from gross production to peat. Transformities increase in like fashion. An energy transformation is a conversion of one kind of energy to another kind. As required by the second law of thermodynamics, the input energies (sun, wind, rain, etc) with available potential to do work are partly degraded in the process of generating a lesser quantity of each output energy. With each successive step, a lesser amount of higher-quality resources are developed.

When the output energy of a process is expressed as a percentage of the input energy, an efficiency results. Lindeman (1942) efficiencies in ecological systems are an expression of the efficiency of transfer of energy between trophic levels. Table 14.3 lists transformities of trophic levels in the Prince William Sound of Alaska calculated from a food web and using Lindeman efficiencies of about 10% (Brown et al., 1993). The transformity, which is a ratio of the emergy input to the available energy output, is an expression of quality of the output energy; the higher the transformity, the more emergy is required to make it.

Table 14.2 Summary of transformities in terrestrial ecosystems

Ecosystem	Transformity (seJ/J)	Reference
Gross primary production		
Subtropical mixed hardwood forest, Florida	1.03E+03	Orrel, 1998
Subtropical forest, Florida	1.13E+03	Orrel, 1998
Tropical dry savannah, Venezuela	3.15E+03	Prado-Jutar and Brown, 1997
Salt marsh, Florida	3.56E+03	Odum, 1996
Subtropical depressional forested wetland, Florida	7.04E+03	Bardi and Brown, 2001
Subtropical shrub-scrub wetland, Florida	7.14E+03	Bardi and Brown, 2001
Subtropical herbaceous wetland, Florida	7.24E+03	Bardi and Brown, 2001
Floodplain forest, Florida	9.16E+03	Weber, 1996
Net primary production		
Subtropical mixed hardwood forest, Florida	2.59E+03	Orrel, 1998
Subtropical forest, Florida	2.84E+03	Orrel, 1998
Temperate forest, North Carolina (*Quercus* sp.)	7.88E+03	Tilley, 1999
Tropical dry savannah, Venezuela	1.67E+04	Prado-Jutar and Brown, 1997
Subtropical shrub-scrub wetland, Florida	4.05E+04	Bardi and Brown, 2001
Subtropical depressional forested wetland, Florida	5.29E+04	Bardi and Brown, 2001
Subtropical herbaceous wetland, Florida	6.19E+04	Bardi and Brown, 2001
Biomass		
Subtropical mixed hardwood forest, Florida	9.23E+03	Orrel, 1998
Salt marsh, Florida	1.17E+04	Odum, 1996
Tropical dry savannah, Venezuela	1.77E+04	Prado-Jutar and Brown, 1997
Subtropical forest, Florida	1.79E+04	Orrel, 1998
Tropical mangrove, Ecuador	2.47E+04	Odum and Arding, 1991
Subtropical shrub-scrub wetland, Florida	6.91E+04	Bardi and Brown, 2001
Subtropical depressional forested wetland, Florida	7.32E+04	Bardi and Brown, 2001
Subtropical herbaceous wetland, Florida	7.34E+04	Bardi and Brown, 2001
Wood		
Boreal silviculture, Sweden (*Picea aibes, Pinus silvestris*)	8.27E+03	Doherty, 1995
Subtropical silviculture, Florida (*Pinus elliotti*)	9.78E+03	Doherty, 1995
Subtropical plantation, Florida (Eucalyptus & *Malaleuca* sp.)	1.89E+04	Doherty, 1995
Temperate forest, North Carolina (*Quercus* sp.)	2.68E+04	Tilley, 1999
Peat		
Salt marsh, Florida	5.89E+03	Odum, 1996
Subtropical depressional forested wetland	2.52E+05	Bardi and Brown, 2001
Subtropical shrub-scrub wetland	2.87E+05	Bardi and Brown, 2001
Subtropical wetland	3.09E+05	Bardi and Brown, 2001

14.3.2 Hierarchy

A hierarchy is a form of organization resembling a pyramid where each level is subordinate to the one above it. Depending on how one views a hierarchy, it can be an organization whose components are arranged in levels from a top level (small in number, but large in influence) down to a bottom

Table 14.3 Summary of transformities in a marine ecosystem, prince william sound, alaska (after Brown et al., 1993)

Item	Transformity (seJ/J)
Phytoplankton	1.84E+04
Zooplankton	1.68E+05
Small nekton (molluskans, artropods, small fishes)	1.84E+06
Small nekton predators (fish)	1.63E+07
Mammals (seal, porpoise, belukha whale, etc)	6.42E+07
Apex predators (killer whale)	2.85E+08

level (many in number, but small in influence). Alternatively, one can view a hierarchy from the bottom where one observes a partially ordered structure of entities in which every entity but one is a successor to at least one other entity; and every entity except the highest entity is a predecessor to at least one other. In general, in ecology we consider hierarchical organization to be a group of processes arranged in order of rank or class in which the nature of function at each higher level becomes more broadly embracing than at the lower level. Thus we often speak of food chains as hierarchical in organization.

Most, if not all, systems form hierarchical energy transformation series, where the scale of space and time increases along the series of energy transformations. Many small-scale processes contribute to fewer and fewer larger-scale ones (Figure 14.4). Energy is converged from lower to higher order processes, and with each transformation step, much energy loses its availability (a consequence of the second law of thermodynamics), while only a small amount is passed along to the next step. In addition some energy is fed back,

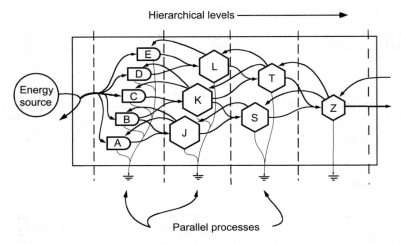

Figure 14.4 Diagram of the organization of systems showing the convergence of energy and matter into higher and higher levels via parallel and hierarchical processes.

reinforcing power flows up the hierarchy. Note in Figure 14.4 the reinforcing feedbacks by which each transformed power flow feeds backward so that its special properties can have amplifier actions.

14.3.3 Transformities and Hierarchy

Transformities are quality indicators, by virtue of the fact that they quantify the convergence of energy into products and account for the total amount of energy required to make something. Quality is a system property, which means that an "absolute" scale of quality cannot be made, nor can the usefulness of a measure of quality be assessed without first defining the structure and boundaries of the system. For instance, quality as synonymous with usefulness to the human economy is only one possible definition of quality — a "user-based quality." A second possibility of defining quality is one where quality increases with increased input. That is, the more energy invested in something, the higher its quality. We might describe this type of quality as "donor-based quality."

Self-organizing systems (be they the biosphere or an ecosystem) are organized with hierarchical levels (Figure 14.4) and each level is composed of many parallel processes. This leads to two other properties of quality: (a) parallel quality; and (b) cross quality.

In the first kind — parallel quality — quality is related to the efficiency of a process that produces a given flow of energy or matter within the same hierarchical level (comparison among units in the same hierarchical level in Figure 14.4). For any given ecological product (organic matter, wood, herbivore, carnivore etc) there are almost an infinite number of ways of producing it, depending on surrounding conditions. For example, the same tree species may have different gross production and yield different numbers and quality of fruit depending on climate, soil quality, rain, etc. Individual processes have their own efficiency, and as a result the output has a distinct transformity. Quality as measured by transformity in this case relates to the emergy required to make similar products under differing conditions and processes. Note in Table 14.2 where several transformities are given for each of the ecosystem products listed.

The second definition of quality — cross quality — is related to the hierarchical organization of the system. In this case, transformity is used to compare components or outputs from the different levels of the hierarchy, accounting for the convergence of emergy at higher and higher levels (see the comparison of transformity between different hierarchical levels in Figure 14.4). At higher levels, a larger convergence of inputs is required to support the component (a huge amount of grass is needed to support an herbivore, many kg of herbivore are required to support a predator, many villages to support a city, etc). Also, higher feedback and control ability characterize components at higher hierarchical levels, so that higher transformity is linked to higher control ability on lower levels. Therefore higher transformity, as equated with higher level in the hierarchy, often

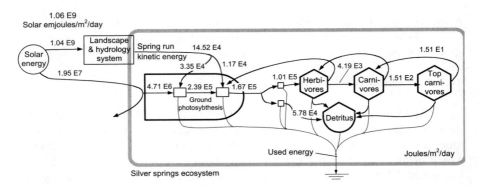

Figure 14.5   Aggregated systems diagram of the ecosystem at Silver Springs, Florida, showing decreasing energy with each level in the metabolic chain (after Odum, 2004). Table 14.5 gives the transformities that result from the transformations at each level.

Table 14.4   Solar transformities of ecosystem components of the Silver Springs

Item	Transformity (seJ/J)
Solar energy	1
Kinetic energy of spring flow	7170
Gross plant production	1620
Net plant production	4660
Detritus	6600
Herbivores	127,000
Carnivores	4,090,000
Top carnivores	40,600,000

means greater flexibility and is accompanied by greater spatial and temporal effects.

Figure 14.5 and Table 14.4 give energy and transformity values for an aggregated system diagram of Silver Springs, Florida. The data were taken from H.T. Odum's earlier studies on this ecosystem (Odum, 1957). Solar energy drives the system directly (i.e., through photosynthesis) and indirectly through landscape processes that develop aquifer storages, which provide the spring-run kinetic energy. Vegetation in the spring run uses solar energy and capitalizes on the kinetic energy of the spring, which brings a constant supply of nutrients. Products of photosynthesis are consumed directly by herbivores and are also deposited in detritus. Herbivores are consumed by carnivores who are, in turn, consumed by top carnivores. With each step in the food chain, energy is degraded.

14.3.4   Transformity and Efficiency

Transformities can sometimes play the role of efficiency indicators and sometimes the role of hierarchical position indicator. This is completely true in systems selected under maximum power principle constraints (Lotka, 1922a, 1922b; Odum, 1983) and is therefore true in untouched and healthy ecosystems. Things are different in an ecosystem stressed by an excess of outside pressure. Relations among components are likely to change, some components may disappear, and the whole hierarchy may be altered. The efficiency of given processes may change (they may decrease or increase) and some patterns of hierarchical control of higher to lower levels may diminish or disappear due to a simplified structure of the system. These performance changes translate into different values of the transformities, the variations of which become clear measures of lost or decreased system integrity.

When an ecological network is expressed as a series of energy flows and transformation steps where the transformation steps are represented as Lindeman efficiencies, the resulting transformities represent trophic convergence and a measure of the amount of solar energy required to produce each level in the hierarchy.

14.4 EMERGY, TRANSFORMITY AND BIODIVERSITY

In practice, the conservation of biodiversity suggests sustaining the diversity of species in ecosystems as we plan human activities that affect ecosystem health. Biodiversity has no single standard definition. Generally speaking, biodiversity is a measure of the relative diversity among organisms present in different ecosystems. "Diversity" in this case includes diversity within species (i.e., genetic diversity), among species, and among ecosystems. Another definition, is simply the totality of genes, species, and ecosystems of a region. Three levels of biodiversity have been recognized:

- Genetic diversity: diversity of genes within a species
- Species diversity: diversity among species
- Ecosystem diversity: diversity among ecosystem

A fourth level of biodiversity, cultural diversity, has also been recognized.

A main problem with quantifying biodiversity, especially in light of the definition above, is that there is no overall measure of biodiversity since diversity at various levels of an ecological hierarchy cannot be summed up. If they were summed up, bacteria and other small animals and plants would dominate the resulting diversity to the total neglect of the larger species. It therefore may be possible to develop a quantitative evaluation of total biodiversity within regions or ecosystems by weighting biodiversity at each hierarchical level by typical trophic level transformities (see, for instance, Table 14.5). In this way quantitative measures of biodiversity can be compared and changes resulting from species loss can be scaled based on transformities. A more realistic picture of total biodiversity may emerge and allow quantitative comparison of losses and gains that result from changes in ecological health.

Table 14.5 Transformities of information in forest components and the emergy to generate global biodiversity (after Odum, 1996; Ager, 1965)

Item	Solar transformity	Units
Forest scale		
DNA in leaves	1.2E+07	seJ/J
DNA in seeds	1.9E+09	seJ/J
DNA in species	1.2E+12	seJ/J
Generate a new species	8.0E+15	seJ/J
Global scale		
Generate global biodiversity	2.1E+25	seJ/species

14.5 EMERGY AND INFORMATION

Ecosystems create, store, and cycle information. The cycles of material driven by energy are also cycles of information. Ecosystems, driven by a spectrum of input resources generate information accordingly and store it in different ways (seeds, structure, biodiversity). The emergy cost of the generated information can be measured by a transformity value and may be a measure of healthy ecosystem dynamics. Odum (1996) suggested transformities for various categories of information within ecosystems and these are given in Table 14.5.

In healthy ecosystems (as well as in healthy human-dominated systems such as a good university) suitable emergy input flows contribute to generating, copying, storing, and disseminating information. In stressed ecosystems such as those where some simplification occurs due to improper loading from outside, the cycle of information is broken or impaired. In this case, the ecosystem exhibits a loss of information, which may manifest itself in simplification of structural complexity, losses of diversity, or decreases in genetic diversity (reduced reproduction).

There are two different concepts of information shown in Table 14.6. The first aspect refers to the emergy required to maintain information, as in the

Table 14.6 Number of species and average transformities of generalized compartments in the Everglades cypress ecosystem

Compartment	No. species*	Avg. transformity
Bacteria[a]	?	1–10
Primary producers[b]	250(est)	3.39E+04
Invertebrates	48	3.24E+05
Fishes	24	9.87E+05
Amphibians	14	1.16E+06
Mammals	20	3.35E+06
Reptiles	19	3.42E+06
Birds	59	3.76E+06

*after Ulanowicz et al. (2001).
[a]Jorgensen, Odum and Brown (2004).
[b]Lane et al. (2003).

maintenance of DNA in leaves (i.e., copying), and the maintenance of information of the population of trees (emergy in seed DNA which is the storing and disseminating information). The second concept is related to generating new information. When a species must be generated anew, the costs are associated with developing one from existing information sources such as trees within the same forest. However, the emergy required to generate biodiversity at the global scale, that is to generate all species anew, required billions of years and a huge amount of total emergy. Table 14.6 provides very average data for tropical forest ecosystems and of course represents only "order of magnitude" estimates of the costs of information generation, copying, storing, testing, and disseminating.

14.6 MEASURING CHANGES IN ECOSYSTEM HEALTH

Changes in ecosystem health can result from alterations in driving energy signature, inflows of a high-quality stressor such as pollutants, or unsustainable activity like overharvesting. In each case there is a consequent change in the pattern of energy flows supporting the ecosystem organization. An energy signature (see Figure 14.6) could change, resulting in ripples that could

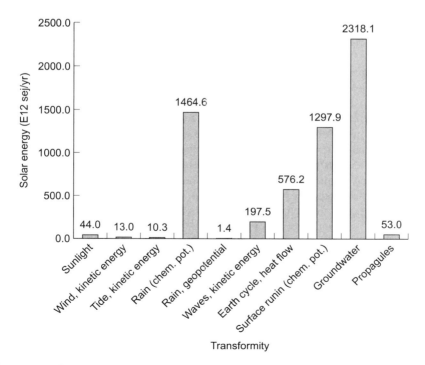

Figure 14.6 Emergy signature of driving energies for 1 ha of typical mangrove ecosystem in Florida.

Table 14.7 Transformities of selected metals as global flows to atmosphere and storages within a river ecosystem

	Annual releases to atmosphere[a] (seJ/J)	River ecosystem[b] (seJ/J)
Aluminum	9.65E+06	3.30E+07
Iron	8.46E+07	6.19E+07
Chromium	2.59E+10	1.99E+10
Arsenic	8.56E+11	—
Lead	2.39E+12	3.59E+10
Cadmium	1.52E+13	8.78E+10
Mercury	6.85E+14	—

[a]Not including human release.
[b]Genoni et al. (2003).

propagate through the ecosystem. If the change in signature is outside the normal range of fluctuations in the driving energy pattern, the effect is a change in the flows of energy and material throughout the ecosystem. Significant change in system organization might be interpreted as a change in ecosystem health. In general, chemicals (including metals, pollutants and toxins) have a high transformity (see Table 14.7) and as a result of an excess concentration, they are capable of instigating significant changes in ecosystem processes, which often result in a decline in ecosystem health. As transformities (emergy intensities) increase, their potential effect within ecosystem increases. Effects can be both positive and negative: transformity does not suggest the outcome that might result from the interaction of a stressor within an ecosystem, only that with high transformity, the effect is greater.

The ultimate effect of a pollutant or toxin is not only related to its transformity, but more importantly to its concentration or empower density (emergy per unit area per unit time; i.e., seJ/m^2 per day) in the ecosystem. Where empower density of a stressor is significantly higher than the average empower density of the ecosystem it is released into, one can expect significant changes in ecosystem function. For instance because of the very high transformities of most metals like those at the bottom of Table 14.7, their concentrations need only be in the parts per billion range to still have empower densities greater than most natural ecosystems. For instance, using the transformity of mercury in Table 14.7, and the exergy of mercury (Szargut et al., 1988), one can convert the transformity to a specific emergy of 3.7 E17 seJ/g. Using this specific emergy, and a mercury concentration of 0.001 ppb (the level the EPA considers to have chronic effects on aquatic life) the emergy density of the mercury in a lake would be 3.7 E12 seJ/m^2. This emergy density is about two orders of magnitude greater than the empower of renewable sources driving the lake ecosystem. Genoni et al. (2003) measured concentrations of 25 different elements in trophic compartments and in the physical environment of the Steina River in Germany (Table 14.6). They calculated transformities of each element based on global emergy supporting river ecosystem, which cycles the elements and their Gibbs energy. They

Table 14.8 Empower density of selected land use categories (after Brown and Vivas, 2004)

Land use	Empower density (E14 seJ/ha/yr)
Natural land/open water	7.0
Silviculture and pasture	10–25
High intensity pasture and agriculture	26–100
Residential and recreational uses	1000–3500
Commercial, transport, and light industrial	3700–5200
High intensity residential, commercial and business	8000–30,000

suggested that the tendency to bioaccumulate was related to transformity of the elements and the transformity of accumulating compartments (i.e., metals and heavy elements accumulated in high transformity compartments).

Empower density has been used as a predictor of impact of human dominated activities on ecosystems. In recent studies of the Florida landscape, Brown and Vivas (2004) showed strong correlations between empower density of urban and agricultural land uses with declines in wetland ecosystem health and pollutant loads in streams. Table 14.8 shows general empower densities of urban and agricultural land uses with natural wildlands for comparison. The empower densities of urban and agricultural land uses are from two to four orders of magnitude greater than the empower density of the natural environment.

A change in ecosystem health is manifested in changes in structural and functional relationships within the system of interest (region, landscape, ecosystem). Often the signs are subtle enough that change is difficult to detect. In other circumstances indicators are not sensitive enough to detect change or to discern changes in health from "normal variability." Network analysis of the flows of emergy on pathways of ecological systems may add insight into changes in ecosystem health. Using the data from Silver Springs in Figure 14.5, a network analysis of changes in emergy flows and cycling that results from removing the top carnivores (Table 14.9) shows changes in overall cycling emergy of about 15% at the top end of the food chain and diminishing effect cascading back downward toward the bottom. The analysis uses a matrix technique to assign emergy to pathways and includes cycling so that feedbacks within the system are accounted for. Evaluation of the changes in pathway emergy may provide a tool that can help in measuring changes in overall ecosystem health with alterations of components or elimination of trophic levels within the system.

14.7 RESTORING ECOSYSTEM HEALTH

Restoration of ecosystems falls within the sphere of ecological engineering. Ecological engineering is the design and management of self-organizing

Table 14.9 The effect of changes in system organization resulting from loss of top carnivore (Silver Springs, Florida data)

Item	Transformity (seJ/J)	Pathway emergy with top carn.[a] (seJ/m²/day)	Pathway emergy without top carn.[b] (seJ/m²/day)	Percent change
Solar energy	1	NC	NC	NC
Kinetic energy of spring flow	7170	NC	NC	NC
Gross plant production	1620	3.87E+08	3.84E+08	0.8%
Net plant production	4660	4.71E+08	4.68E+08	0.6%
Detritus	6600	6.67E+08	6.58E+08	1.4%
Herbivores	127,000	5.32E+08	5.20E+08	2.3%
Carnivores	4,090,000	6.13E+08	5.20E+08	15.2%
Top carnivores	40,600,000	6.13E+08	0	100.0%

[a]Emergy on pathways of the system depicted in Figure 14.5. Emergy is calculated using a network analysis method (Odum, 2002).

[b]Emergy on pathways of the system depicted in Figure 14.5 when the top carnivore is excluded. Emergy is calculated using a network analysis method (Odum, 2002).

ecosystems that integrate human society with its natural environment for the benefit of both. The restoration of damaged ecosystems, while resulting in benefits for humanity (increased ecological services) is also necessary to maintain landscape-scale information cycles and ultimately, biodiversity. The value of active restoration can be measured as the decrease in the time required to restore ecosystem functions to levels characteristic of levels prior to disturbance. The graph in Figure 14.7 illustrates the concept of a net benefit from ecological restoration. The difference between the upper and lower lines in the graph is the benefit of restoration. If the benefit is divided by the costs of restoration, a cost/benefit ratio results.

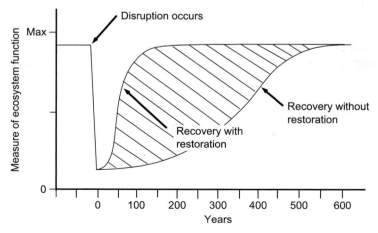

Figure 14.7 Graph illustrating the net benefit from ecological restoration. The net benefits can be calculated as the difference between recovery of ecosystem function with and without restoration efforts.

Table 14.10 Emergy costs for restoration of forested wetland in Florida (after Bardi, 2001)

Item	Data*	Units	Unit emergy values (seJ/unit)	Emergy (E15 seJ)
Environmental flows				
Sunlight	4.2E+13	J/yr	1	2.10
Wind	3.0E+09	J/yr	2.5E+03	0.38
Rain, chemical potential	6.4E+10	J/yr	3.1E+04	97.60
Total				97.70
Construction flows				
Planting material	8.4E+07	J	6.7E+04	0.01
Services	8.7E+02	$	1.7E+12	1.48
Fertilizer	6.7E+03	g	4.7E+09	0.03
Services	1.0E+02	$	1.7E+12	0.17
Labor (unskilled)	3.1E+07	J	4.2E+07	1.30
Labor (skilled)	5.4E+07	J	1.2E+08	6.61
Services	4.1E+03	$	1.7E+12	7.01
Total				16.61
Management				
Chemicals (herbicides)	1.9E+04	g	2.5E+10	0.47
Labor (unskilled)	2.3E+07	J	4.2E+07	0.96
Labor(skilled)	4.6E+07	J	1.2E+08	5.63
Total				7.06

*based on assumption of 50-year recovery time.

Stressed or damaged ecosystems may be rejuvenated or restored by removal of stresses or in the case of significant losses, by reconstruction. Table 14.10 gives data for the construction of a forested wetland system in Florida. The data are given for a 50-year time period assuming that 50 years are required to develop a relatively mature forested wetland. While the inputs of nonrenewable and human-dominated resources are significant, over the 50-year time frame of the restoration effort, renewable emergy dominates.

14.8 SUMMARY AND CONCLUSIONS

Emergy and transformity are useful measures that may be applied to concepts of ecosystem health. Transformity measures the convergence of biosphere work into processes and products of ecosystems and as such offers the opportunity to scale ecosystem and their parts based on the energy required to develop and maintain them. Ecosystems are composed of physical structure (i.e., wood, biomass, detritus, animal tissue, etc) and information found in both its genetic makeup, as well as relationships and connections between individuals and groups of individuals. Declines in ecosystem health are manifested in changes in the quality and quantity of relationships and connections between individuals. Stressors may change driving energies pathways, and connections.

When one component in a system is affected, the energy and matter flows in the whole system change, which may translate into declines in ecosystem health. We suggest in this chapter that changes in ecosystem structure and functions are reflected in changes of emergy flows and the corresponding transformities of system components. We also suggest that there may be a relationship between the empower density of urban and agricultural lands and their effects on ecosystem health. The effect of a stressor may be predicted by its empower density. Changes in ecosystem structure translate into changes in pathway empower and thus quantifying changes on networks may provide quantitative evaluation of changes in ecosystem health.

REFERENCES

Ager, D.U. *Principles of Paleontology*. McGraw-Hill, New York, 1965.

Brandt-Williams, S. "Emergy of Florida agriculture," in *Handbook of Emergy Evaluation*, folio no. 4. Center for Environmental Policy, University of Florida, Gainesville, FL, 2002, 93 p. (See http://www.ees.ufl.edu/cep/).

Brown, M.T. and Bardi, E. "Emergy of ecosystems," in *Handbook of Emergy Evaluation*, folio no. 3. Center for Environmental Policy, University of Florida, Gainesville, FL, 2002, 93 p. (See http://www.ees.ufl.edu/cep/).

Brown, M.T. and Vivas, M.B. A landscape development intensity index. *Ecological Monitoring and Assessment* (in press).

Brown, M.T. and Ulgiati. S. Emergy based indices and ratios to evaluate sustainability: monitoring technology and economies toward environmentally sound innovation. *Ecological Engineering* 9, 51–69, 1997.

Brown. M.T., Woithe, R.D., Montague, C.L., Odum, H.T., and Odum, E.C. *Emergy Analysis Perspectives of the Exxon Valdez Oil Spill in Prince William Sound, Alaska*. Final report to the Cousteau Society. Center for Wetlands, University of Florida, Gainesville, FL, 1993, 114 p.

Collins, D. and Odum, H.T. Calculating transformities with an eigenvector method. *Emergy Synthesis: Proceedings of the Emergy Research Conference*, Brown, M.T., Ed. Center for Environmental Policy, University of Florida, Gainesville, FL, 2001, pp. 265–280.

Doherty, S.J. Emergy Evaluations of and Limits to Forest Production. PhD dissertation, Department of Environmental Engineering Sciences University of Florida. Gainesville, FL, 1995.

Doherty, S.J., Scatena, F.N., and Odum, H.T. Emergy Evaluation of the Luquillo Experimental Forest and Puerto Rico. Final report submitted to the International Institute of Tropical Forestry, cooperative project 19-93-023. University of Florida, Gainesville, FL, 1994, 98 p.

Ehrenfeld, D. *From Beginning Again: People and Nature in the New Millennium*. Oxford University Press, New York, 1993.

Genoni, G.P., Meyer, E.I., and Ulrich, A. Energy flow and elemental concentrations in the Steina River ecosystem (Black Forest, Germany). *Aquatic Sciences* 6, 143–157, 2003.

Huang, S.L. and Odum, H.T. Ecology and economy: emergy synthesis and public policy in Taiwan. *Journal of Environmental Management*, 32, 313–333, 1991.

Huang, S.-L. and Shih, T.-H. The evolution and prospects of Taiwan's ecological economic system. *Proceedings of the Second Summer Institute of the Pacific Regional Science Conference Organization*. Chinese Regional Science Association, Taipei, 1992.

Kangas, P.C. "Emergy of landforms," in *Handbook of Emergy Evaluation*, folio no. 4. Center for Environmental Policy, University of Florida, Gainesville, FL, 2002, 93 p. (See http://www.ees.ufl.edu/cep/).

Keitt T.H. Hierarchical Organization of Energy and Information in a Tropical Rain Forest Ecosystem. MSc thesis, University of Florida, Gainesville, FL, 1991, 72 p.

Lane, C., Brown, M.T., Murray-Hudson, M., and Vivas, B. Florida Wetland Condition Index (FWCI): Biological Indicators for Wetland Condition of Herbaceous Wetlands in Florida. Final report to Florida Department of Environmental Protection. Center for Wetlands, University of Florida, Gainesville, FL, 2003.

Leopold, A. *A Sand County Almanac, and Sketches Here and There*. Oxford University Press, New York, 1949.

Lindeman, R.L. The trophic-dynamic aspects of ecology. *Ecology* 23, 399–418, 1942.

Lotka, A.J. Contribution to the energetics of evolution. *Proceedings of the National Academy of Sciences U.S.A* 8, 147–151, 1922a.

Lotka, A.J. Natural selection as a physical principle. *Proceedings of the National Academy of Sciences U.S.A.* 8, 147–151, 1922b.

Odum, H.T. Trophic structure and productivity of Silver Springs, Florida. *Ecol. Monogr.* 27, 55–112, 1957.

Odum, H.T. *Systems Ecology: An Introduction*. Wiley, New York, 1983.

Odum, H.T. Self organization, transformity, and information. *Science* 242, 1132–1139, 1988.

Odum, H.T. *Environmental Accounting. Emergy and Environmental Decision Making*. Wiley, New York, 1996.

Odum, H.T. "Emergy of global processes," in *Handbook of Emergy Evaluation*, folio no. 2. Center for Environmental Policy, University of Florida, Gainesville, FL, 2002, 93 p. (See http://www.ees.ufl.edu/cep/).

Odum, H.T. *Environment, Power and Society*, 2nd ed. University of Columbia Press, New York, 2004.

Odum, H.T. and Arding, J. *Emergy Analysis of Shrimp Mariculture in Ecuador*. Coastal Resources Center, University of Rhode Island, Narragansett, RI, 1991.

Odum, H.T., Brown, M.T., and Williams, S.B. "Introduction and global budget," in *Handbook of Emergy Evaluation*, folio no. 1. Center for Environmental Policy, University of Florida, Gainesville, FL, 2002, 93 p. (See http://www.ees.ufl.edu/cep/).

Orrell, J.J. Cross Scale Comparison of Plant Production and Diversity. MSc thesis, University of Florida, Gainesville, 1998.

Prado-Jartar, M.A. and Brown, M.T. Interface ecosystems with an oil spill in a Venezuelan tropical savannah. *Ecological Engineering* 8, 49–78, 1996.

Szargut, J., Morris, D.R., and Steward, F.R. Exergy analysis of thermal, chemical and metallurgical processes. Hemisphere Publishing Corporation, London, 1988.

Tilley, D.R. Emergy Basis of Forest Systems. PhD dissertation, University of Florida, Gainesville, 1999.

Ulanowicz, R., Heymans, S., Bondavalli, C., and Egnotovich, M.S. *Network Analysis of Trophic Dynamics of South Florida Ecosystems*. University of Maryland Center for Environmental Science, Chesapeake Biological Laboratory, 2001. (See http://www.cbl.umces.edu/~atlss/ATLSS.html).

Ulgiati, S. and Brown, M.T. "Emergy accounting of human-dominated, large scale ecosystems," in *Thermodynamics and Ecology*, Jørgensen, S.E. and Kay, Eds. Elsevier, 1999.

Ulgiati S., Odum, H.T., and Bastianoni, S. Emergy use, environmental loading and sustainability: an emergy analysis of Italy. *Ecological Modelling* 73, 215–268, 1994.

Ulgiati S., Brown, M.T., Bastianoni, S., and Marchettini, N. Emergy based indices and ratios to evaluate the sustainable use of resources. *Ecological Engineering* 5, 519–531, 1995.

Weber, T. Spatial and Temporal Simulation of Forest Succession with Implications for Management of Bioreserves. MSc thesis, University of Florida, Gainesville, 1994.

CHAPTER 15

Mass Accounting
and Mass-Based Indicators

S. Bargigli, M. Raugei, and S. Ulgiati

The accounting of material flows, which are diverted from their natural pathways to support modern societal metabolism, is of key importance for the evaluation of the related impacts on the environment, both on a local and a global scale. In fact, there is a close relationship between resource use and environmental impacts and therefore the evaluation of resource use (and the related hidden flows) can be considered an aggregated indirect measure of ecosystem disturbance. Among the several different mass-based methods and indicators, MAIA (material intensity analysis), for local scale evaluations, and nationwide material flow analysis (on the national and international levels) represent in the opinion of the authors the most relevant examples, because of their widespread use in Europe. By means of indirect mass-flow accounting, reference is made to the extent to which the technological choices of industrialized countries affect the environmental integrity of primary producing countries. Possible synergies from increased integration with other methods for natural-resource accounting are highlighted.

1-56670-665-3/05/$0.00 + $1.50

15.1 INTRODUCTION

An economic system is *environmentally sustainable* only as long as it is physically in a (dynamic) steady state — that is, the amount of resources utilized to generate welfare is permanently restricted to a size and a quality that does not overexploit the storage of resources, or overburden the sinks, provided by the ecosphere.

As current experience with several environmental issues indicates, we are already at or even beyond the limits of the Earth's *carrying capacity*, mainly due to the exploitation of large fractions of biophysical storages. For instance the 2003 State of The World Survey published by the Worldwatch Institute points out the huge environmental consequences of the ongoing overexploitation of mineral resources. In addition, the report underlines that for some metals the amount already extracted exceeds the estimated amount of existing underground reserves (Sampat, 2003).

Due to the technical skills of humankind and the material growth of the anthroposphere, an infinite number of ever-changing, disruptive interactions can occur at the boundaries of the ecosphere. Moreover, these impacts are characterized by *nonlinear relationships* between stresses and responses. An unknown quantity of these effects cannot usually be detected within human time horizons, and even if this would be the case, they could not be easily attributed to distinct causes. This precludes the observation or even the theoretical calculation — and thus quantification — of the totality of the actual consequences of human activities on ecosystems.

Since neither the carrying capacity nor the critical load can ever be precisely determined, the political application of these natural science-based concepts must necessarily take into account the *precautionary principle*. Decision makers should adopt this approach and keep the economy within a sustainable framework. Economists and scientists should provide proper tools of evaluation.

A widely accepted theoretical framework for explaining the physical relationship of society and nature is the so-called *socio-economic metabolism*, a concept applied to investigate the interactions between social and natural systems. It is the socio-economic metabolism (see Fischer-Kowalski, 1997) that exerts pressure upon the environment. It comprises the extraction of materials and energy, their transformation in the processes of production, consumption, and transport and their eventual release into the environment. Other different frameworks have been suggested (emergy synthesis by H.T. Odum, 1996; cumulative exergy accounting by Szargut and Morris, 1987; ecological footprint by Wackernagel and Rees, 1996), but will not be dealt with in this chapter.

The basis of socio-economic metabolism approach is the accounting of the material flows of resources. This material flow analysis (MFA) accounts for the overall material input which humans use, move or take away while generating products and services. Consequently, it can be used as a direct measure of the exploitation of natural resources (soil excavation, water withdrawal, biotic material degradation, etc.) and, from a precautionary principle point of view, an indirect measure of environmental impact (ecosystem stress, alterations of local climate, loss of biodiversity). This method has gained wide acceptance due to its simplicity and straightforwardness.

15.1.1 Targets of Material Flow Accouting

Two main approaches for assessing indirect flows associated to human activities can be identified. In most studies carried out so far, the calculation of indirect flows was limited to a simplified life-cycle analysis (LCA) of products or product groups (MAIA or material intensity analysis, Ritthof et al., 2002). A second approach applies input-output analysis on the national level (macroscale), extended to the environmental dimension (nationwide MFA, also known as bulk MFA, Eurostat, 2001). Table 15.1 shows the main differences between the two types of analysis.

15.2 MAIA: GENERAL INTRODUCTION TO THE METHODOLOGY

15.2.1 Historical Background

Material flow analysis builds on earlier concepts of material and energy balancing, as presented by Ayres, for example (Ayres and Kneese, 1969). The so-called material intensity analysis was originally developed at the Wuppertal Institute (WI) in Germany in 1992. Although the principles that form the basis of this methodology have gained wide acceptance (ecological rucksack, Schmidt-Bleek, 1992, 1994; factor 4 and factor 10, Von Weinsacker et al., 1998), the methodology itself remained confined to northern Europe, especially German-speaking countries, for almost a decade. Today MAIA is finally "crossing the border" and is applied by many LCA analysts, mainly throughout Europe. LCA is in fact a comprehensive framework that comprises a thorough "inventory" of input and output flows as well as the

Table 15.1 Types of MFA analysis

Type	Target	Description	Objective
MAIA	Intermediate or final products and services.	Analyzes the direct and indirect material inputs, including energy, which are required to produce a product.	To calculate the ecological rucksack of a product.
Nationwide MFA (bulk MFA)	National economy or economy sector.	Analyzes the flows which constitute the basis of an economy or a sector.	To find out which sectors and economies have the highest material basis and to find out the relation between material basis and imports/exports.

"determination of the environmental impacts." MFA can provide a useful tool of evaluation in both LCA stages.

15.2.2   The MAIA Method

The method is based on a careful inventory of material flows to a process. Since a crucial aspect of the method is the classification of such flows as well as the boundary of the analyzed system, it is of paramount importance to clarify what kind of flows are considered:

15.2.2.1   *Used versus Unused*

The category of used materials is defined as the amount of extracted resources, which enters the process or an economic system for further processing or consumption. All used materials become (part of) products exchanged within the economic system. Unused extraction refers to materials that never enter the economic system and thus can be described as physical market externalities (Hinterberger et al., 1999). This category comprises, for example, overburden and parting materials from mining, by-catch, wood-harvesting losses from biomass extraction, soil excavation and dredged materials from construction activities.

15.2.2.2   *Direct versus Indirect*

Direct flows refer to the actual weight of the products and thus do not take into account the life-cycle dimension of production chains. Indirect flows, however, indicate all materials that have been required for manufacturing (upstream resource requirements) and comprise both used and unused materials.

Each material and energy flow to a process or a system is multiplied by a suitable intensity factor, which accounts for all the indirect hidden material flows over its whole production chain. The sum of all direct and indirect flows calculated yields an estimate of the total amount of matter moved and processed for this purpose. Intensity factors are calculated separately or taken from literature.

This method of calculating (used and unused) indirect material flows required in the life cycle of a product leads to the so-called "ecological rucksack" (Schmidt-Bleek, 1992, 1994). Tish can be defined as "the total sum of all materials which are not physically included in the marketable output under consideration, but which were necessary for production, use, recycling and disposal. Thus, by definition, the ecological rucksack results from the life-cycle-wide material input (MI) minus the mass of the product itself," (Spangenberg et al., 1998).

MAIA is currently widely used to quantify the life-cycle requirement of primary materials for products and services. Analogously to the quantification of the embodied energy requirements, MAIA provides information on basic environmental pressures associated with the magnitude of resource extraction and the subsequent material flows that end up as waste or emission.

According to the concept of the ecological rucksack, a set of distinct indicators have been developed (Ritthof et al., 2002). These are:

- Material input (MI): the sum (measured in physical units; e.g., ton) of all the resources used to produce a given amount of product
- Material intensity (MIT): the material input (MI) expressed per unit of product (e.g: t/t, t/kWh, t/tkm)
- Material input per service unit (MIPS): the material input (MI) referred to the amount of product which is able to provide a given final service to the user. The service ability of a product is very variable and has to be defined case by case

MI, MIT and MIPS are generally differentiated in five main categories:

- Abiotic raw materials
- Biotic raw materials
- Soil removal
- Water
- Air

These categories are presented in details in Table 15.2.

Table 15.2 MAIA categories

Category	Description
Abiotic raw materials	This category covers all minerals and ores extracted in mining operations but also the overburden and other earth movements. It also includes all kinds of fossil fuels used, expressed in mass units. This category becomes particularly relevant for metals and other industrial products as their processing usually implies a considerable use of fossil fuels and a significant amount of overburden to be produced.
Water	All actively extracted or diverted water flows are accounted for in this category. This also includes the extraction of ground and surface water, cooling water in power generation and industries, water for irrigation in agriculture, but also rivers diverted to other places and water running off from sealed areas (controlled landfills). This category indicates the influence on ecosystems due to changes in water flows rather than the direct water pollution.
Air	All air chemically or physically processed or converted into another physical state is measured in this category. This is strongly correlated with fossil fuel combustion and the CO_2 emissions as principal gaseous output of processes. Thus, this category indirectly reflects the potential mobilization of atoms which are up to now bound in the lithosphere in the "reserve pool" (e.g., carbon in fossil fuels).
Biotic raw materials	It covers all biomass which is altered but not used during any economic activity, sometimes called "unused extraction" (e.g., a forest clear-cut for mining purposes). Some authors also include in this category all the products of modern agriculture and forestry.
Soil removal	It accounts for human-induced erosion especially due to agricultural and mining activities. This category as well as the biotic one are not yet widely used because of the intrinsic higher difficulty in calculating the intensity factors.

15.2.3 Calculation Rules

When only one product is produced in a given process, all the material inputs (A, B, . . .,) used along the whole production chain (as well as their ecological rucksacks, if any) are assigned to that product (see Figure 15.1). If two or more co-products are produced in the same process, the ecological rucksack is generally distributed to those products according to their mass fraction (see Figure 15.2). However, when energetic outputs and services are considered, other parameters like energy content or price might serve as the basis for the allocation.

Figure 15.1 MAIA calculation rule I.

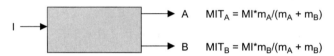

Figure 15.2 MAIA calculation rule II.

Waste or nonmarketable byproducts are not assigned any ecological rucksack by definition (Stiller, 1999). They only bear the additional inputs needed for their further processing. This implies that if a sufficiently efficient recycling process exists for them, secondary materials will have lower material intensities than the primary ones and this will make them favorable as alternative inputs in other industrial processes. Other methods such as emergy synthesis assign an emergy content to any byproduct still characterized by available energy (exergy) different to zero, in so recognizing that it is a potential resource, no matter whether it is presently marketable or not.

If two or more products and byproducts (A and B), produced in the same production chain, converge into a further new process for the production of a product (P), both of them will contribute to the MI of the product P (i.e., their ecological rucksacks are summed like if they were originated by two independent processes). This does not imply double counting because of calculation rule II (see Figure 15.3)).

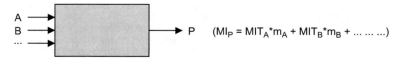

Figure 15.3 MAIA calculation rule III.

It is also common practice to keep the cumulative electricity input separate in the calculation of the final material intensity factors, since its contribution in terms of material intensities is often overwhelmingly large and would easily hide all the other contributions, thus causing the loss of useful information.

Moreover, the material intensity factors of electricity are highly dependent on the kind of technology and fuels used to produce it and thus very variable. Therefore, it is recommendable to specify which "kind" of electricity has been considered in the analysis before presenting aggregated data on the ecological rucksack of a given product. Comparison among several different products on an MIT basis (as defined above) should instead be made by using aggregated factors, which include the electricity rucksacks calculated in the same way. Some analyses have focused on the evaluation of the ecological rucksack of fossil fuels and electricity mix of several European countries (Frischknecht et al., 1996; Manstein et al., 1996; Hacker, 2003). This is very useful for the analysts in order to take into proper consideration the influence of the energy sources (i.e., fossil fuels, renewables or nuclear power) used to produce the electricity on the material intensity of the final products.

15.2.4 MAIA Database

The Wuppertal Institute for Climate, Environment and Energy in Germany has been one of the central institutions in the development of a standardized methodology for MFA and today is one of the most important sources for material flow data. Ongoing research projects, tutorials, as well as spreadsheets with a number of "rucksack factors," mostly for abiotic raw materials, building and construction materials, and selected chemical substances, can be downloaded from the website of the Wuppertal Institute (see http://www.wupperinst.org/Projekte/mipsonline/). However, these downloadable files do not contain any detailed description concerning the calculation procedure used. Thus an important piece of information is not available to the user. Very few MFA papers have been published in international journals up to date so that the interested reader can only refer to the Wuppertal working papers, most of which are in German.

Apart from the Wuppertal Institute, other research groups have investigated the material and energy requirements of resource extraction and processing. In particular, the study series "Material flows and energy requirements in the extraction of selected mineral raw materials," published by the German Federal Geological Institute (see Kippenberger, 1999, for an executive summary) provides detailed information on the resource inputs for the extraction, processing and transportation of eight of the most important mineral resources.

15.2.5 Selected Case Studies: Fuel Cells and Hydrogen

Calculating indirect flows for semi-manufactured and finished products by applying MAIA requires the collection of an enormous amount of data for

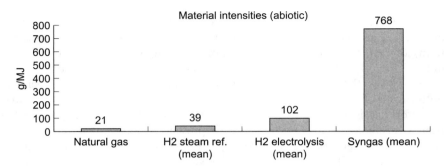

Figure 15.4 Comparison among the abiotic factors of natural gas, hydrogen and syngas.

every product under consideration. Thus rucksack factors have only been published for a very small number of finished products (Stiller, 1999; Bargigli et al., 2003; Raugei et al., 2003). Two case studies have been selected in order to show the kind of information provided by MAIA:

1. A comparison among selected fuels (hydrogen, syngas, and natural gas) (Bargigli et al., 2003)
2. A 500-kW molten carbonate fuel cell (MCFC) plant production and operation and its comparison to other plants (Raugei et al., 2003).

Figure 15.4 shows the comparison among the abiotic factors of natural gas, hydrogen (produced via steam reforming and water electrolysis) and syngas produced via coal gasification.

The three energy carriers are comparable in terms of possible use for many applications; for example, in fuel cells, but the abiotic material intensity factor of syngas is considerably higher than the others (the other factors are not shown). This is partly due to the high abiotic factor of coal compared to natural gas and oil and indicates that syngas production from coal has a higher load on the environment. The precautionary principle allows us to consider that the more material flows are diverted from their natural pathways, the higher environmental impact may result. The evaluation of process emissions confirms the above consideration: 78.2 g of solid emissions are produced per MJ of syngas while the others have negligible amounts of local solid emissions. It is also important to note that these solid emissions are mainly composed of ashes and coal tars, which are rich in carcinogenic polynucleated aromatic hydrocarbons (PAHs), and can cause serious ecotoxicological problems to the environment in the area surrounding the plant if they are simply dumped in a landfill.

Another interesting example is provided by the application of MAIA to MCFCs. The MCFC production implies the use of rare metals such as nickel, chromium, and lithium for fuel cells components, which require the excavation of large amounts of overburden and provide considerable disturbance to the ecosystems at the mining sites. Furthermore, the processing of these metals is generally energy intensive and therefore indirectly requires a large use of natural resources.

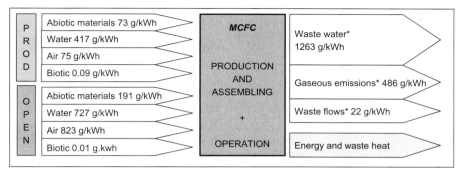

Figure 15.5 Material Flows to and from a 500-kw MCFC pilot module. Note: The imbalance between the total outputs and the total inputs in Figure 15.5 is due to the very theory of MFA. In fact, inputs are calculated as total embodied material flows (ecological rucksacks), whereas output are those physically released on all the production sites of the components of the finished product. It is possible that some of the emissions have not been fully accounted for due to incomplete available data. A further source of imbalance is due to the materials stored in the plant infrastructure, which will only be released at the end of the structure life cycle.

Figure 15.5 shows the input and output flows to and from a 500-kW MCFC pilot module, expressed per kWh of electricity delivered along its whole life cycle. These material intensity factors (classified in abiotic, water, air and biotic factors) are presented separately for the production phase and the operation phase. It is apparent that, at least in mass terms, the production and assembling phase is not the one that causes the major environmental load, due to the dominant role of the fuel in the operation phase. The mass of the module is 45 tons, but it required an indirect material flow equal to 1120 tons, including structure steel and NG for start up operations. This translates into a total mass processed of 1165 tons, 69% of which is indirect input from outside of Italy. If we disregard structure and start up inputs, and only focus on the materials used to manufacture the active components of the fuel cells, it emerges that 99% of their indirect material flows are generated abroad. Since the latter involve over their whole life cycle non-negligible amounts of potentially toxic substances, which need to be dealt with carefully, it is very likely that these emissions occur far from the assembling and use sites.

In general, this is particularly true for all the materials used in advanced technologies. Their extraction and primary processing stages usually take place in countries different than those of final users (HIPCs (Heavily Indebted Poor Countries) for ore extraction, developing countries for assembling and preprocessing), where this is more economically profitable and where environmental laws are less strictly enforced. They are then processed and assembled in developed countries. This implies that if we only look at the production chain that "physically" takes place within a developed nation's territory, it may appear that the analyzed process causes only minor environmental problems, due to the fact that the major environmental impacts are located abroad. To evaluate the life cycle environmental impact of a product or service, requires that reliable data are available concerning the first part of the production chain as well as its transportation costs and routes. If these data are lacking, the applicability and the meaning of the approach are limited, especially as far as highly technological industrial products are concerned.

15.3 NATIONWIDE MFA: GENERAL INTRODUTION TO THE METHODOLOGY

15.3.1 Historical Background of Bulk MFA

The first material flow accounts on the national level have been presented at the beginning of the 1990s for Austria (Steurer, 1992) and Japan (Japanese Environment Agency Japan, 1992). Since then, MFA has become a rapidly growing field of scientific interest, and major efforts have been undertaken to harmonize the methodological approaches developed by different research teams. The Concerted Action "ConAccount" (Bringezu et al., 1997; Kleijn et al., 1999), funded by the European Commission, was one of these milestones in the international harmonization of MFA methodologies. A second cooperation led by the World Resources Institute (WRI), brought together MFA experts for the investigation of the material basis of several industrialized countries. In their first publication (Adriaanse et al., 1997) the material inputs of four industrial societies (U.S., Germany, The Netherlands, Japan) have been assessed and guidelines for resource input indicators have been defined. Their second study (Matthews et al., 2000) focused on material outflows and introduced emission indicators.

Finally, with the publication of a methodological guide "Economy-Wide Material Flow Accounts and Derived Indicators" by the European Statistical Office (Eurostat (2001)), an officially approved harmonized standard was reached.

15.3.2 The Bulk MFA Model

Monitoring the transition of modern societies towards a path of sustainable development requires comprehensive information on the relationships between economic activities and their environmental impacts. Physical accounting systems fulfill these requirements by (a) describing these relationships in biophysical terms, and (b) by being compatible with the standard system of local and national economic accounting. Resource use indicators derived from physical accounts play a major role in environmental and sustainability reporting (Spangenberg et al., 1998). A substantial reduction of the resource throughput of societies by a factor of 10 or more (also referred to as a strategy of "dematerialization" (Hinterberger et al., 1996) was suggested as a requirement for achieving sustainability (Schmidt-Bleek, 1994). Resource flow-based indicators help monitoring progress towards this goal.

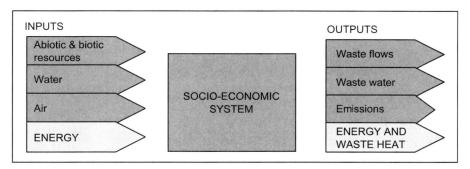

Figure 15.6 The basic model bulk MFA.

The basic principle is that the economy is physically embedded into the environment that is, the economy is an open system with regard to matter and energy (see Figure 15.6).

In economy-wide MFA the whole economy including production and consumption activities is a single black box. Only flows that cross the system boundary of the economy are recorded.

15.3.3 The System Boundaries and System Stock

Two main boundaries for resource flows can be defined for the accomplishment of an MFA on the national level. The first is the interface between the economy and the domestic natural environment, from which resources (materials, water, air) are extracted and to which they are released after being used. The second kind of boundary is the political border of a country, crossed by imported and exported commodities.

15.3.3.1 Boundary between the Economy and the Natural Environment

For a consistent compilation of an economy-wide material flow account, it is at first necessary to define exactly where the boundary between the economic and the environmental system is to be set (i.e., which elements of the material world belong to society and which to nature) as only resources crossing this border will be accounted for.

Every part of the material world produced by (or periodically maintained by) human labor are part of the material components of society. This implies that human bodies, livestock, and all man-made infrastructures along with all their complete metabolism has to be included in society's metabolism. As a consequence, products from livestock, like meat and milk, as well as the waste generated in the process, are not to be treated as inputs but as internal transfers within the socio-economic system.

However, experience suggests that these stocks are very small compared to other stocks such as buildings, machinery, or consumer durables and also do not change much over time. In practice, therefore, human bodies and livestock and their changes may be ignored unless there is evidence that these stocks change rapidly.

Same theoretical considerations could be raised about whether to include plants as a component of the socio-economic system, as they are maintained by labor in agriculture and forestry. For pragmatic reasons it was suggested not to consider plants as a component of the socio-economic system (Fischer-Kowalski, 1997). Therefore, plant harvest can be seen as an input to the socio-economic system whereas manure and fertilizers are an output to nature. Eurostat (2001) recommends treating forests and agricultural plants as part of the environment in economy-wide MFA and the harvest of timber and other plants as material inputs. This correspond to the economic logic of the System of National Accounts (SNA) and to economic statistics. As described in the System of Environmental and Economic Accounts (SEEA) (United Nations, 1993, 2001), the economic sphere is defined in close relation to the flows covered by the conventional System of National Accounts (SNA). Thus all flows related to the three types of economic activities included in the SNA (production, consumption, and stock change) are referred to as part of the economic system.

Once these components are recognized (human bodies, livestock and artifacts) every material *flow* that is needed to sustain these components is considered to be an input to society's metabolism. These material flows are set in motion via society's activities to produce and maintain society's material *stock*.

Stocks of materials that belong to the economy are mainly man-made fixed assets as defined in the national accounts such as infrastructures, buildings, vehicles, and machinery as well as inventories of finished products. Durable goods purchased by households for final consumption are not considered fixed assets in the national accounts but should be included in economy-wide MFA and balances (Eurostat (2001)). Of course the lifetime of goods plays a role in determining to which category (stock, durable goods, etc.) the product itself can be assigned.

There are some material stocks for which compilers have to determine whether they should be treated as part of the economy or of the environment. Cases in point are controlled landfills and cultivated forests. These decisions have an impact on the input and output flows that are recorded in the accounts. When controlled landfills are included within the system boundary, the emissions and leakages from landfills rather than the actual waste landfilled must be recorded as an output to the environment. For cultivated forests, the nutrients taken up by the trees rather than the timber harvested would be recorded as an input.

In Eurostat (2001), landfilled waste is considered an output to the environment but compilers are free to choose the treatment they prefer. If controlled landfills are included within the system boundary, the classification of outputs and stock changes and must be adapted. Showing waste landfilled as a separate category of stock changes so as to facilitate international data comparison is recommended.[1] Clearly, there is a close link between stocks and flows and a positive feedback as well. The bigger the material stocks are, the bigger the future material flows needed to reproduce and maintain the material stock.

15.3.3.2 Frontier to Other Economies (the Residence vs. Territory Principle)

Economy-wide material flow accounts and balances should be consistent with national accounts. The national accounts define a national economy as the activities and transactions of producer and consumer units that are resident (i.e., have their center of economic interest) on the economic territory of a country. Some activities and transactions of these units may occur outside the economic territory and some activities and transactions on the geographical territory of a country may involve nonresidents. Standard examples for illustrating this difference are tourists or international transport by road, air, or water. Due to such activities the environmental pressures generated by a national economy may differ from the environmental pressures generated on a nation's geographical territory. Transboundary flows of emissions through natural media (e.g., emissions to air or water generated in one country but which are carried by air or rivers and impact on another country) are not part of economy-wide MFA.

In order to make physical accounts consistent with the national economic accounts it is necessary to apply the residence (rather than territory) principle. Hence, in principle, materials purchased (or extracted for use) by resident units abroad would have to be considered as material inputs (and emissions abroad as material outputs) of the economy for which the accounts are made. Likewise, materials extracted or purchased by nonresidents on a nation's territory (and corresponding emissions and wastes) would have to be identified and excluded from that nation's economy-wide MFA and balances. Current knowledge suggests that the most important difference between residence and territory principle results from fuel use and corresponding air emissions related to international transport including bunkering of fuels and emissions by ships and international air transport as well as to fuel use and emissions of tourists.

Framed like this, MFA accounts for the overall material throughout, (i.e., the overall metabolism of a given socio-economic system).

15.3.4 Classification of Flows

In the MFA methodological guide, Eurostat (2001), various types of material flows are distinguished according to the abovementioned "direct vs.

---

[1]For a more detailed discussion on this topic please refer to Eurostat (2001).

Table 15.3 Categories of material inputs for economy-wide MFA

Product chain	Economic fate (used/unused)	Origin (domestic/ROW)	Term to be used
Direct	Used	Domestic	Domestic extraction (used)
(Not applied)	Unused	Domestic	Unused domestic extraction
Direct	Used	Rest of the world	Imports
Indirect (up stream)	Used	Rest of the world	Indirect input flows associated to imports
Indirect (up stream)	Unused	Rest of the world	

Source: modified from Eurostat, 2001.

Table 15.4 Categories of material outputs for economy-wide MFA

Product chain	Economic fate processed or not	Destination (domestic/ROW)	Term to be used
Direct	Processed	Domestic	Domestic processed output to nature
(Not applied)	Unprocessed	Domestic	Disposal of unused domestic extraction
Direct	Processed	Rest of the world	Exports
Indirect (up stream)	Processed	Rest of the world	Indirect output flows associated to exports
Indirect (up stream)	Unprocessed	Rest of the world	

Source: modified from Eurostat, 2001.

indirect" and "used vs. unused" classification. When economic systems are investigated a further category applies (i.e., domestic vs. rest of the world (ROW)) which refers to the origin or destination of the flows. Combining the three dimensions leads to five categories of inputs relevant for economy-wide MFA, as summarized in Table 15.3.

The output categories relevant for economy-wide MFA are summarized in Table 15.4. For output flows the column "used vs. unused" is called "processed vs. nonprocessed," thus referring to their stemming from an economic system or not, and the distinction "domestic vs. ROW" refers to the destination (rather than the origin) of the flows. A more detailed classification of the output based on their final environmental fate and harm has yet to be developed.

15.3.5 Categories of Materials

A standard classification of materials, which should be applied in the preparation of material flow accounts on the national level is summarized in Table 15.5. A very detailed material classification can be found in the annex of Eurostat (2001).

Table 15.5 Classification of input and output flows in economy-wide MFA, broad categories

Inputs	Outputs
Domestic extraction (used)	Emissions and wastes
Fossil fuels	Emissions to air
Minerals	Waste landfilled
Biomass	Emissions to water
Imports	Dissipative use of products and dissipative losses
Raw materials	Dissipative use of products
Semi-manufactured products	Dissipative losses
Finished products	Exports
Other products	Raw materials
Packaging material imported with	Semi-manufactured products
products	Finished products
Waste imported for final treatment	Other products
and disposal	Packaging material exported with products
Memorandum items for balancing	Waste exported for final treatment and disposal
O_2 *for combustion* (*of C, H, S, N* ...)	Memorandum items for balancing
O_2 *for respiration*	*Water vapor from combustion*
Nitrogen for emission from combustion	*Water evaporation from products*
Air for other industrial processes	*Respiration of humans and livestock* (*CO_2 and*
(*liquefied technical gases.* ..)	*water vapor*)
Unused domestic extraction	Disposal of unused domestic extraction
Unused extraction from mining	Unused extraction from mining and quarrying
and quarrying	Unused extraction from biomass harvest (discarding
Unused biomass from harvest	of by-catch, harvesting losses and wastes)
Soil excavation and dredging	Soil excavation and dredging
Indirect flows associated to imports*	Indirect flows associated to exports

Source: modified from Eurostat, 2001.
*The ecological rucksacks of all imported products equal the indirect flows associated to imports on the national level. Descriptions of indirect flows for imported products can be obtained from various publications of the WI (Bringezu, 2000; Bringezu and Schütz, 2001; Schütz, 1999). In several studies, these "rucksack-factors," which have been calculated for Germany, have been used in other country studies in order to estimate indirect flows (for example, Chen and Qiao (2001) for China; Hammer (2002) for Hungary; Mündl et al. (1999) for Poland).

As water flows in most cases exceed all other material inputs by a factor of ten or more (especially if water for cooling is also accounted for, see Stahmer et al. (1997) for example), Eurostat recommends presenting a water balance separately from solid materials. Thus in the standard accounts, water should only be included when becoming part of a product.

In order to close the overall material balance, the input of air has to be considered as corresponding to air emissions on the output side. In this respect, the most relevant processes are the combustion of fossil energy carriers (O_2 on the input side as a balancing item corresponding to CO_2 emissions), air for other industrial processes, and air for respiration of humans and livestock. The consideration of balancing items is not only an accounting trick because, for example, airflow through a system carries the heat produced by combustion processes. This affects a system performance positively (by cooling) or negatively (by losing still useful energy).

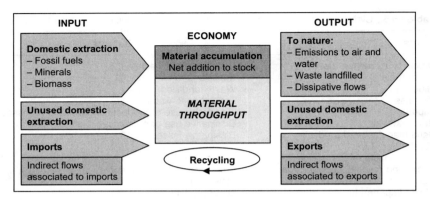

Figure 15.7   Diagram of a nationwide MFA system.

15.3.6   The Final Scheme and Material Balance

A general balance scheme including all the relevant input and output flows, but water and air, is given below (see Figure 15.7).

The *law of conservation of matter* states that matter is neither created nor destroyed by any physical transformation (production or consumption) process. This material balance principle provides a logical basis for the physical book keeping of the economy environment relationship and for the consistent and comprehensive recording of inputs, outputs, and material accumulation. The material balance principle can be applied from either a systems perspective or from a flow perspective.

For any given *system* such as production or consumption processes, companies, regions or national economies, the material balance principle leads to the following identity:

Total inputs = total outputs + net accumulation

That is, any input to the system is either accumulated in the system itself or exits the system again as an output.

For any given physical *flow* the material balance identity can be expressed as:

Origin = destination (other terms used are supply

 = demand, or resources = uses).

That is, any input flow must have an origin and a destination, and a breakdown by origin must be exhaustive in the sense that the sum of masses by origin must be equal to the sum of masses by destination, although matter may change form and state during production and consumption processes. When this identity is used to establish economy-wide balances for specific material groups (e.g., fossil fuels or biomass), the raw materials must be related to, for example, the emissions or wastes that are the final destinations of these materials.

Table 15.6 Material balance in an economy-wide MFA

Inputs	Outputs
Domestic extraction	Emissions and waste
Fossil fuels	Emissions to air
Minerals	Waste landfilled
Biomass	Emissions to water
Imports	Dissipative use of products and losses (fertilizers, manure, seeds, corrosion)
DMI (direct material inputs)	DPO (domestic processed output to nature)
Unused domestic extraction	Disposal of unused domestic extraction
From mining/quarrying	From mining/quarrying
From biomass harvest	From biomass harvest
Soil/rock excavation	Soil/rock excavation
TMI (total material input)	TDO (total domestic output to nature)
Indirect flows associated to imports	Exports
TMR (Total material requirements)	TMO (Total material output)
	Net addition to stock
	Infrastructure
	Building
	Other (machinery, durable goods)
	Indirect flows associated to exports
+ Balancing items (air and water inputs and related outputs)	

Note: In addition to the items shown in the table there are items (on both sides) for input and output balance, they are called balancing items. An insight on them is given in Section 15.3.6.1.

Moreover, the previous considerations can be summarized in a composite material balance (see Table 15.6).

15.3.6.1 Memorandum Items for Balancing

The full material balance, material inputs and outputs must be measured consistently. There are different options to ensure consistency of the material balance and to allow a meaningful interpretation of differences between inputs and outputs. Eurostat (2001) recommends the introduction of memorandum items for balancing purpose. The most important are listed in the classifications of inputs and outputs in the Eurostat guide in sections 3.5 and 3.6. For example, for the air emissions to balance with the fuels used in combustion, the oxygen must be included as a memorandum item on the input side. Alternatively, CO_2 emissions and water vapor could be described only in terms of their carbon and hydrogen content. Also, memorandum items for the water content of materials should be introduced. These memorandum items, however, are not to be included in the indicators derived from the accounts.

15.3.7 Indicators

The material balance also allows the derivation of several aggregate material-related indicators (see Table 15.7 below). They can be classified into input, output, and consumption indicators.

Table 15.7   Main material-related indicators

Input indicators
  DMI (Direct material inputs) = Domestic extraction + imports ☺
  TMI (Total material inputs) = DMI + Unused domestic extraction ☺
  Domestic TMR = TMI − Imports ☺
  TMR (Total material requirements) = DMI + imports + unused domestic extraction + indirect
    flows associated to imports ☺

Output indicators
  DPO (Domestic processed output to nature) = Emissions and waste + Dissipative use of
    products and losses ☺
  DMO (Direct material output) = DPO + exports ☺
  TDO (Total domestic output to nature) = DPO + Disposal of unused domestic extraction ☺
  TMO (Total material output) = TDO + exports ☺

Consumption indicators
  DMC (Domestic material consumption)= DMI − exports ☺
  TMC (Total material consumption) = TMR − exports — indirect flows associated to exports ☺
  NAS (Net addition to stock) = TMR − TMO ☺
  PTB (Physical trade balance) = Imports − exports ☺

  Memorandum items for balancing are not to be included when compiling indicators.
  ☺ Not additive across countries.
  ☺ Additive across countries.

When import-export foreign trade is included, directly or indirectly, in the calculation of indicators, the latter becomes not additive across countries. This is due to an unavoidable double counting related to foreign trade statistics. For example, as far as the European Union DMI (direct material input) is concerned, the intra-EU foreign trade flows must be netted out of the DMIs of member states.

15.3.7.1   The Physical Trade Balance

Concerning the trade and environment issue, the physical trade balance (PTB) is the most important indicator derivable from economy-wide MFA. The PTB expresses whether economies of countries or regions are dependent on resource inputs from other countries/regions as well as to what extent domestic material consumption is based on domestic resource extraction and imports of resources from abroad, respectively.

A physical trade balance is compiled in two steps. First, a PTB for direct material flows is calculated, which equals imports minus exports of a country or region. Second, a PTB can also be calculated including indirect flows associated to imports and exports, which include both used resource flows and unused resource flows.

In fact, for economy-wide MFA, two components of indirect flows are distinguished:

    1. Upstream indirect flows expressed as the raw material equivalents (RME) of the imported or exported products (less the weight of the imported or

exported product). The RME is the used extraction that was needed to provide the products

2. Upstream indirect flows of unused extraction (e.g., mining overburden) associated to this RME.

The first step is to compile the RME of imports or exports — that is, the vector of raw materials needed to provide the product at the border. In a second step, the unused extraction associated to this RME is compiled.

When imports and exports are converted into their RME, the weight of the RME includes the weight of the imports or exports. For the purpose of economy-wide MFA and balances, the indirect flows of type 1 (i.e., those based on the RME) are calculated by subtracting the weight of the imports or exports from the RME associated to these imports or exports so as to ensure additivity. This methodology of calculating direct and indirect material flows required in the life cycle of a product has been developed at the Wuppertal Institute in Germany.

Some of the indirect flows associated to exports may consist of the indirect flows associated to products previously imported. This effect would be particularly pronounced for countries with important harbors where a substantial part of imports is direct transit to other countries (the "Rotterdam effect"). It is recommended to show direct transit as a separate category of imports and exports in the accounts and to leave out direct transit when compiling indicators. For further discussion on this topic see Eurostat (2001).

15.3.8 Data Sources

An extensive description of indirect flows for imported products can be obtained from the report "Total Material Requirement of the European Union" (Bringezu and Schütz, 2001c). Detailed lists with "rucksack-factors" for minerals and metals as raw materials and semimanufactured products as well as some factors for biotic resources are provided. Good summaries for the calculation of indirect flows with the LCA-based approach have also been published by Schütz (1999) and Bringezu (2000). The annexes in both publications present comprehensive compilations of all available "rucksack factors," both for abiotic and biotic products, for domestic extraction as well as imports to Germany. This calculation methodology is mainly suitable for the calculation of indirect flows associated to biotic and abiotic raw materials and products with a low level of processing. To calculate indirect flows for semi-manufactured and finished products by applying this methodology requires the collection of an enormous amount of data for every product under consideration. A more convenient methodology for calculating the indirect flows on the macro level therefore is to apply input-output analysis. This allows quantifying the overall amount of material requirements stemming from inter-industry interrelations along the production chain (what is similar to the indirect effects in input-output analysis). The input-output technique is presented in Eurostat (2001).

15.3.9   State of the Art at a National Level

National material flow accounts are readily available for a number of national economies. Economy-wide material flow analyses have recently been published or are in progress for a number of countries, including Germany, Japan, The Netherlands and the U.S. (Adriaanse et al., 1997; Matthews et al., 2000), Australia (Durney, 1997), Austria (BMUJF, 1996; Schandl, 1998; Wolf et al., 1998; Gerhold and Petrovic, 2000; Schandl et al., 2000; Eurostat, 2000; Matthews et al., 2000), China (Chen and Qiao, 2000; Chen and Qiao, 2001), Finland (Ministry of the Environment, 1999; Juutinen and Mäenpää, 1999; Muukkonen, 2000; Mäenpää and Juutinen, 2000), Italy (Femia, 2000; De Marco et al., 2001), Japan (Moriguchi, 2001), Poland (Mündl et al., 1999; Schütz and Welfens, 2000), Sweden (Isacsson et al., 2000), the U.K. (Schandl and Schulz, 2000; Bringezu and Schütz, 2001d; Schandl and Schulz, 2002; Sheerin, 2002), France (Chabannes, 1998), Brazil (Machado, 2001; Amann et al., 2002), Venezuela (Castellano, 2001; Amann et al., 2002), Bolivia (Amann et al., 2002), and the European Union (Bringezu and Schütz, 2001a, 2001b, 2001c; Eurostat, 2002).

Several countries have integrated material flow statistics into their official statistics or are planning to do so (Austria, Denmark, Finland, France, Germany, Italy, Japan, The Netherlands and Sweden, according to Fischer-Kowalski and Hüttler, 1999). The United Nations integrated physical flow accounts into its System of Environmental and Economic Accounting (SEEA) (United Nation, 2000c).

15.3.10   Limits and Needed Improvements of MFA

As stated above MFA indicators reflect environmental pressures stemming from the societal metabolism. This is a good start for the understanding of the material basis of a process or an economy. However, several theoretical aspects are still weak and not yet investigated to the needed extent (Eurostat, 2001).

One of these aspects is the noninclusion into the MFA of the environmental services provided by the environment (e.g., dilution of pollutants, cooling, local microclimate maintenance, water cycling, soil buffering and filtering capacity, etc). These services are free but essential at the national level, especially for densely populated countries with significant internal agricultural and industrial activities (e.g., EU countries).

A further shortcoming of MFA is closely related to one of its main strengths, namely its simplicity and straightforwardness. The very fact that in MFA all material inputs are accounted for on a common mass basis potentially leads to the possible underestimation of the ecological impacts connected with the use of specific substances which may have a great impact on the environment (e.g., because of their ecotoxicity), despite their small or even negligible contribution to the overall mass balance.

A promising approach is to link MFA to other physical accounting methods (Bargigli et al., 2002; Ulgiati, 2002, 2003). The connection to ongoing research about land-use accounting and land-use change is particularly important, in order to integrate spatial aspects in the interpretation of MFA results. Another interesting link is among MFA and energy-based accounting methods (e.g., emergy, H.T. Odum, 1996), and downstream environmental impact evaluation methods (e.g., CML2). It should be underlined that MFA shares with emergy synthesis (Odum, 1988, 1996), cumulative exergy accounting (Szargut and Morris, 1987) and ecological footprint (Wackernagel and Rees, 1996) the characteristics of assessing the embodiment of some fundamental physical quantity (matter, time, exergy, land), as a measure of the support received from larger scales. This translates into similarities in their definitions of boundaries, algebra, and indicators, although significant differences have not yet been removed. A potential synergy could be achieved if researchers displayed a larger effort towards their integration and complementarity.

Moreover, direct connections between resource use, type of activity, and specific environmental impacts are still lacking in the MFA, both on the national and on the local scale, along with a detailed classification of outputs to the environment according to their final environmental fate and potential harm to humans and ecosystems. Ayres et al. (1998) proposed to evaluate the potential harm of outputs according to their residual exergy content. The World Resources Institute (WRI), when compiling the national physical accounts for the U.S. for 1975 to 1996, developed categories to characterize material flows (see annex 2 of Matthews et al., 2000). The characterization is made on the basis of quantity, mode of first release, quality, and velocity (expressed as residence time). The method also allows the estimation of outflows to the environment and net additions to stock from the input flows based on the velocity. Nevertheless, an internationally standardized procedure for considering qualitative differences in the quantitative concept of MFA is so far still missing.

In particular, there is an urgent need for careful investigation of the environmental fate and the ecotoxicological impact of the chemicals released by the societal metabolism. Societal catabolites can be very harmful for the environment on a national scale too.

This topic is especially crucial if we consider that many of the materials used in industrialized countries are very material- and energy-intensive, and are based on the exploitation of rare minerals very often concentrated in small regions of developing countries, where environmental protection norms are lacking or less strictly enforced. This consideration applies to some extent to all kinds of industrial processes, but is very often disregarded, and calls for a careful assessment of the consequences that the technological choices of a country may have on other areas of the world (Ulgiati et al., 2002). This point should be carefully considered while observing dematerialization paths of modern societies.

REFERENCES

Adriaanse, A., Bringezu, S., Hammond, A., Moriguchi, Y., Rodenburg, E., Rogich, D., and Schütz, H. *Resource Flows. The Material Basis of Industrial Economies.* World Resource Institute, Washington, D.C., 1997.

Amann, C., Bruckner, W., Fischer-Kowalski, M., and Grünbühel, C.M. Material Flow Accounting in Amazonia. A Tool for Sustainable Development. Social Ecology Working Paper. 63. Institute for Interdisciplinary Studies of Austrian Universities, Vienna, 2002.

Ayres, R.U. and Masini, A. "Waste exergy as a measure of potential harm," in *Advances in Energy Studies. Energy Flow in Ecology and Economy*, Ulgiati, S., Brown, M.T., Giampietro, M., Herendeen, R.A., and Mayumi, K., Eds. Musis, Rome, 1988, pp. 113–128.

Ayres, R.U. and Kneese, A.V. Production, consumption and externalities. *American Economic Review* 59, 282–297, 1969.

Bargigli, S., Raugei, M., Ulgiati, S. Thermodynamic and environmental profile of selected gaseous energy carriers. Comparing natural gas, syngas and hydrogen, in *ECOS 2002, Book of Proceedings of the 15th International Conference on Efficiency, Costs, Optimization, Simulation and Environmental Impact of Energy Systems*; Tsatsaronis, G., Moran, M.J., Cziesla, F., and Bruckner, T., Eds. Berlin, Germany, 2–5 July 2002, pp. 604–612.

Bargigli, S., Raugei, M., and Ulgiati, S. Comparison of thermodynamic and environmental indexes of natural gas, syngas and hydrogen production processes. *Energy. The International Journal*, 2004.

BMUJF. Materialfluβrechnung Österreich. Gesellschaftlicher Stoffwechsel und nachhaltige Entwicklung. Schriftenreihe des BMUJF Band 1/96. Vienna: Bundesministerium für Umwelt, Jugend und Familie, 1996.

Bringezu, S., Fischer-Kowalski, M., Klein, R., and Palm, V. in Analysis for Action. Support for Policy towards Sustainability by Material flow Accounting. *Regional and National Material Flow Accounting: From Paradigm to Practice of Sustainability.* Proceedings of the ConAccount, Leiden, 1997.

Bringezu, S. Ressourcennutzung in Wirtschaftsräumen. Stoffstromanalysen für eine nachhaltige Raumentwicklung. Springer, Berlin, 2000.

Bringezu, S. and Schütz, H. The Material Requirement of the European Union. Technical report no. 55. EEA (European Environment Agency), Copenhagen, 2001a.

Bringezu, S. and Schütz, H. Material Use Indicators for the European Union, 1980–1997. Eurostat Working Papers 2/2001/B/2: Eurostat, 2001b.

Bringezu, S. and Schütz, H. Total Material Requirement of the European Union. Technical Part. Technical report no. 56. EEA (European Environment Agency), Copenhagen, 2001c.

Bringezu, S. and Schütz, H. Total Material Resource Flows of the United Kingdom. Wuppertal Institute for Climate, Environment, Energy, Wuppertal, 2001d.

Castellano, H. Material Flow Analysis in Venezuela. Internal report (unpublished). Caracas, 2001.

Chabannes, G. Material Flows Analysis for France. (Unpubl. man)

Chen, X. and Qiao, L. Material flow analysis of the Chinese economic-environmental system. *journal of Natural Resources* 15, 17–23, 2000.

Chen, X. and Qiao, L. A preliminary material input analyses of China. *Population and Environment* 23, 117–126, 2001.

De Marco, O., Lagioia, G., and Mazzacane, E.P. Materials flow analysis of the Italian economy. *Journal of Industrial Ecology* 4, 55–70, 2001.

Durney, A. Industrial Metabolism. Extended Definition. Possible Instruments and an Australian Case Study. Wissenschaftszentrum Berlin für Sozialforschung, Berlin, 1997.

Eurostat. Material Flow Accounts — Material Balance and Indicators, Austria 1960–1998. Eurostat Working Papers 2/2000/B/7: Eurostat, Luxembourg, 2000.

Eurostat. Economy-wide Material Flow Accounts and Derived Indicators. A Methodological Guide. European Communities, Luxembourg, 2001.

Eurostat. Material Use Indicators for the European Union 1980–2000. Background document. Eurostat, Luxembourg, 2002.

Femia, Aldo. A Material Flow Account for Italy, 1988. Eurostat Working Papers 2/2000/B/8: Eurostat, Luxembourg, 2000.

Fischer-Kowalski, Marina. "Methodische Grundsatzfragen," in *Gesellschaftlicher Stoffwechsel und Kolonisierung von Natur*, Fischer-Kowalski, M., Haberl, H., Hüttler, W., Payer, H., Schandl, H., Winiwarter, V., and Zangerl- Weisz, H., Eds. G+B Verlag Facultas, Amsterdam, 1997, pp. 57–66.

Fischer-Kowalski, M. and Hüttler, W. Society's metabolism. the intellectual history of materials flow analysis, part II, 1970–1998. *Journal of Industrial Ecology* 2, 107–136, 1999.

Frischknecht, R., Bollens, U., Bosshart, S., Ciot, M., Ciseri, L., Doka, G., Hischier, R., Martin, A. (ETH Zürich), Dones, R., and Gantner, U. (PSI Villigen). *Ökoinventare von Energiesystemen, Grundlagen für den Ökologischen Vergleich von Energiesystemen und den Einbezug von Energiesystemen in Ökobilanzen für die Schweiz*, 3rd ed. CD-ROM and hard copy, Gruppe Energie-Stoffe-Umwelt, ETH Zürich, Sektion Ganzheitliche Systemanalysen, PSI Villigen, 1996.

Gerhold, S. and Petrovic, B. Materialflussrechnung: Bilanzen 1997 und abgeleitete Indikatoren 1960–1997. *Statistische Nachrichten* 4, 298–305, 2000.

Giljum, S. Trade, material flows and economic development in the south: the example of Chile. *Journal of Industrial Ecology*, 2003.

Hacker, J. Diplomarbeit. Bestimmung des lebenszyklusweiten Naturverbrauches für die Elektrizitätsproduktion in den Ländern der Europäischen Union. Technische Universität Wien, 2003.

Hinterberger, F., Luks, F., and Stewen, M. Ökologische Wirtschaftspolitik. Zwischen Ökodiktatur und Umweltkatastrophe. Birkhäuser, Berlin, 1996.

Hinterberger, F., Renn, S., and Schütz, H. Arbeit-Wirtschaft-Umwelt. Wuppertal Papers. 89. Wuppertal Institut für Klima, Umwelt, Energie, Wuppertal, 1999.

Isacsson, A., Jonsson, K., Linder, I., Palm, V., and Wadeskog, A. Material Flow Accounts, DMI and DMC for Sweden 1987–1997, Eurostat Working Papers, No. 2/2000/B/2. Statistics Sweden, 2000.

Japanese Environment Agency. *Quality of the Environment in Japan 1992*, Tokyo, 1992.

Juutinen, A. and Mäenpää, I. Time Series for the Total Material Requirement of the Finnish Economy. Summary, interim report, 15 August 1999. Thule Institute, University of Oulu, Oulu, 1999.

Kippenberger, C. Stoffmengenflüsse und Energiebedarf bei der Gewinnung ausgewählter ineralischer Rohstoffe. Auswertende Zusammenfassung, No. Heft SH 10. Bundesanstalt fuer Geowissenschaften und Rohstoffe, Hannover, 1999.

Kleijn, R., Bringezu, S., Fischer-Kowalski, M., and Palm, V. Ecologizing Societal Metabolism. Designing Scenarios for Sustainable Materials Management, No. CML Report 148, Leiden, 1999.

Machado, J.A. Material Flow Analysis in Brazil. Internal report (unbublished). Manaus, 2001.

Mäenpää, I. and Juutinen, A. Explaining the Material Intensity in the Dynamics of Economic Growth: The Case of Finland. *Proceedings of the ISEE Conference*, Canberra, Australia, 5–8 July, 2000.

Manstein, C. Das Elektrizitaetsmodul im MIPs-Konzept, Wuppertal Paper no. 51, Wuppertal Institute, Wuppertal, 1996.

Matthews, E., Amann, C., Bringezu, S., Fischer-Kowalski, M., Hüttler, W., Kleijn, R., Moriguchi, Y., Ottke, C., Rodenburg, E., Rogich, D., Schandl, H., Schütz, H., van der Voet, E., and Weisz, H. *The Weight of Nations. Material Outflows from Industrial Economies*. World Resources Institute, Washington, D.C., 2000.

Ministry of the Environment. Material Flow Accounting as a Measure of the Total Consumption of Natural Resources. The Finnish Environment 287. Helsinki, 1999.

Moriguchi, Y. Rapid socio-economic transition and material flows in Japan. *Population and Environment* 23, 105–115, 2001.

Mündl, A., Schütz, H., Stodulski, W., Sleszynski, J., and Jolanta Welfens, M. Sustainable Development by Dematerialisation in Production and Consumption. Strategy for the New Environmental Policy in Poland. Institute for Sustainable Development, Warsaw, 1999.

Muukkonen, J. Material Flow Accounts. TMR, DMI and Material Balances, Finland 1980–1997. Eurostat Working Papers 2/2000/B/1: Eurostat, Luxembourg, 2000.

Odum, H.T. Self organization, transformity and information. *Science* 242, 1132–1139, 1988.

Odum, H.T. *Environmental Accounting: Emergy and Environmental Decision Making*. Wiley, New York, 1996.

Organisation for Economic Co-operation and Development. Working Group on the State of the Environment 30th Meeting. Special Session on Material Flow Accounting. History and Overview. Room Document — MFA 1. Agenda Item 2a. OECD, Paris, 2000.

Ritthof, M., Rohn, H., and Liedtke, C. MIPS Berechnungen — Ressourceproduktivität von Produkten und Dienstleistungen Wuppertal Spezial 27. Wuppertal Institute publications, 2002.

Raugei, M., Bargigli, S., and Ulgiati, S. A multi-criteria life cycle assessment of Molten Carbonate Fuel Cells (MCFC). A comparison to natural gas turbines. *International Journal of Hydrogen Energy*, 2003 (submitted).

Sampat, P. "Scrapping mining dependence," in State of The World 2003, A Worldwatch Institute Report on Progress toward a Sustainable Society. Washington, D.C., 2003.

Schandl, H. Materialfulß Österreich. Wien, Interuniversitäres Institut für Interdiszilpinäre Forschung und Fortbildung (IFF). IFF Paper, 1998.

Schandl, H. and Schulz, N. Using Material Flow Accounting to Operationalize the Concept of Society's Metabolism. A Preliminary MFA for the United Kingdom for the Period of 1937–1997. ISER Working Papers. 2000–2003. University of Essex, Colchester, 2000.

Schandl, H. and Schulz, N. Changes in the United Kingdom's natural relations in terms of society's metabolism and land-use from 1850 to the present day. *Ecological Economics* 41, 203–221, 2002.

Schandl, H., Weisz, H., and Petrovic, B. Materialflussrechnung für Österreich 1960 bis 1997. *Statistische Nachrichten* 2, 128–137, 2000.

Schmidt-Bleek, F. MIPS — A Universal Ecological Measure. Fresenius Environmental Bulletin 2, 407–412, 1992.

Schmidt-Bleek, F. Wieviel Umwelt braucht der Mensch? MIPS — Das Ma für ökologisches Wirtschaften. Birkäuser, Berlin, 1994.

Schmidt-Bleek, F., Bringezu, S., Hinterbegrer, F., Liedtke, C., Spangenberg, J., Stiller, H., Welfens, M., MAIA. Einführung in die Material-Intensitäts-Analyse nach dem MIPS-Konzept. Berlin, Basel, Boston, Birkhäuser, 1998.

Schütz, H. Technical Details of NMFA (Inputside) for Germany. Wuppertal Institute, Wuppertal, 1999.

Schütz, H. and Welfens, M.J. Sustainable Development by Dematerialisation in Production and Consumption — Strategy for the New Environmental Policy in Poland. Wuppertal Papers No. 103. Wuppertal Institut für Klima, Umwelt, Energie, Wuppertal, 2000.

Sheerin, C. UK material flow accounting. *Economic Trends* 583, 53–61, 2002.

Simonis, U.E. "Industrial restructuring in industrial countries," in *Industrial Metabolism. Restructuring for Sustainable Development*, Ayres, R.U. and Simonis, U.E., Eds. United Nations University Press, Tokyo, 1994.

Spangenberg, J.H., Hinterberger, F. "Material Flow Analysis, TMR and the mips — Concept: A Contribution to the Development of Indicators for Measuring Changes in Consumption and Production Patterns", *International Journal of Sustainable Development* 1(2), 1999.

Stahmer, C., Kuhn, M., and Braun, N. Physische Input-Output-Tabellen 1990. Wiesbaden: Statistisches Bundesamt. UCTAD and WTO. Trade Analysis System on CD-ROM (PC-TAS). United Nations Conference on Tariffs and Trade (UNCTAD), World Trade Organisation (WTO). UN (1997a). Energy Statistics Yearbook 1995. Department of Economic and Social Affairs. Statistics Division, United Nations, New York, 1997.

Steurer, A. Stoffstrombilanz Oesterreich, 1988, Schriftenreihe Soziale Oekologie, No. Band 26. IFF/Abteilung Soziale Oekologie, 1992.

Stiller, H. Material Intensity of Advanced Composite Materials. Wuppertal Papers No. 90, 1999.

Szargut, J., Morris, D.R., and Steward, F.R. Exergy analysis of thermal, chemical and metallurgical processes. Hemisphere, London, 1988.

Ulgiati S. "Energy flows in ecology and in the economy," in *Encyclopedia of Physical Science and Technology*, 3rd ed, Vol. 5. Academy Press, New York, 2002, pp. 441–460.

Ulgiati, S., Bargigli, S., Raugei, M., and Tabacco, A.M. Analisi Energetica e Valutazione di Impatto Ambientale della Produzione ed Uso di Celle a Combustibile a Carbonati Fusi. Report to ENEA (in Italian), Contract 1033/TEA, 2002.

Ulgiati, S. Raugei, M., and Bargigli, S. Overcoming the inadequacy of single-criterion approaches to Life Cycle Assessment. *Ecological Modeling* 2003 (submitted).

United Nations. *Energy Statistics Yearbook 1997*. Department of Economic and Social Affairs. Statistics Division, United Nations, New York, 2000a.

United Nations. *Statistical Yearbook 1997*. Department of Economic and Social Affairs. Statistics Division, United Nations, New York, 2000b.

United Nations. *System of Environmental and Economic Accounting (SEEA) 2000*. United Nations, Voorburg draft, 2000c.

United Nations. *Handbook of National Accounting: Integrated Environmental and Economic Accounting. Studies in Methods*. New York, United Nations, 1993.

United Nations. *System of Environmental and Economic Accounting*. SEEA 2000 Revision United Nations, New York, 2001.

Wackernagel, M. and Rees, W. *Our Ecological Footprint*. New Society Publisher, 1996.

Wolf, M.E., Petrovic, B., and Payer, H. Materialfluβrechnung Österreich 1996. *Statistische Nachrichten* 11, 939–948, 1998.

CHAPTER 16

The Health of Ecosystems: The Ythan Estuary Case Study

D. Raffaelli, P. White, A. Renwick, J. Smart, and C. Perrings

16.1 INTRODUCTION

The last ten years have seen a call to refocus the sustainability research agendas of the environmental, social, and economic sciences, so that they can be integrated within the overarching concept of ecosystem health (e.g., Costanza et al., 1992; Rapport et al., 1999). Such integration will not be easy. Like all novel approaches, ecosystem health has its critics harping from the sidelines, as well as having significant conceptual and operational issues to overcome (discussed throughout this volume). One of the most productive ways to present the area of ecosystem health to research scientists and environmental managers alike is through well-documented and understood case studies. Here, we describe agriculture-induced changes in an extremely well-documented system, the Ythan Estuary, Aberdeenshire, Scotland, from the early 1960s to date.

The Ythan is one of the smallest rivers in Europe, but one of the most significant in the present context, due to:

- A long history of ecological research on the system carried out continuously over the last 40 years by staff and students at the

University of Aberdeen's Culterty Field Station located on the banks of the estuary

- The small spatial extent of the Ythan, especially the estuary, which has facilitated detailed descriptions of the physical and biological elements, as well as scale-appropriate manipulative experiments for testing hypotheses about function
- Detailed historical data on agricultural land-use within the catchment and on river water quality, available on an annual basis
- The U.K. national and wider European interest in the Ythan's nature conservation status and as a test-bed for implementation of European environmental legislation, policy, and restoration

In this chapter, we review the temporal changes in the major physical and biological components of the Ythan Estuary and link these to historical trends in land use and agricultural practice in the catchment driven by U.K. and European agriculture policies. We also evaluate changes in the health of the ecosystem using a number of indicators and explore the potential for different policy instruments for restoring ecosystem health in this system.

16.1.1 The Physical Context

The Ythan River drains a catchment of approximately $640 \, \text{km}^2$ lying north of the city of Aberdeen, northeast Scotland. The river rises only a few hundred meters in altitude near the Wells of Ythan and runs east to the North Sea coast where it enters the sea at the village of Newburgh. Several major tributaries join the Ythan towards the lower reaches, some almost as large as the main river itself, which is never more than a few tens of meters wide, even at the estuary. The estuary is basically salt-wedge — that is, with a saline intrusion on the flood tide running beneath a surface wedge of freshwater (Leach, 1971), although wind-induced wave action can easily destratify the vertical salinity gradient because of the estuary's shallow depths (1 to 5 m). The salinity intrusion reaches approximately 8 km upstream from the estuary mouth. The area experiences two tides per day ranging from around 2 m (neap) to around 4 m (spring). The upper and middle reaches of the estuary are characterized by significant areas of mudflat with a high silt content, while the lower reaches are much sandier. The lower reaches support extensive intertidal and shallow subtidal mussel (*Mytilus edulis*) beds. Whilst the Ythan is at a relatively high latitude (57° 20′ N), the river and estuary are always ice-free, because of the modifying influence of the North Atlantic drift on this region of Europe. A full account of the physical characteristics of the estuary can be found in Baird and Milne (1981).

16.1.2 Long-Term Data Sets

There is a wealth of relevant data on the Ythan (Raffaelli et al., 1999), including:

- Annual digests of land use integrated at the level of each of the seven parishes that fall entirely within the catchment boundary; river water

quality measurements made by Culterty staff and the Scottish Environ-
mental Protection Agency (SEPA)
- Topographic data from charts and maps from local sources, Her Majesty's
  Admiralty and the Ordnance Survey of Great Britain
- Aerial photography (irregularly)
- Biological surveys of macroalgal mats, benthic invertebrates and
  shorebirds.

16.2 CHANGES IN AGRICULTURE

The period after World War II saw dramatic changes in agricultural
production across Europe, and these accelerated with the adoption of the
Common Agricultural Policy (CAP) in the early 1970s. The CAP provided
incentives for particular kinds of land use through a series of subsidies and
interventions. In Aberdeenshire, a field rotation system was still common
practice in the 1960s, with oats, spring-sown barley and wheat, potatoes, neaps
(swedes), cattle, and sheep, the mainstay of farming. The next 10 to 20 years
saw major changes in land use within the Ythan parishes (Figure 16.1). These
changes reflect:

1. The preferential growing of subsidized cereals, such as wheat and barley,
   at the expense of the more traditional oats
2. The introduction of novel crops, such as oil-seed rape
3. An increase in the total land area under fertilizer-hungry cereals and
   rape, at the expense of grassland, especially rough grassland
4. A shift towards winter- and autumn-sown cereals, such that land is tilled
   at a time of high precipitation and run-off
5. An increase in pig production

In addition, intensive agriculture now extends right to the edge of the river
bank over large sections of the water course, creating significant soil erosion at
many locations. Over this 40-year period, application of chemical fertilizer,
especially nitrogen-based, increased by a factor of two to three (Domberg et al.,
1998, 2000), independently of the spreading of livestock waste to the land.
In addition, about 95% of the land in the catchment is under agriculture, much
of it arable.

16.3 CHANGES IN WATER QUALITY

Land-use changes are unambiguously reflected in the water quality of the
Ythan River over the same period (Figure 16.2). Since 1958, there has been a
two- to three-fold increase in the concentration of total oxidized nitrogen
(almost entirely nitrate) in the river water (Figure 16.2a) and a similar pattern
is seen within the estuary, once dilution by seawater is accounted for
(Figure 16.2b). Above the town of Ellon, representing the tidal limit of the
system and where salinities are typically less than 2 psu, the total load of

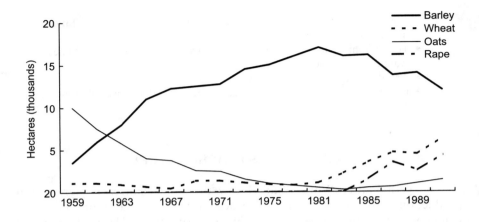

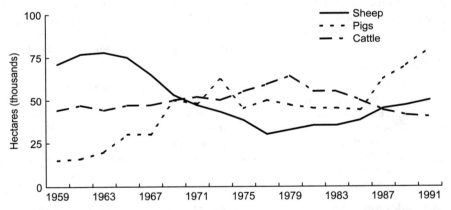

Figure 16.1 Land-use change in the Ythan catchment over a 30-year period. Data from Scottish Records Office, after Raffaelli (1997).

nitrogen entering the estuary from the river has increased by a similar factor (Figure 16.3). The only sewage-associated nutrient inputs to the Ythan are at small towns of Ellon (population 15,000) and Newburgh (population 1500). These inputs are negligible in terms of the overall nutrient budget (Raffaelli et al., 1989) There is negligible import of nitrogen from the sea (Figure 16.2b).

The emphasis above has been on nitrogen rather than phosphorus, because while phosphorus is a limiting nutrient for plant growth in freshwater, it is much less important than nitrogen in marine and estuarine systems. In addition, phosphorus shows at best only a weak temporal signal (Figure 16.3). However, it should be noted that the significance of phosphorus remains a lively and contentious issue in this debate, probably because sewage works at Ellon contributes about half of the estuarine phosphorus budget.

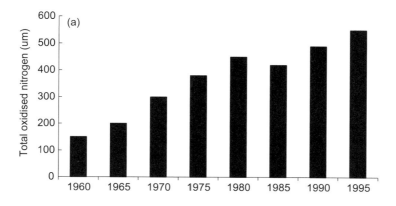

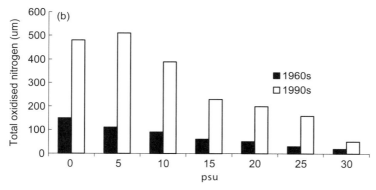

Figure 16.2 Nitrogen (mostly nitrate) recorded: (a) in the Ythan River over a 40-year period by SEPA (after Raffaelli et al., 1989); (b) in the Ythan Estuary pre- and posteutrophication (1960s and 1990s, respectively). Data from Leach (1971) and Balls et al. (1995), after Raffaelli (1999).

16.4 CHANGES IN BIOLOGY

The biology of the tributaries and the main river has changed consistent with a picture of nutrient increase (see below). In addition to the routine invertebrate monitoring carried out by SEPA, macrophyte surveys reveal evidence of eutrophication in several stretches, filamentous algal growth is an increasingly common feature of the stream bed and spawning areas (redds) for salmonids have become silted up and hence unsuitable.

By far the best-documented changes in biology are for the estuary itself. The area of intertidal flat covered by mats of opportunistic macroalgae (*Enteromorpha*[1], *Ulva* and *Chaetomorpha*) has increased substantially

---

[1]Molecular studies of the genus Enteromorpha on the Ythan has recently lead to its revision into the genus *Ulva* (Maggs et al., 2003). However, to retain consistency with previous texts in this review, and to avoid confusion with the *Ulva* described in those texts, the name *Enteromorpha* is used here.

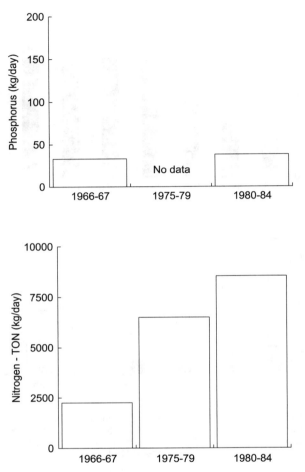

Figure 16.3 Nutrient loads (kg/day) in the Ythan River upstream of the town of Ellon and therefore representing run-off from the catchment, 95% of which is agricultural. Data from SEPA and Leach (1971), after Raffaelli et al. (1989).

(Figure 16.4) so that now around 40% of the intertidal area is regularly covered by mats of biomass in excess of $1 \, \text{kg} \, \text{m}^{-1}$. Nitrogen is implicated in this algal growth (Fletcher, 1996; Raffaelli et al., 1998) and other competing explanations, such as changes in estuarine topography and hydrography, have been evaluated and discounted (Raffaelli et al., 1999).

Algal mats have a major impact on the underlying sediment invertebrate communities. The most important is a decline in the amphipod *Corophium volutator* and an increase in the opportunistic polychaete worm *Capitella* sp., although many other taxa are also affected (Raffaelli, 2000). These effects are dose-dependent (algal biomass) and their causality has been confirmed through a series of manipulative field experiments (Figure 16.5; Hull, 1987; Raffaelli et al., 1998, 2000). Algal mats create a hostile physical and

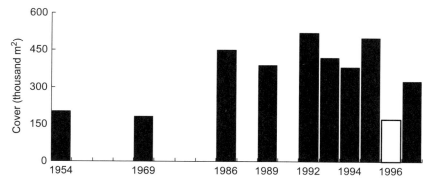

Figure 16.4   Coverage of intertidal area by macroalgal mats (*Enteromorpha, Chaetomorpha* and *Ulva*) digitized from aerial photographs taken during the late summer when the bloom is at its maximum. Only macroalgal biomass in excess of $1\,kg\,ww/m^{-2}$ can be detected and the contribution from individual algal taxa cannot be separated. After Raffaelli et al. (1999).

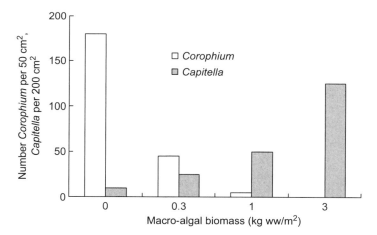

Figure 16.5   The response of *Corophium* and *Capitella* to controlled addition of algal biomass in a field experiment carried out on the Ythan Estuary by Hull (1987). A series of similar experiments have all confirmed these effects (Raffaelli, 2000).

redox environment for *Corophium*, which can be exploited by the more tolerant *Capitella*. The effects on *Corophium* are dramatic, with densities reduced from around $60,000\,m^{-1}$ to zero (Figure 16.5). *Corophium* is the main species in the diet of many of the Ythan's shrimp, crabs, fish, and shorebirds, including eider chicks (*Somateria mollissima*), a flagship conservation species. The increase in *Capitella* in areas affected by algal mats does not compensate for the decline in *Corophium* because this polychaete is not preferred by consumers and is protected by the presence of the overlying mats anyway.

The decline in the prey base for shorebirds is reflected in changes in the distribution and abundance of wildfowl and waders on the Ythan (Raffaelli et al., 1999). Redshank (*Tringa totanus*) and shelduck (*Tadorna tadorna*) appear to have shifted their distributions to upstream and downstream areas less affected by algal mats. Analysis of shorebird counts (Figure 16.6) reveal an initial increase at the start of the eutrophication period, consistent with an overall increase on the carrying capacity of the system (areas not affected by algal growth became more productive through enrichment), followed by a decline over time as these once algal-free and productive areas are reduced in extent through continued algal spread (Figure 16.6).

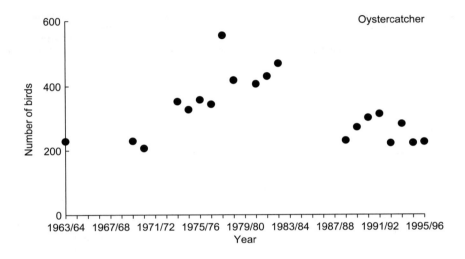

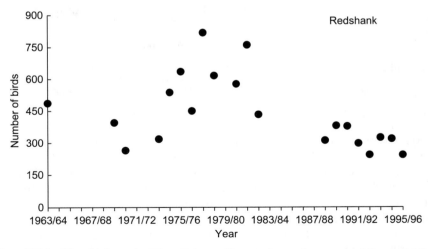

Figure 16.6   Shorebird counts, Ythan Estuary. The counts are the means of the period from December to February. After Raffaelli et al. (1999).

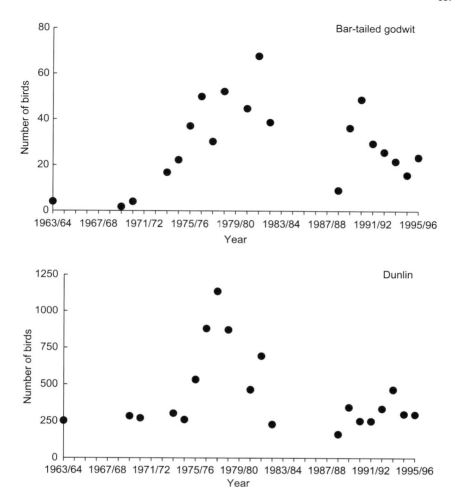

Figure 16.6 Continued.

16.5 MEASURES OF ECOSYSTEM HEALTH

A number of measures of ecosystem health have been made on the Ythan, partly in response to an awareness of the nutrient enrichment issue, but also because agencies such as SEPA have statutory responsibilities to carry out such monitoring.

16.5.1 Water Quality Index (WQI)

The SEPA Water Quality Index (Scottish Development Department, 1976) integrates measurements of around 15 chemical determinants over a scale of zero (poorest quality) to 100 (best quality). Data collected from 1980 to 1991 show a very slight decline in quality over that period (only a few points on the scale) and a more pronounced decline downstream, reflecting the progressively greater contribution of nitrate nitrogen (Pugh, 1997).

16.5.2 Macroinvertebrate Indices of Water Quality

As for many other parts of the world, various biotic indices exist for U.K. rivers which integrate score measures of abundance of the benthic invertebrates taken by kick sample. For northeast Scotland, the Biological Monitoring Working Party (BMWP, 1978) scoring method has been adopted by SEPA. Data from 1982 to 1991 (Pugh, 1997) reveal the following. For the main river, there is a reduction in water quality downstream, reflecting increasing inputs as the catchment area increases, with a further depression in the BMWP scores downstream of the main settlement of Ellon, probably due to localized sewage inputs. In addition, there is evidence of depression of water quality due to agricultural inputs in the major primary feeder streams (Pugh, 1997).

16.5.3 Estuary Quality Indicators

Estuary water quality in the U.K. is classified as "good," "fair," "poor," or "bad," based on the U.K. National Water Council Scheme. This scheme is the summation of scores for biological quality (presence of migratory fish, healthy benthic communities), esthetic quality (evidence of esthetic pollution) and chemical quality (dissolved oxygen concentration). The scores for the Ythan pre- and posteutrophication are shown in Table 16.1. Two features emerge from this analysis: (1) the index is relatively insensitive to major changes in benthic community structure (range available, 0 to 2), which is the most noticeable ecological impact on the Ythan Estuary; and (2) esthetic changes have a much greater effect on the score total than biological changes.

16.5.4 Ecosystem Indicators

A number of indicators of ecosystem health derive from ecological and ecosystem theory and several of these have been explored for the Ythan.

Table 16.1 Estuary water quality in the Ythan, pre- and posteutrophication (1970s and 1990s, respectively). Data from personal observations and various sources

	1970s	1990s
Chemistry (dissolved O_2, range 0–10)	10	10
Biology (migratory fish, salmon and flounder, range 0–2)	2	2
Good resident fish population? (range 0–2)	2	2
Good benthic communities (range 0–2)	2	1
Absence of toxins in biota (range 0–4)	3*	3**
Aesthetics (sewage associated material, range 0–10)	8	5
Maximum score possible: 30	27	23

*copper originating from pig food.
**PCBs detected by Plymouth Marine Laboratory Mussel Watch program.

A detailed account will be presented elsewhere (Raffaelli et al., in prep.) and only illustrative examples are provided here.

One indicator associated with ecosystem maturity and stability is the proportion of r- and K-trait taxa selected for by the environment (Odum, 1969). The Ythan has seen a major increase in at least two r-trait taxa which dominate the benthic assemblage: the green alga *Enteromorpha* (more r-selected than other macroalgae) and the polychaete *Capitella* sp. (more r-selected than other macroinvertebrates). Changes in the abundance of both taxa are intimately linked, because of the dose-dependency in the cause-effect relationship (Figure 16.5). The population dynamics of *Corophium* are in marked contrast to *Capitella* and considerably further along the r–K continuum. It is clear that by this criterion, the maturity (sensu Odum, 1969; Christensen, 1995) of the Ythan system has declined markedly with increasing nutrient enrichment.

Various metrics from the field of network analysis are thought to have potential as indicators of ecosystem health (Contanza, 1992 and references therein). Of these, only ascendancy will be considered here. Ascendancy is defined as the average mutual information content of the system scaled by the total system throughput (Ulanowicz, 1986). Ascendancy has potential as an indicator of ecosystem health because average information content increases with system maturity, and there is empirical evidence to support this from Christensen's (1995) analysis of real aquatic systems.

Ascendancy can be derived from mass-balance models of ecosystems, such as ECOPATH (Christensen and Pauly, 1992). Previous studies of the Ythan by Baird and Ulanowicz (1993) and Buchan (1997) give ascendancy values for the pre-eutrophication state of the Ythan of 34.4% and 27%, respectively. Formal exploration of mass-balance models of Ythan Estuary system at different stages of the eutrophication process is ongoing, but it is clear that the biomass and productivity of primary producers, *Enteromorpha* in the case of the Ythan, will contribute significantly to ascendancy estimates. Pre-eutrophication average biomasses of *Enteromorpha* were typically in the order of a few $g\,ww\,m^{-2}$, whereas now 40% of the estuary is covered by algal mats of biomass of at least $1\,kg\,m^{-2}$ (Raffaelli et al., 1999). In contrast, the biomasses and productivities of most of other elements of the Ythan ecosystem are probably similar now to those in the pre-eutrophic period because of the nonmonotonic (hump-backed) nature of their response to eutrophication (e.g., Figure 16.6). Thus it is likely that ascendancy will increase with eutrophication in estuaries such as the Ythan, rather than decline as expected from ecosystem theory, although this remains to be confirmed.

Surprisingly, despite the major ecological changes which have taken place on the Ythan over the study period, no species have disappeared. Furthermore, the abundances of focal species of conservation interest, shorebirds, are probably similar today as in the pre-eutrophication period. This does not necessarily reflect a high system resilience, rather the nonlinear dynamics of the eutrophication process (Figure 16.6) which only become apparent in long-term data sets. Similarly, the compensatory increase in invertebrate production in algal-free areas in the 1980s, and the consequent maintenance of the estuary's carrying capacity for shorebirds, were only revealed by working at the scale of the entire estuary. If the Ythan study had been restricted to timescales typical of present research grant funding and to only part of the estuary, the dynamics of the system would have been misrepresented.

16.6 A COUPLED HUMAN–ECOLOGICAL SYSTEM?

The relationship between human activity in the Ythan catchment and ecological processes is all too obvious. Changes in the pattern of farming activity in the catchment have resulted in significant spillover effects downstream at the estuary. However, there are few obvious direct feedbacks between ecological change and human activity at the catchment scale: increased cover of mudflats by macroalgal mats is unlikely to affect farming practice directly. This is only achievable through the development of new policy instruments, participatory or legislative. Indirect linkages can be identified, but these are tenuous. For example, a decline in estuarine and river water quality may reduce angling revenues to large landowners who own the rights to such recreation. These landowners could then enforce changes in agricultural practice amongst their tenant farmers, as well as adopting better practice on the home estate, in order to reduce nutrient run-off. Similarly, reduction in tourist income, especially from birdwatchers, could theoretically feedback into more sympathetic land use.

In reality none of these feedbacks operate at present, probably because it is unclear to stakeholders what the relative economic and social advantages would be of such actions, given that the ecological effects are largely spillover (without the site of impact production) and that changes in catchment land use initiated today would probably not be translated into an improved estuarine environment for several decades. Predicting the future behavior of this coupled human-ecological system will therefore be challenging, but success would provide a model for many similarly impacted catchments worldwide.

16.7 POLICY, DEBATE, AND THE BURDEN OF SCIENTIFIC PROOF

The concomitant changes in land use, nutrients, and biology of the Ythan seen over the past 40 years provide compelling arguments for direct downstream (spillover) effects of agriculture on estuarine and coastal ecology. Making a persuasive case for causal linkages between agriculture and ecological change was given impetus in 1991 through the adoption of the EC Nitrates Directive (91, 676/EEC) by the U.K. government. Prior to this, substantial material existed in the public domain arguing that a eutrophication problem existed in the Ythan (e.g., Raffaelli et al., 1989, and references therein), including the identification of the Ythan as one of only a few eutrophic areas in the U.K. by the Oslo and Parish Commission (OSPARCOM).

In 1991, the Environment Agency (EA) and SEPA were requested by the U.K. government to examine the status of all rivers and estuaries with respect to the criteria laid out in the EC directive. Waters not meeting those criteria would be considered for designation under the directive. The criteria for designation on the basis of eutrophication include higher concentrations of nitrogen in river water than would be expected from comparison of similar catchments, evidence of algal blooms, and changes in the overall ecology of the river and the estuary. (The criteria for nitrates permissible in drinking water refer only to potable water and do not concern the Ythan River). The Ythan met all these criteria and was consequently proposed for designation as a nitrogen vulnerable zone (NVZ) under the directive. At the time, the Ythan was the only catchment in the U.K. deemed to meet the directive's criteria, and was thus suitable for designation, a somewhat surprising outcome given that the whole of Denmark was proposed by the Danish government.

The consequences of possible designation for the thousand-plus farms within the catchment were not well understood at the time. This uncertainty, together with the independent nature of the Aberdeenshire farmer and their distrust of, and resistance to, externally imposed policy (especially from Brussels) led to considerable public debate and political lobbying. One of the consequences of this was a major three-year research program on the Ythan carried out by a range of agencies and institutions. The aims of this program were to confirm the scientific case for designation by carrying out more detailed research on agricultural practice within the catchment (Domberg et al., 1998, 2000) nutrient relationships of the major mat-forming algae (Taylor and Raven, 2003) and exploring competing hypothesis for the drivers of ecological change, particularly changes in estuarine topography (Raffaelli et al., 1999). Publication of the results of this program did not change SEPA's original proposal for designation, which was imposed by the European Commission in May 2000.

The implications for the farmers within the catchment are still not clear. According to the terms of the directive, the U.K. is merely required to take steps to restore the system. There are no formal restoration targets or restrictions on fertilizer or manure and slurry use, other than adhering to a voluntary code of good agricultural practice. Fears that land prices (and hence financial borrowing power) within the Ythan catchment might be reduced compared to nondesignated catchments have not been realized.

Given the contentious nature of designation, there is considerable pressure on the scientific community to rigorously establish causal linkages and to discriminate between competing explanations for the observed ecological changes. The most persuasive arguments in science (but not necessarily in society) come from experimental tests of hypotheses (e.g., Raffaelli and Moller, 2000), and the Ythan Estuary lends itself well to this approach. Since the mid-1980s, a series of rigorously designed manipulative field experiments have confirmed the proposition that the observed changes in invertebrates and shorebirds are due to the increased coverage by algal mats (Hull, 1987;

Raffaelli et al., 1998; Raffaelli, 1999, 2000). Competing hypotheses concerning the nonagricultural origins of the nutrients entering in the Ythan, long-term climate shifts and suspected changes in the bathymetry of the estuary have been ruled out (Raffaelli et al., 1999). Despite this, the case of NVZ designation of the Ythan remains contentious and considered unproven for many of the farmer stakeholders. However, a participatory approach is being developed by the different stakeholders within the Ythan catchment, which shows considerable promise for dealing with the nutrient issue at a community level (Morris and Morris, 2005).

REFERENCES

Baird, D. and Milne, H. Energy flow in the Ythan estuary, Aberdeenshire, Scotland. *Estuarine and Coastal Marine Science* 13, 443–455, 1981.

Baird, D. and Ulanowicz, R.E. Comparative study on the trophic structure, cycling and ecosystem properties of four tidal estuaries. *Marine Ecology Progress Series* 99, 221–237, 1993.

Balls, P.W., MacDonald, A., Pugh, K., and Edwards, A.C. Long-term nutrient enrichment of an estuarine system: Ythan, Scotland (1958–1993). *Environmental Pollution* 90, 311–321, 1995.

Buchan, R.G. A Comparative Study of Two East Coast Scottish Estuaries, the Lower Forth and Ythan, using the ECOPATH Trophic Interaction Model. MSc thesis, University of Napier, Edinburgh, 1997.

Biological Monitoring Working Party. Final Report: Assessment and Presentation of the Biological Quality of Rivers in Great Britain. Water Data Unit, Department of the Environment, London, 1978.

EC Nitrates Directive. Council Directive of 12 December 1991 concerning the protection of waters against pollution caused by nitrates from agricultural sources. *Official Journal of the European Community* 91/676/EEC, 1991.

Christensen, V. Ecosystem maturity — towards quantification. *Ecological Modelling* 77, 3–32, 1995.

Christensen, V. and Pauly, D. ECOPATH II. A software for balancing steady-state ecosystem models and calculating network characteristics. *Ecological Modelling* 61, 169–185, 1992.

Costanza, R. (1992). "Towards an operational definition of ecosystem health," in *Ecosystem Health*, Costanza, R., Norton, B.G., and Haskel, B.D., Eds. Island Press, Washington, D.C., 1992, pp. 239–256.

Costanza, R., Norton, B.G., and Haskel, B.D. *Ecosystem Health*. Island Press, Washington, D.C., 1992.

Domburg, P., Edwards, A.C., Sinclair, A.H., Wright, G.G., and Ferrier, R.C. Changes in fertilizer and manurial practices during 1960–1990: implications for N and P inputs to the Ythan catchment, NE Scotland. *Nutrient Cycling and Agroecosystems* 52, 19–29, 1998.

Domburg, P., Edwards, A.C., and Sinclair, A.H. A comparison of N and P inputs to the soil from fertilizers and manures summarized at farm and catchment scale. *Journal of Agricultural Science* 134, 147–158, 2000.

Hull, S. Macro-algal mats and species abundance: a field experiment. *Estuarine and Coastal Shelf Science* 25, 519–532, 1987.

Leach, J. Hydrology of the Ythan estuary with reference to distribution of major nutrients and detritus. *Journal of the Marine Biological Association of the United Kingdom* 51, 137–157, 1971.

Morris, T. and Morris, R. The Ythan Project: a case study on improving catchment management through community involvement. *Journal of Environmental Planning and Management* (in press), 2004.

Odum, E.P. The strategy of ecosystem development. *Science* 104, 262–270, 1969.

Pugh, K. "Water quality and aquatic biology of the River Ythan," in *The Ythan*, Gorman, M.L., Ed. University of Aberdeen, 1997.

Raffaelli, D. "The community ecology of the Ythan Estuary," in *The Ythan*, Gorman, M.L., Ed. University of Aberdeen, 1998.

Raffaelli, D. Nutrient enrichment and trophic organisation in an estuarine food web. *Acta Oecologia* 20, 449–461, 1999.

Raffaelli, D. Interactions between macro-algal mats and invertebrates in the Ythan estuary, Aberdeenshire, Scotland. *Helgoland Marine Research* 54, 71–79, 2000.

Raffaelli, D., Hull, S., and Milne, H. Long-term changes in nutrients, weed mats and shorebirds in an estuarine system. *Cahiers de Biologie Marine* 30, 259–270, 1989.

Raffaelli, D., Raven, J., and Poole, L. Ecological impact of green macroalgal blooms. *Annual Review of Marine Biology and Oceanography* 36, 97–125, 1998.

Raffaelli, D., Balls, P., Way, S., Patterson, I.J., Hohmann, S., and Corp, N. Major long-term changes in the ecology of the Ythan estuary, Aberdeenshire, Scotland; how important are physical factors? *Aquatic Conservation: Marine and Freshwater Ecosystems* 9, 219–236, 1999.

Raffaelli, D. and Moller, H. Manipulative experiments in animal ecology — do they promise more than they can deliver? *Advances in Ecological Research* 30, 299–330, 2000.

Rapport, D. et al. Ecosystem health: the concept, the ISEH and the important tasks ahead. *Ecosystem Health* 5, 82–88, 1999.

Scottish Development Department. Development of a Water Quality Index. Applied Research and Development Report No. ARD 3. Scottish Development Department, Edinburgh, 1976.

Taylor, R. and Raven, J.A. "Survival strategies of Enteromorpha from eutrophic sites," in *The Estuaries of North-East Scotland*, Raffaelli, D.G., Solan, M., and Paterson, D., Eds. *Coastal Zone Topics* 5, 31–40, 2003.

Ulanowicz, R.E. *Growth and Development: Ecosystem Phenomenology*. Springer-Verlag, New York, 1986, 203 p.

Ulanowicz, R.E. and Norden, J.S. Symmetrical overhead in flow and networks. *International Journal of Systems Science*, 21, 429–437, 1990.

CHAPTER 17

Assessing Marine Ecosystem Health — Concepts and Indicators, with Reference to the Bay of Fundy and Gulf of Maine, Northwest Atlantic

P.G. Wells

17.1 INTRODUCTION

The Gulf of Maine and Bay of Fundy are located in the northwest Atlantic, bounded by the states of Massachusetts, New Hampshire and Maine, the provinces of New Brunswick and Nova Scotia, and in the coastal waters by various oceanic sub-sea banks such as Stellwagen, Georges, and Brown. It is a highly productive coastal area, noted for its abundant fisheries and marine wildlife. It has sustained coastal peoples and communities before and after European settlement. The Gulf of Maine has been studied for over 100 years, both inshore and offshore, benefiting from the presence of numerous marine institutes and universities ringing its shores. It has an exceptionally long coastline, many islands, and receives numerous large rivers, especially the Penobscot, Kennebec, and Merrimack. The Bay of Fundy is a large macrotidal embayment, forming the northeastern arm of the larger Gulf of Maine. It is

1-56670-665-3/05/$0.00 + $1.50

closely linked oceanographically to the Scotian Shelf and northwestern Atlantic, and receives the inputs of 44 major rivers such as the Saint John Petitcodiac, Avon, and Annapolis countless smaller ones. It is bounded by two provinces with 1.5 to 2 million people, is extensively used by the fishing, shipping, forestry, aquaculture and ecotourism industries, and has two moderately sized cities and many towns and villages along its shores.

The Bay of Fundy is faced with a number of major issues, 38 as counted at the first Fundy workshop in 1996 (Percy et al., 1997). The issues included contaminants and pathogens, barriers on rivers, sediments and coastal erosion, climate change, impacts of fisheries and aquaculture, invasive species, and species and habitat loss. The first workshop has led to five others (Burt and Wells, 1997; Ollerhead et al., 1999; Chopin and Wells, 2001; Wells et al., 2004; Percy et al., 2005 in press), all of them contributing to the information on Fundy and assisting with plotting a path forward for research, monitoring, assessment, community action and management. In particular, the workshop volumes will contribute to an eventual state of environment or SOE report on the Gulf of Maine and Bay of Fundy.

For such SOE or health assessments, data and information are required from a large number of indicators used for monitoring and describing the health of the environment — in this case, the Bay of Fundy. Also required is a process (an outline or framework) for producing periodic, carefully prepared, peer-reviewed reports on the Bay's health and quality (see definitions below), in the context of the greater Gulf of Maine and Northwest Atlantic (Sherman et al., 1996; Wallace and Braasch, 1996; Sherman, 2004). Such monitoring, analysis, and reporting involves the contributions of many people and organizations (a consideration of many perspectives on the approach and necessary information), understanding of some key concepts, and the incorporation of the experiences and knowledge of people who know the bay and the region. It also requires commitment, time, and money.

The task of assessing the bay's health is not simple. Many other "state of the environment" reports have shown the task of objective analysis and synthesis to be challenging. Ecosystems are complex, incompletely known, and constantly changing. Also the measures of health (of organisms other than humans), ecosystem health, and environmental quality are in their infancy. It is important to avoid the pitfalls of the recent tome *The Skeptical Environmentalist* (Lomborg, 2001), where oversimplification, incomplete knowledge, and bias colored the analysis, according to most reviewers. Once a credible approach is chosen for the Bay of Fundy, there is a very large body of literature and experience to distill. The northwest Atlantic and the Gulf of Maine benefit from more than 100 years of oceanographic study, from the works of Bigelow and Schroeder (1953) and Huntsman (1922) to Plant (1985), Backus and Bourne (1987), and Percy et al. (1997), among many other sources. The challenges notwithstanding, we should try to produce a current assessment and a series of reports on the Bay of Fundy and the greater Gulf of Maine, building upon the growing knowledge of the ecosystem and its indicators of health.

Given these objectives, this chapter presents a brief review of current concepts of ecosystem health, environmental quality, and ecosystem integrity; a summary of what is required in health measurements; e.g., the indicators of ecosystem health for the Bay of Fundy; and a framework for the process for assessing and reporting on the health of the Bay of Fundy and its key issues.

17.2 CONCEPTS OF MARINE ECOSYSTEM HEALTH

17.2.1 Conceptual Framework

Various conceptual frameworks have been presented for assessing ecosystem health and environmental quality. Rapport (1986) described a stress-response framework and used it as the basis for Canada's early SOE reports. The Marine Environmental Quality (MEQ) Working Group of Environment Canada (Wells and Rolston, 1991) built on Rapport's stress-response model and presented a framework with four components:

1. Characteristics and uses
2. Stress factors
3. Ecosystem responses (using indicators)
4. Health or condition of the environment

Harding (1992) presented the MEQ model as including stressors, characteristics of exposure, measurement of effects, and indicators of quality. Most recently, Smiley et al. (1998) presented a modified MEQ framework, with the components being condition, stress, effects and response (indicators), and he is applying this in the Strait of Georgia, BC, area. Finally, Fisheries and Oceans Canada (2000a, 2000b) succinctly described MEQ in the new federal Oceans Act as involving guidelines and objectives, indicators, and assessment.

The various approaches show some common needs in coastal and ocean assessments: (1) understanding of the habitats and ecosystem(s) under consideration, and recognition of what we do not know well or at all, given ecosystem complexity; (2) identifying indicators of "health" and "quality" that can be developed by research and used in monitoring; and (3) installing a feedback loop via assessments, monitoring, and management and societal action (through a range of mechanisms, including regulations). Other mechanisms to complete the loop, such as the Fundy Science Workshops, and periodic reports and report cards on progress, are essential as they keep interested parties focussed on the key issues.

By necessity, various terms are used in this field, but not always in the same way. After all, the field is interdisciplinary and evolving. However, this currently causes confusion both in conceptualizing the issue of "ecological" or "ecosystem" health (EH) and in its application — that is, conducting assessments and prioritizing issues. One solution, a major part of this chapter, is to discuss the key terms and concepts including health, marine ecosystem

Table 17.1 Relationship of the terms an concepts on health and ecosystem health (EH), across time, space and levels of biological organization (adapted from Wells, 2003).

Components and Levels of an Ecosystem			
Time/space scale	Individuals	Pop.[1]	Comm./Ecosys.[2]
Short term, current state or condition, generally local	Health	Health	EH*, Integrity**
Long term, status/trends, generally regional	N/A	Quality, Change	Change, EQ***
			Integrity**

* ecosystem health.
** ecological or ecosystem integrity.
*** environment quality.
N/A – not applicable.
1 – population.
2 – communities and ecosystems, including habitats.

health, ecological ecosystem integrity, ecological change, and (marine) environmental quality.

One view of the relationship between the various health concepts, time and space, and level of biological organization is shown in Table 17.1. Health and integrity are a description of the current state, condition or status, a view over the short term (one to a few generations, varying with organism, lasting from hours to decades). Quality and change refer to trends from the "baseline" or original, undisturbed (by humanity) conditions, a view over the longer term (many generations and lifespans, covering decades to centuries). In practice, as shown below, ecological or ecosystem integrity has been used to describe both short- and long-term conditions. Importantly, the terms are used precisely in relation to the level of biological organization (i.e., an individual organism is healthy or unhealthy, whereas a biological community has or lacks ecological integrity). The distinctions are not trivial as they reflect the need to choose quite different indicators across the structural and functional components of ecosystems e.g., biomarkers describe health, whereas species diversity describes ecological integrity.

17.2.2 Health

Health, as in ocean health or health of the oceans, is a commonly used and publicly accepted term referring to the condition or state of the seas (see Goldberg, 1976; Kullenberg, 1982; Wells and Rolston, 1991; McGinn, 1999; Knap et al., 2002). Curiously however, its users usually avoid exact definitions of the word or phrase. Health is defined in the Oxford dictionary as "soundness or condition of body (good, poor, bad, ill, health)" (Sykes, 1976). Health means freedom from or coping with disease on the one hand (the medical view), and the promotion of well-being and productivity on the other (the public health view). There are two dimensions of health — the capacity for maintaining organization or renewal, and the capacity for achieving reasonable human goals or meeting needs (Nielsen, 1999). Importantly, Nielsen states that "health is not a science per se; it is a social construct and its defining characteristics will evolve with time and circumstance." Earlier, Rapport et al. (1980) considered the concept of health and the need to recognize vital signs, a topic explored below. Finally, health is usually defined by what it is *not*, such as "the occurrence of disease, trauma or dysfunction" (Webster's Dictionary, 1993). Therefore, a healthy marine environment requires individuals of each species (ecologically, individual organisms) with signs of wellness and productivity, based on vital signs, and the absence of obvious disease or lack of function. Health as a concept is readily understood, has social capital (healthy is preferred to unhealthy), is transferable to ecosystems (as shown below), and in practice is measurable (though with great difficulty for most marine organisms).

17.2.3 Ecosystem Health

Ecosystem health, as a concept and practice, has been discussed at length for at least two decades. Papers and reports include Rapport et al. (1980, 1998a, 1999); Kutchenberg (1985); Rapport (1989, 1992, 1998); International Joint Commission (1991); Calow (1992, 1993, 1995, 2000); Costanza et al. (1992); Suter (1993); Sherman (1994b, 2000c); Environment Canada (1996); Schaeffer (1996); Jørgensen (1997); Vandermeulen (1998); Fairweather (1999); Tait et al. (2000); Wood and Lavery (2000); Sutter (2001); and Wilcox (2001). Some of these publications are discussed below.

Rapport and his colleagues are leaders in exploring the field of ecosystem health. Rapport et al. (1980) discussed early warning indicators of disease, hypersensitivity, epidemiological models, the crucial role(s) of certain parts of a living system, Selye's concept of stress without distress (Selye, 1974), and immune antibody responses. These topics have advanced further due to research in medicine, epidemiology, toxicology, and environmental toxicology since the 1980s. Rapport et al. (1980) stated: "The corresponding ecological concept to health might be ecosystem persistence, or ecological resilience. Presumably this property can be assessed using a range of indicators ... candidate vital signs include primary productivity, nutrient turnover rates, species diversity, indicator organisms, and the ratio of community production to community respiration." A very important observation was that "once ecosystems are adequately characterized in terms of vital signs, the development of more comprehensive diagnostic protocols might be the next logical step." Developing and standardizing such protocols has been at the heart of applied ecotoxicology and environmental monitoring for years now. In addition, this step has been taken by groups such as Health Ecological and Economic Dimensions (HEED), of Global Change Program (Center for Health and the Global Environment at Harvard University) (B.H. Sherman, 2000) and Kenneth Sherman of National Oceanic and Atmospheric Administration

(NOAA) (USA) with his internationally recognized work on large marine ecosystems (LMEs) (K. Sherman, 2000, 2004). Much work is continuing with molecular biomarkers (Depledge, M., pers. comm.; Galloway et al., 2004), and the connections across levels of biological organization to populations and communities (e.g., Downs and Ambrose, 2001; Livingston, 2003).

A review of the core studies of ecological health and ecosystem health reveals some key observations. On indicators and indices: "Ecosystem health is a characteristic of complex natural systems ... defining it is a process involving (a) the identification of important indicators of health; (b) the identification of important endpoints of health; and (c) the identification of a healthy state incorporating our values. Historically, the health of an ecosystem has been measured using indices of a particular species or component." (Haskell et al., 1992). It is clear that we need to choose indicators and monitor ecosystems with them, and then summarize and interpret the responses using indices. This in fact is being done, for example, in the U.S. Environmental Protection Agency's (EPA's) estuarine programs, in larger comprehensive coastal programs such as in Chesapeake Bay and other U.S. mid-Atlantic estuaries (Kiddon et al., 2003), and in a multitude of community-led monitoring programs in places around the Gulf of Maine (Chandler, 2001; Pesch and Wells, 2004).

On the components of ecosystem health, Schaeffer et al. (1988) gave ten guidelines for assessing ecosystem health (Haskell et al., 1992). At two 1991 workshops, participants developed a working definition of ecosystem health, defining health in terms of four characteristics applicable to any complex system — sustainability, which is a function of activity, organization, and resilience. Sustainability implies that the ecosystem can maintain its structure and function over time and space, maintaining its dynamic nature and changing slowly. The conclusion was that "an ecological system is healthy and free of 'distress syndrome' if it is stable and sustainable — that is, if it is active and maintains its organization and autonomy over time, and is resilient to stress." This, of course, implies that activity, organization and resilience can be measured for each, at least major, component of each marine ecosystem under scrutiny, a daunting task indeed.

One problem of defining and describing ecosystem health is choosing the appropriate geographic and biological scales (i.e., defining which ecosystem we are managing and what level we are focusing on). For example, for Chesapeake Bay, are the bay and its many estuaries being considered, or is it the whole Chesapeake watershed? The same question applies to the Bay of Fundy — do we focus on the whole bay (there are already 38-plus issues), the extensive watersheds (the proposed approaches of the GOMCME, and the Minas Basin Working Group of (Bay of Fundy Ecosystem Partnership, BOFEP), or just one part (e.g., Passamaquoddy Bay). Choosing the appropriate spatial scale has implications for a wide range of activities associated with describing, managing and maintaining ecosystem health; issues of policy, governance, research, assessment, management tools, monitoring, communication, and stakeholder involvement ultimately have to be considered.

The human health assessment model was described by Haskell et al. (1992). It has six parts:

1. Identify symptoms
2. Identify and measure vital signs
3. Make a provisional diagnosis
4. Conduct tests to verify the diagnosis
5. Make a prognosis
6. Prescribe a treatment

For a large marine ecosystem, in this case the Bay of Fundy (part of the greater Gulf of Maine), this model of health assessment could work as below. A compendium for items 1 and 2 is as yet incomplete, so examples come from Bay of Fundy Science Workshops, ongoing projects, and the literature.

17.2.3.1 *Identify Symptoms*

What are the first signals that the system is "unhealthy"? From the physical to the biotic environment, some are:

- Physical changes to shorelines (e.g., barriers, such as causeways and dykes; increased coastal development especially homes along the shorelines)
- Changed sediment patterns in estuaries (e.g., Avon R. estuary)
- Contaminants in sediments and tissues (e.g., mussels, salmon)
- Increased numbers of aquaculture sites in bays
- Abundant debris on shorelines
- Reduced fisheries catches or failing fisheries (e.g., cod, salmon)
- The requirement to open new fisheries on new or previously "under-utilized" species (e.g., sea urchins, seaweeds)
- Reduced numbers of seabirds (phalaropes), and marine mammals (right whales)
- Increased small boat traffic in bays and inlets (noise, water and air pollution).

17.2.3.2 *Identify and Measure Vital Signs*

There are a number of critical changes in key attributes of the ecosystem that collectively show the system is under stress or change. For example:

- Loss of/reduced fisheries (cod)
- Loss or reduction of species (wild Atlantic salmon, *Salmo salar*)
- Changed distributions of seabirds such as the red-necked phalarope, *Phalaropus lobatus*
- High levels of some chemicals in biota (e.g., Cu in crustaceans, and PAHs and PCBs in birds and mammals)
- Changed water flows or hydrologies in estuaries (e.g., Petitcodiac, Avon)
- Overall reduced salt marsh acreage in the upper bay.

17.2.3.3 Provisional Diagnosis

At the 1996 Fundy Science Workshop (Percy et al., 1997), the participants concluded that the Bay of Fundy was showing a number of signs of poor health and lowered quality, and a list of 38 key issues was made. Many of these have been discussed at subsequent Fundy Science Workshops, and many other recent meetings around the Gulf of Maine (e.g., RARGOM Conference, Wallace and Braasch, 1997; Rim of the Gulf Conference 1997; habitat conferences; further RARGOM meetings, e.g., Pesch, 2000). The Conservation Council of New Brunswick has recently expressed concerns for coastal habitats throughout the bay, with a careful record of habitat loss or modification (Harvey et al., 1998) and there are marked changes in fisheries over 200 years and the presence of chemical burdens in Passamaquoddy Bay species (Lotze and Milewski, 2002; Mills, 2004).

17.2.3.4 Tests to Verify Diagnosis

Diagnostic tests include:

- Monitoring for trace contaminants in mussels and in the food chain (e.g., levels, biomarkers, effects)
- Monitoring for algal toxins (e.g., domoic acid)
- Monitoring for bacterial pathogens (e.g., at all shellfish beds)
- Monitoring for effects of salmon aquaculture wastes on benthic species and communities
- Assessment of condition of remaining salt marsh habitat
- Assessment of effects of tidal barriers (e.g., the 2000 to 2002 tidal restriction audits being completed in New Brunswick and Nova Scotia).

This important verification step ensures that the ecological health issue is real and important (economically and ecologically). It is also more tractable to address single issues than the whole system at once. With multiple issues, the potential for cumulative effects, and the potential for confounding with natural variables in ways difficult to predict, diagnosing the whole system is the goal but it can be achieved most successfully with "bite-sized" efforts.

17.2.3.5 Make a Prognosis for the Bay

This is the "ecosystem health report" and a statement of the future for the bay's habitats, and natural and living resources. Is there a good chance of "recovery," or "maintaining the status quo" if we continue to act through protection, conservation, and remediation efforts? This prognosis is probably most effective if looked at by sector — fisheries, marine mammals, wildlife, sediments, coastlines, estuaries, etc. — and by regions within the bay, from Passamaquoddy Bay around to Annapolis Basin, St. Mary's Bay and the coast to Yarmouth. The prognosis is best captured in the periodic "State of the Bay of Fundy" and "State of the Gulf of Maine" reports (e.g., Pesch and Wells, 2004; GPAC, 2004).

17.2.3.6 Treatment

This step describes the actions required to restore ecosystem health, in this case to the Bay of Fundy. For example, recent positive actions include working with the IMO (UN) to select sea lanes away from the northern right whale feeding areas (a real success); remediation of unused salt marsh in Shepody Bay; implementation of better management plans for specific fisheries species such as bait worms (polychaetes) and green sea urchins (*Strongylocentrotus droebachiensis*); gradually improved sewage treatment, such as at Saint John; efforts to remediate a tidal river, for example, Petitcodiac River; improved aquaculture practices in southwestern NB; and identifying the potential for opening selected causeways and restoring tidal flows in estuaries, for example, Windsor, NS. Actions are small and large, most are opportunistic, but all contribute to the momentum of addressing issues confronting the bay.

What is our capacity to conduct ecosystem health assessments? Many traditional "state of the environment" reports have been prepared, some very thoroughly (e.g., Arctic Monitoring and Assessment Programme, AMAP, 1997), but we may only be in the early stages of being able to do actual ecosystem health assessments because we lack the "medical encyclopedia" for ecosystems (Haskell et al., 1992). As Norton et al. (1991) and Haskell et al. (1992) point out that medicine deals with the individual (i.e., the person), whereas "ecosystems exist on many (biological) levels, can be described on many scales, and require a consensus of public goals on the road to having diagnostic tests for ecosystem stress." A framework for starting to evaluate an ecosystem is to assemble a table of (1) types of stress; (2) response variables or symptoms of ecosystem distress; (3) monitoring, fiscal resource, management, and other needs. The table (6.1) in Percy et al. (1997), pages 140–141, is an excellent tool but requires an update. We have to move from the traditional environment report, a valued but disconnected tally of characteristics of the system, to an actual health assessment, an integrated analysis of how well the system is functioning or not. The Gulf of Maine and Bay of Fundy offer an opportunity few other places (if any) have to prepare such an ecosystem health assessment. Encouragingly, EPA (e.g., EPA, 2001) and K. Sherman (e.g., Sherman and Skjoldal, in press) are moving in this direction.

There are limitations to the concept of ecosystem health, and especially to putting the concept into practice. The concept has a recent history in western-based science, medicine, and conservation. It first formally "emerged in the mature thought of Aldo Leopold as a bridge between technical management and formulation of management goals," (Leopold, 1949; Haskell et al., 1992), hence it is not just a scientifically based concept. This is very important because there is great value, indeed crucial value, in the link to environmental management goals (see section 17.3.4). Metaphorically, the concept and term has the strength of communicating the problem to a wide audience. The term "ecosystem health" was used in the 1960s and 1970s in the context of the Great Lakes, especially Lake Erie, once considered "dead," rather than having impaired "ecosystem health". Lake Erie survives, with impaired but improving health (it is alive and productive) and a lower quality (its current condition compared to the original state of the lake) (see IJC 1991). Likewise, in the U.S., the ecosystem health concept has been applied to important estuaries and coastal bays, such as Chesapeake and San Francisco (see numerous EPA reports, such as EPA, 1998, 1999, 2000a, 2000b; Kiddon et al., 2003).

However, several important limitations with the ecosystem health concept should be kept in mind, as we consider the Bay of Fundy and the greater Gulf of Maine. First, "no longer are communities (natural) considered normative. Disturbance is common; communities and ecosystems are in constant flux. Knowing what is natural is difficult," (Ehrenfeld, 1992, in Costanza et al., 1992). That is, the normal range for a variable may be quite wide (note in particular Schindler, 1987), and in this age of marked climate change, even more so (e.g., air temperature, storm events, and levels of precipitation). The so-called "baseline" for normal ecosystem health fluctuates. See Pauly (1995) for discussion in context of fisheries. Second, "a determination of ecosystem health can be a function of which process you are looking at, which in turn is determined by your own values," (Ehrenfeld, 1992). Ecosystem health has a social context, as does the science behind it. Third, the word "health", as well as the concept of "quality", should not be defined or applied too rigorously because communities of plants and organisms comprising ecosystems vary greatly in their state of equilibrium. Hence the term 'ecosytem health' is best used as a bridging concept between the scientists and nonscientists (Ehrenfeld, 1992), a starting place for dialogue on issues, priorities, and choice of indicators.

Systems ecologists have views as to what is meant by ecosystem health. These views shed light on the selection of suitable indicators for the Bay of Fundy. For example, Ulanowicz (1992) in Costanza et al. (1992) states: "A healthy ecosystem is one whose trajectory toward a climax (referring to ecological succession) is relatively unimpeded and whose configuration is homeostatic to influences that would displace it back to early successional stages. Assessing the health of ecosystems requires a pluralistic approach and a number of indicators of system status," (also see Schaeffer et al., 1988; Karr, 1991). Ulanowicz uses the approach of network ascendancy, an index that captures four key properties of quantified networks of trophic interactions: greater species richness, more niche specialization, more developed cycling and feedback, and greater overall activity, in healthy systems. This approach could be usefully applied to the Bay of Fundy, and its various ecosystems and regions; one could hypothesize that in some places (e.g., near aquaculture sites, in urbanized harbors, and near industrial effluent locations) these properties have been diminished, and can be investigated (as in recent aquaculture studies in the lower Bay of Fundy, (G. Pohle, HMSC, pers. comm.)).

Table 17.2 A hypothetical health status of the Bay of Fundy (based on approach in Karr, 1992 and earlier papers). Arrows denote direction of change. Adapted from Wells, 2003.

	Health status		
System criterion	**Good**	**Fair**	**Poor**
Inherent potential	+>	+>	
Condition			
Self-repair	+>	+>	
Management support			

Karr, a fisheries biologist, stated "A biological system can be considered healthy when its inherent potential, whether individual or ecological, is realized, it's condition is stable (meta-stable), its capacity for self-repair when perturbed is preserved, and minimal external support for management is needed," (Karr et al., 1986). One can analyze the Bay of Fundy using this approach (Table 17.2), based on personal judgment. Hence, when treated as a system, the Bay of Fundy's condition could be considered as deteriorating and in need of enlightened integrated management.

Finally, and following from above, one should further consider which additional components of ecosystem health are required. Costanza (1992) discusses ecosystem health in the context of a system's overall performance: "To understand and manage complex systems, we need some way of assessing the system's overall performance (its relative health)." He summarizes the components of ecosystem health as:

- Homeostasis
- The absence of disease
- Diversity or complexity
- Stability or resilience
- Vigor or scope for growth
- Balance between system components.

"Systems are healthy if they can absorb stress and use it creatively, rather than simply resisting it and maintaining their former configurations. An ecological system is healthy and free from distress syndrome if it is stable and sustainable — that is, if it is active and maintains its organization and autonomy over time, and is resilient to stress." (Costanza, 1992) To be healthy and sustainable, the system must maintain its metabolic activity level, maintain its internal structure and organization, and be resilient to outside stresses.

Costanza (1992) has attempted to quantify estimates of ecosystem health:

HI = overall health index

$$HI = V \times O \times R$$

where V is the system vigor, O is the system organization index (0 to 1), and R is the system resilience (to stress) (0 to 1) (see Box 1, Rapport et al., 1998a). This approach, producing an overall health index, could be tried for the various regions, habitats and trophic levels of the Bay of Fundy. In fact,

Table 17.3 Hypothetical assessment of ecosystem health (EH) of the Bay of Fundy, using Rapport's indicators of stress (Rapport et al., 1998) (from Wells, 2003).

Stress results in	Bay of Fundy	Passamaquody Bay	Minas Basin	Saint John Harbor
Biotic impoverishment	yes	yes	?	yes
Impaired productivity	?	?	yes	yes
Altered biotic composition	yes	yes	yes	yes
Reduced resilience	?	?	?	yes
Increased disease prevalence	yes	yes	?	yes
Reduced economics	yes	yes	yes	yes
Risks to human/org.health	yes	yes	?	yes

Rapport et al. (1998a) have taken this approach and advanced upon it. They stated: "Many ecosystems are unhealthy — their functions have become impaired." They looked at the literature and presented several case studies (e.g., Central Rio Grande Valley, the Great Lakes Basin). Using their choice of indicators of ecosystem health, a hypothetical assessment of the Bay of Fundy might appear as Table 17.3, although this requires quantitative verification.

17.2.4 Marine Ecosystem Health

In an early paper on ecological terms for large lakes, Pamela Stokes' description (Stokes, 1981) of 'healthy,' in the context of aquatic ecosystem health, was: "It includes (1) stability — gross structure unchanged over many years; (2) balance; and (3) functioning. In the context of lakes, an example of 'unhealthy' would be a condition caused by the addition of toxins; if algae are killed and bacteria increase out of balance, the lake is not healthy."

Since that time, for coastal and open ocean systems, the term "marine ecosystem health" or MEH has come into common usage, at least in North America (amongst others: Wells and Rolston, 1991; Wells, 1996, 1999a; Smiley et al., 1998; Vandermeulen, 1998; Sherman, B.H. 2000; Sherman, K. 2000). Paul Epstein's definition of marine ecosystem health is: "To be healthy and sustainable, an ecosystem must maintain its metabolic activity level, its internal structure and organization, and must be resistant to stress over a wide range of temporal and spatial scales" (Epstein, 1999, 2000). This approach to MEH was comprehensively presented in HEED (1998).

B.H. Sherman (2000d) states: "Ecosystem health is a concept of wide interest for which a single precise scientific definition is problematical." He described the HEED approach of Harvard University: "A marine health assessment is a rapid global survey of possible connections and costs associated with marine disturbance types." This program was initiated in 1995; it published its first survey in 1998 (Epstein and Rapport, 1996; HEED, 1998). Eight disturbance types are described, shown above (Table 17.4) in the context of Bay of Fundy. "The 8 general types of disturbance may provide a first approximation of the comparative health of coastal marine ecosystems.

Table 17.4 Disturbances in the Bay of Fundy: a preliminary list, following from Epstein and Rapport (1996) and HEED (1998). Adapted from Wells (2003).

Type of disturbance*	Y/N	Bay of Fundy occurrence/comments
Biotoxin and exposure	Y	Toxic algal blooms
		Beach closures
Anoxic/hypoxic	N	Unlikely due to tidal exchange
Trophic-magnification	Y	Toxic algal blooms
		Contaminants e.g., Hg
Mass lethal mortality	N?	None recorded
Physically-forced (climate/ocean)	Y	Severe storms
Disease	Y/N?	Imposex in snails; none in birds
New, novel occurrences and invasives	Y	Crabs (green, Japanese shore)
Keystone endangered and	Y	Algal blooms; beach closures;
chronic cyclical		fisheries closures; invert. declines;
		bivalve contamination

*As per HEED (1998) and B.H. Sherman (2000).

Mortality, disease and chronic disturbances are the three major variables or changes reported across a wide spectrum of taxonomic groups," (B.H. Sherman, 2000). Assessments using such a marine epidemiological approach can track these changes in ecosystem health (Epstein and Rapport, 1996), providing that these indicators are included routinely in monitoring programs, which currently they are not.

Kenneth Sherman's large marine ecosystem (LME) approach (Sherman et al., 1996; Sherman and Skjoldal, 2002; other LME publications) is applicable to the Gulf of Maine and Bay of Fundy. Indeed, the Gulf of Maine is part of an LME and is identified as the Northeast Shelf Ecosystem (unfortunately, the boundaries are jurisdictional rather than ecological as the Gulf in part of the greater Northwest Atlantic). Methods to assess the health of LMEs are being developed from modifications to a series of indicators and indices described by several investigators (see Costanza and Mageau, 1999); the methods form part of the pollution and ecosystem health module of the LME approach (Sherman, 2004). Recent workshops sponsored by NOAA (USA) and the Nordic Council of Environment Ministers (Europe) considered the concepts of ecosystem health and marine ecosystem health (K. Sherman, 2000, 2004). The Pollution and Ecosystem Health module consists of eutrophication, biotoxins, pathology, emerging disease, and health indices. Five health indices (biodiversity, productivity, yield, resilience, stability) are included as experimental measures of changing ecosystem states and health. Hence, the module emphasizes stresses and indices (based on numerous indicators), consistent with current thought on how to approach marine ecosystem health (MEH).

In practice, the pollution and ecosystem health module of the LME uses benthic invertebrates, fish, and other biological indicator species, and accepts the following set of measures (Sherman et al., 1996; K. Sherman, 2000, 2004):

- Bivalves (Musselwatch) (similar to Gulfwatch employed in Bay of Fundy, Chase et al., 2001; Jones et al., 2001)

- Patho-biological examination of fish
- Estuarine and nearshore monitoring of contaminants and contaminant effects in water, sediments and organisms (note NOAA's programs, Wade et al., 1998; for Gulf of Maine, see Jones, 2004.)
- Routes of bioaccumulation and trophic transfer of contaminants
- Examination of critical life stages and food chain organisms to demonstrate exposure
- Impaired reproductive capacity
- Organ disease
- Impaired growth
- Impacts on individuals and populations.

Sherman et al. (1996) adopted a holistic approach inherent in the LME concept. It encourages agencies and other stakeholders to address issues of overfishing, habitat loss, pollution, and recreation needs from a multidisciplinary ecosystems perspective. This approach is finally being incorporated into the Gulf of Maine Program, as the linkages are made between indicators, monitoring and reporting, across the Gulf (Jones and Wells, 2002; GOMCME, 2002) .

Vandermeulen (1998), in a Canadian summary, stated that MEH indicators are being identified in five categories:

1. Contaminants
2. Biotoxins, pathogens, and disease
3. Species diversity and size spectrum
4. Primary productivity and nutrients
5. Instability or "regime shifts."

These were adopted from the literature, and hence are similar to MEH indicators chosen by the many expert groups involved in the LME and GOMCME approaches (K. Sherman, 2000; Jones and Wells, 2002).

The terms "marine ecosystem health" (MEH) and "marine environmental quality" (MEQ — see below) are often used interchangeably in the literature and in common practice when communicating about the health of coastal seas and the oceans (Wells, 1991; chapter 6 in Wells and Rolston, 1991). The case made in this chapter, however, is that the terms health and quality are not the same (section 17.2.1), and that there are benefits from using them more precisely in an assessment of the health of coastal waters, in this case the Bay of Fundy and Gulf of Maine — that is, we want to maintain and sustain a healthy bay and gulf of high quality.

17.2.5 Ecological or Ecosystem Integrity

Ecological integrity (also called ecosystem integrity) is "the dimension of health that reflects the capacity to maintain organization; it is akin to the term 'integrity', especially when used at the scale of ecosystems," (Karr, 1992). It incorporates the ideas of resilience, vigor, and homeostasis. "Many regard integrity, when used in a purely ecological sense, to refer to the evolution of the ecosystem without human disturbance," (Nielsen, 1999). Key papers include

Karr (1981, 1991, 1992, 1993); Harris et al. (1990); Kay (1991); Holling (1992); Woodley et al. (1993); Noss (1995); Nielsen (1999); and Campbell (2000).

Karr (1992) and Campbell (2000) discussed the concept in detail. Integrity "implies an un-impaired condition, or the quality or state of being complete or undivided," (Karr, 1992). It also means "un-impaired when compared with the original condition," (Campbell, 2000).

Systematic assessments of the status of ecological resources (i.e., ecological integrity) have three requirements: they must be based in biology and biological processes; there must be a selection of measures of health or integrity appropriate for the place and biological attributes of concern; and there must be biological benchmarks or reference conditions (Karr, 1992). On the second requirement, an array of attributes of biological/ecological integrity is used. These equate to the indicators used to measure ecosystem health and environmental quality in monitoring programs. Collectively, the attributes must be very diagnostic of local conditions. For example, five attributes (individual health, species richness, relative abundance of species, population age structure, genetic diversity) measured together can describe local conditions. Efforts to assess ecological integrity are more likely to detect degradation if those efforts are conceptually diverse — from the use of individuals to populations to assemblages to landscapes, for the measurement of attributes (Karr, 1992).

Karr uses the terms "conditions," "ecological health," and "integrity" interchangeably — not surprising given the similarity of the components, but unhelpful to using the concepts in practice with precision. Campbell (2000) agrees with previous writers (e.g. Suter, 1993) that we need operational definitions of concepts such as ecosystem health, and by analogy, ecological integrity. Campbell concludes that "ecological integrity is an ecosystem property that is greatest when all the structural components of a system that should be there, are there (i.e., structure is complete), and all the processes operating within the system are functioning optimally (i.e., the ecosystem is 'healthy')." Perhaps what is important is that condition, ecological health, and ecological/ecosystem integrity refer to the current state of a system, how well it is composed and functioning now. Semantic arguments fall prey to practical needs.

In this context, the important question is: Can we assess the Bay of Fundy with the more operationally useful definition of Campbell (2000) — are all the structures/parts and functions of the Bay of Fundy's Gulf of Maine's ecosystems present and operating optimally? Are we monitoring enough of the ecosystem to be able to describe the integrity of natural communities and ecosystems in the Bay of Fundy as a whole, or will we do such monitoring of key attributes (i.e., indicators) only for selected sites, for example, salmon aquaculture sites, tidal barriers, harbors, points of industrial discharge, mudflats, seabird breeding sites, etc.?

A second important question is: Which of the terms, "ecosystem health" or "ecological/ecosystem integrity," has more social capital associated with it? It is noteworthy that the GOMCME, in Goal 2 of its third five-year action plan

(GOMCME, 2002), addresses "human health and ecosystem integrity," with three social, management-oriented objectives: (1) increasing awareness and improving management of priority contaminants; (2) identifying reduction strategies for priority contaminants; and (3) enhancing citizen stewardship. Protecting and assessing the ecological or ecosystem integrity of the Gulf of Maine and the Bay of Fundy has been identified as a long-term social goal of institutions around the Gulf. The challenge will be to put the supporting monitoring programs into place for many decades to achieve the goal.

17.2.6 Ecological Change

Change is constant (i.e., continual), in Earth's ecosystems. Marine ecosystems are no exception to this rule. What is important is to distinguish between natural ecological change, important anthropogenically driven change, and the two combined (Schindler, 1987; Spellerberg, 1991; Ollerhead et al., 1999; Wells, 1999b; Jackson, 2001; Jackson et al., 2001), and to identify the important adverse change(s) that can be ameliorated (e.g. ozone depletion due to CFCs; contamination of food supplies and ecosystems by other synthetic chemicals; climate change if we control CO_2 emissions). Changes should be observed or measured over the long term, and compared to measurements of, or approximations of, the original conditions (set at some arbitrary time). The choice of appropriate indicators (see section 17.3.3.2), the monitoring design, and modeling (Jakeman et al., 1993) are critical to success. "The question is how to better identify, monitor, anticipate, and respond to the network of changes in the ecosystem," (Zelazny, 2001), and how best to periodically report on and interpret such change for the Bay of Fundy and the Gulf of Maine.

Change and ecological change has been studied and discussed recently by Schindler (1987); Rapport (1990); Duarte et al. (1992); Spellerberg (1991); Hall and Wadleigh (1993); McMichael (1993, 2001); Earle (1995); Myers (1995); Epstein and Rapport (1996); HEED (1998); Jickells (1998); McGowan et al. (1998); Harvell et al. (1999); Rapport and Whitford (1999); Wells (1999b); Mann (2000); Rose et al. (2000); Baird and Burton (2001); Clark and Frid (2001); Downs and Ambrose (2001); Jackson (2001); Lotze and Milewski (2002); Martens and McMichael (2002); and others. Ecological change has been taking place over the geological epochs, and in "recent times," for the Gulf of Maine region, since the last ice cover 12,500 to 15,000 years ago (Atlantic Geoscience Society, 2001), sea levels have lowered and the land has been reoccupied by plants and animals. Gradual change, and occasional abrupt occurrences (disturbances, including extinction events), are normal to ecosystems. Organisms, populations, and animal and plant communities adapt in a variety of ways, from physiological to reproductive to distributional patterns. What must be understood is how ecosystems accommodate to the natural change and the changes imposed by human activity at the same time, particularly when the latter includes large perturbations such as biomass removal, habitat destruction or modification, chemical effects (toxicity), and competition from bio-invaders or exotics in coastal waters (Wells, 1999a; GESAMP, 2001a).

Ecological change can be subtle. A change in the health of the system moves to a change in the systems overall quality, often without being noticed or measured. Examples are numerous in the Bay of Fundy: the ecological effects of new fisheries for so-called under-utilized species (e.g., sea urchins, sea cucumbers, rockweed, gastropods, polychaetes); the impacts of barriers on tidal rivers on mudflats and other intertidal zones; the progressive loss of freshwater reproductive habitat (e.g., salmon, striped base); and the potential impact of tourism on migratory shorebirds (i.e., disturbance at critical feeding and roosting areas in the intertidal zones and on islands, affecting piping plovers and semipalmated sandpipers).

Ecological systems are complex and chaotic, many interactions are nonlinear, and some species play pivotal roles in the transfer of energy between trophic layers (e.g., keystone species such as *Corophium volutator*, D. Hamilton et al., pers. comm.) (Myers, 1995; Livingston, 2003). Once disturbed by an anthropogenic stress, the ecosystem may not recover or offer the possibility for remediation (e.g., overfished areas, water bodies with introduced species, areas of coastal development, highly contaminated sites), the system entering a new and different ecological state, possibly forever (Pavly and MacLean, 2003).

Ecological change(s) occurs at different spatial and temporal scales. For example, compare the impact of a single fishery for a keystone benthic species such as sea urchins, the change occurring rapidly over many hectares, to the predicted impacts of global climate change on sea temperatures and sea surface levels, the change occurring gradually by years and decades, over tens of thousands of square kilometers. This range of change is probably very common in ecosystems, though most of it goes undetected and unaccounted for in our assessments of ecosystem health. The challenge is to examine the Bay of Fundy and Gulf of Maine as a system and an ecosystem, and determine the type of change(s) occurring, the causes, the interactions, and the pragmatic actions society should take.

17.2.7 Marine Environmental Quality (MEQ)

Papers covering the concepts and practice of marine environmental quality (MEQ) include Wells and Côté (1988); Côté (1989); Wells and Rolston (1991); Harding (1992b); Buckley (1995); IOC (1996); Lane (1998); NOAA's many programs regarding the long-standing MEQ Status and Trends Program (NOS, 1998; T. O'Connor and D. Wolfe, pers. comm.); DFO (2000); Percy and Wells (2002); and Rapport's many papers (see References). Recent discussions include those of Vandermeulen and Cobb (2001) and Westhead and Reynoldson (2004).

In Canada, MEQ was defined during the 1980s and early 1990s by the Environment Canada MEQ Working Group (Wells and Côté, 1988; Wells and Gratwick, 1988; Wells, 1991, 1996), and accepted by the federal

Interdepartmental Committee on Oceans in 1989. "MEQ is the condition of a particular marine environment measured in relation to each of its intended uses and functions. It can be described subjectively, especially if stresses impinging on the system are large and if the ecosystem or habitat are obviously degraded. However, MEQ is usually assessed quantitatively for each environmental compartment, on temporal and spatial scales. It is measured using sensitive indicators of natural condition and change. Such measures are interpreted using objectives and limits set by environmental, health and resource agencies." (Wells, 1991).

MEQ differs from MEH, in this author's opinion. Quality denotes historical recorded change in the condition, whereas health is the present condition and the direction of change (as discussed above; A. Gaston, CWS, pers. comm.). However, the terms MEQ and MEH are often used synonymously in the literature, especially nontechnical literature, and importantly in day-to-day practice by conservation and protection groups, to mean "ocean health" (e.g., IOC, 1996; Frith, 1999; McGinn, 1999). The social and political currency of the term "health" captured in the context of "oceans" and "coastal" was the compelling reason for calling the 1991 report on Canadian MEQ "Health of Our Oceans," and likely the reason for the earlier report "Health of the Northwest Atlantic" (Wilson and Addison, 1984), whereas both reports assessed both the health and quality of Canadian marine waters.

Harding (1992b) further explored the MEQ concept and measures, developing a framework focused on chemical contaminants and incorporating the ecological risk assessment components of sources, exposure, effects (indicators), and risk estimation. This was an important conceptual advance to understanding the breadth of MEQ and the importance of linking stresses and effects to management action through risk analysis and risk management. Chang (1999) and Chang and Wells (2001) then developed an MEQ framework for the Bay of Fundy. It shows the linkages between research, monitoring (indicators), objectives/guidelines, assessments, and management response. This framework was very useful at evaluating selected stresses on the ecosystem (e.g., PCBs or polychlorinated biphenyls, mercury, and algal toxins).

As with assessments of health and ecological health, MEQ requires indicators and marine environmental guidelines, objectives and standards. The objective is to take measurements of key ecosystem variables over many years and compare values to original baseline conditions, or their best proxy. Guidelines, objectives, or standards for water, sediments, and tissues are essential. Underpinning both the choice of indicators and guidelines is research. For example, the Gulf of Maine Musselwatch program (Gulfwatch) is an MEQ program, using an indicators species and approach (the mussel), guidelines (largely from human health), and supportive research. Contaminant levels in mussel tissues are measured annually at various stations around the Gulf of Maine; values are compared to earlier ones, and all measurements over time, as part of a status and trends analysis, and the values are compared with environmental and health advisory guidelines (Chase et al., 2001; Jones et al.,

2001). The program has given a picture of trace chemical contamination around and across the Gulf in the 1990s; tissue levels of trace contaminants are elevated, stable or often declining, and largely below health guideline values (Chase et al., 2001; Wells et al., 2004).

Within the new Canadian Oceans Act (1997), MEQ within DFO (or Fisheries and Oceans Canada) focuses on the requirement for objectives and guidelines for protecting ocean health, the latter not being defined. However, MEQ activity under the Oceans Act plans to cover three areas—research on and testing of indicators of ocean health, the development and use of objectives and guidelines, and the production of assessments of ocean health (DFO, 2000). MEQ has also taken on a broader context than just chemical contaminants, although that appears still to be the emphasis in an excellent review of contaminants on the Scotian Shelf and the adjacent coastal waters (Stewart and White, 2001). Most recently, the DFO's Oceans Strategy (DFO, 1997, 2002a, 2000b) considers MEQ under the umbrella of "understanding and protecting the marine environment," with an emphasis on science support for oceans management (which includes "assessing the state of ecosystem health"), MPAs and MEQ guidelines. With DFO and Environment Canada as lead agencies, Canada now has a mandated operational concept of MEQ, which should stimulate cooperative multiagency and multipartner research, monitoring, MPAs, ecosystem-based guideline development, and ecosystem assessments for its oceans, including the Bay of Fundy, Gulf of Maine and their adjacent waters in the Northwest Atlantic.

17.2.8 Sustainability of Marine Ecosystems

Over the long term, measured in at least hundreds of years (not geological time), humans strive to maintain or sustain the ecology and resource values of the sea, especially productive coastal and shelf ecosystems. There is much current discussion of how to achieve this in the context of marine fisheries (Pauly et al., 1998; Jackson et al., 2001; Zabel et al., 2003), marine biodiversity (Norse, 1993), and human population health (Knap et al., 2002; McMichael, 2002). There is a considerable international opinion that marine ecosystems are not being managed sustainably and that coastal seas in particular are being diminished in their natural living resources and their overall quality (e.g., GESAMP 2001a, 2001b; Pauly et al., 1998; Zaitsev and Mamaer, 1997). One of humanity's current greatest challenges is to reverse this trend and put into place an effective international monitoring and assessment program to conserve future sustainability of the seas.

17.2.9. Human Health and Marine Ecosystem Health

There are many connections between human and ecosystem health and integrity, through air, water, sediments, soils, and their ecosystems (e.g., McMichael, 1993; Di Guilio and Monosson, 1996; Epstein and Rapport, 1996; Epstein, 1997; Ealvin et al., 1999; Adey, 2000; Tabor et al., 2001; Wilcox, 2001;

Di Guilio and Benson, 2002; GOMCME, 2002; Knap et al., 2002; Woodwell, 2002).

There are now well-recognized linkages between ecosystem health and human health (McMichael, 1993; di Guilo and Monosson, 1996; Shaw et al. pers. comm.) or ocean health and human health, in its widest context (Knap et al., 2002). For the marine environment, examples abound: algal toxin effects, sewage impacts, trace chemical effects, quality of seafood, reduced use, esthetics, quality of life, etc. Internationally, the linkage between ocean and human health is being pursued actively through the IOC/UNESCO (Knap et al., 2002; Knap, pers. comm.), especially in the context of native subsistence users, and the inhabitants of small island states. For the Bay of Fundy and Gulf of Maine, closed or restricted shellfish beds are the most obvious sign of the social and health impacts of a degraded system (Jones, 2004; also see section 17.2.3).

The various connections are well recognized by environmental scientists, resource managers and policy makers in the context of the Bay of Fundy and greater Gulf of Maine, and they are the imperative for much coordinated and networked action in its coastal waters (GOMCME, 2002; Pesch and Wells, 2004). The second goal of the GOMCME 2001–2006 action plan is to "protect human health and ecosystem integrity" from contaminant exposures to ensure that "contaminants in the Gulf of Maine are at sufficiently low levels to ensure human health and ecosystem integrity." The plan is currently addressing sewage, mercury and nitrogen. Sewage is the one priority 'pollutant' (actually a very complex mixture of chemicals and pathogens) in the Gulf and Bay of Fundy emphasizing the human-ecosystem health connections without doubt (GESAMP, 2001a, 2000b; Hinch et al., 2002), justifying the large expenditures to reduce the inputs.

17.3 INDICATORS FOR ASSESSING MARINE ECOSYSTEM HEALTH

This section describes some of the essential approaches and techniques for acquiring the data and information, and the analyses essential for an assessment of the Bay of Fundy and Gulf of Maine's health. It refers back to how we study, measure, and analyze ecosystem health, and in that context, ecological and ecosystem change (section 17.1). There is a large amount of literature and many active programs pertaining to approaches and techniques, not only relating to the Bay of Fundy and the Gulf of Maine. This section is simply meant to be a checklist of some components to consider while organizing an assessment, and is not exhaustive on the topic of indicators.

17.3.1 Monitoring Approaches

In Canada, the importance of monitoring and the data that it produces has been reiterated recently with the passage of the Oceans Act (1997). The MEQ component of the Oceans Act includes indicators, guidelines, and assessments, but not research and monitoring, explicitly. However, the key role of ocean monitoring is assumed and is slowly being strengthened in the context of programs such as GOOS or Global Ocean Observing System (P. Strain, pers. comm.) and harbor monitoring (e.g., Halifax, J. Hellou, pers. comm.). MEQ was discussed at a 2001 DFO meeting in Victoria on objectives and indicators for ecosystem-based management (Jamieson et al., 2001), and the Coastal Zone Canada Conference 2000 session on ocean health, held in Saint John, NB (S. Courtney, pers. comm.). There is continued monitoring, such as with the Gulf of Maine GOMCME EQ Monitoring Committee, with Gulfwatch (Chase et al., 2001; GOMCME, 2002; Jones et al., 2001; Jones and Wells, 2002; Percy and Wells, 2002). There are offshore EEM programs, using sophisticated methods of monitoring ocean change and contaminant sources, fate and effects (Gordon et al., 2000). There are also the environmental effects monitoring (EEM) programs related to the pulp and paper industry that use current field ecotoxicology techniques.

It is important to mention monitoring activity sponsored by the United Nations system. Some of it goes on under the auspices of GIWA (Global Inland Waters Assessment) and the GPA on LBA land-based activities. There is also Global Environmental Facility (GEF/LME) activity (IUCN, 1998), with LME assessments taking place globally (Hempel and Sherman, 2003; Sherman, 2004). There is one for the U.S. Northeast Shelf, and one for the Scotian Shelf, both of which overlap the Bay of Fundy (Sherman, 2004; Sherman and Skjoldal, 2002). As described above (section 17.2.1.3), the pollution and ecosystem health module of each LME covers: eutrophication, biotoxins, pathology, emerging disease and health indices. "Systematic monitoring data of bio-indicator species including bottom fish and mollusks (Musselwatch) are examined for endocrine disrupters and organ pathology. Water quality and plankton examinations are made for phytoplankton toxicity, eutrophication, persistent organic pollutants, and evidence of emerging disease. Examinations of changing states of health of LMEs are based on indices of ecosystem biodiversity, productivity, yield, resilience, and stability." (K. Sherman, pers. comm.).

The above examples reiterate the role that systematic monitoring plays in providing the data, short- and long-term, essential for any ecosystem health and environmental quality assessment. The challenge is two-fold: agreeing on the issues and indicators, and finding support for the programs. It should also be recognized that effective marine monitoring programs would benefit from having components of data analysis and interpretation, sample and data management, research on new techniques, and communication, in addition to the basic, routine sampling programs (Baird and Burton, 2001; Jones and Wells, 2002; Percy and Wells, 2002b; Simon et al., 2003a; amongst others).

17.3.2 Indicators and Indices

There has been considerable discussion and some agreement on the choice of MEH and MEQ indicators, for short term and long term monitoring, over the past few years. The literature is large. Relevant papers and reports include: O'Connor and Dewling (1986); Salanki (1986); Bilyard (1987); Long and Buchman (1989); International Joint Commission (1991); Kay (1991); Rapport (1991b, 1999); Montevecchi (1993); GESAMP (1994a); Nettleship (1997); Schwaiger (1997); Soule and Kleppel (1998); Vandermeulen (1998); Wichert and Rapport (1998); NOAA (1999); CRMSW (2000); EPA (2000a, 2000b); Pesch (2000); Thompson and Hamer (2000); Van Dolah (2000); Burger and Gochfeld (2001); Adams (2002); Cairns (2003); Busch and Trexler (2003); Yoder and DeShon (2003); Environment Canada (2003); MacDonald et al. (2003); Shaw (2003); Simon (2003a, 2000b); Simon et al. (2003); and Strain and Macdonald (2002). A synthesis of indicators currently used in U.S. NOAA coastal programs has been assembled for GOMCME (Bill O'Bearne, NOAA, pers. comm.), and considerable discussion has taken place by the GOMCME as to the selection of indicators and networking of monitoring programs (Keeley, D. 2000, pers. comm.). A new journal was launched in 2001 on this topic (*Ecological Indicators, Integrating Monitoring, Assessment and Management*). Two relevant monographs, "Biological Response Signatures" and "Biological Indicators of Aquatic Ecosystem Health" also appeared recently (Adams, 2002; Simon, 2003a).

When one views and considers the size, dynamics and various ecologies and species in the Bay of Fundy (and other water bodies), it is very hard to comprehend, without examining the evidence, that their ecosystems are being significantly impaired through human activity (perhaps with the exception of fisheries, with the large biomass removal and physical impact of fishing (Jackson et al., 2001; and others), and the impairment of shellfish beds. The bay simply appears too large, dynamic, biologically diverse, and ever renewed by the tides. However, programs of research and monitoring tell a different story (see the Proceedings of five Fundy Science Workshops, 1996 to 2004: Percy et al. (1997); Burt and Wells (1998); Ollerhead et al. (1999); Chopin and Wells (2001); Wells et al., 2004). Parts of the system can become impacted or impaired, in both the short and long term. So the question becomes: which indicators and indices, used in combination, give the "best" set of measures of the state of the Fundy/Gulf of Maine system and/or its component places and parts?

Schaeffer et al. (1988) described four major measures of ecosystem health, which can be tested here: sustainability, activity, organization, and resilience (also see section 17.2). As stated above, "sustainability implies that the system can maintain its structure and function over time (Schaeffer et al., 1988)." Basically, the authors are saying that the best indicators are structure, function and resilience of each ecosystem, and that there are readily measured endpoints of these indicators to incorporate into a monitoring program. Karr (1992) states: "the development of indicators is, arguably, the most important step needed to mobilize social support for reversal of the trend towards biotic impoverishment." Many ecological concepts, as discussed above throughout this chapter, such as resilience, resistance, connectivity, and ascendancy, may be important indicators of a systems condition, but they have practical difficulties — what do you measure operationally in the real world? The indicators must have practical endpoints, and only those that do will have currency in monitoring and assessment programs.

Hence, it is crucial to use practical, measurable, and current indicators; and where possible, interpret data in the context of formal environmental guidelines (MacDonald et al., 1992; EPA, 2001). There may be "disconnects" between the people discussing the concepts of ecosystem health and indicators (Rapport et al; Costanza et al), and those in the applied aquatic science field who have been deploying techniques in the field and assessing areas as to health and quality (i.e., condition), for several decades (Soule and Kleppel, 1998; Simon, 2003). Immense advances have been made in marine environmental monitoring approaches and techniques over the past three to four decades, and in the production of environmental (e.g., water quality, MEQ) guidelines. We are not starting at the beginning (witness B.H. Sherman's HEED work and Ken Sherman's LME work above; the existing Gulfwatch program, and similar programs worldwide; many others, especially in selected specialized fields such as the fate and effects of oil spills). Dr. John Pearce, who has worked in this area for several decades (note McIntyre and Pearce, 1980; Pearce, 2000, 2004; Pearce and Wells, 2002) has offered valuable insights as to the appropriate set(s) of indicators to use for measuring ecosystem health in the Gulf of Maine. And shown by Butler (1997) on blue herons, and Jones et al., 2001 single well-known species can act as easily monitored "sentinel species."

Smiley et al. (1998), in an internal Canadian review of indicators on MEH (not differentiated from MEQ), reached two conclusions useful to the discussion of appropriate indicators for the Bay of Fundy and Gulf of Maine:

1. The development of indicators to measure MEH generally involves the following key steps:

   a. Scope the issues
   b. Evaluate the knowledge base
   c. Select indicators
   d. Conduct targeted research and monitoring of the indicators

2. The ABC's of selecting indicators are:

   a. Some indicators should be related to ecosystem structure and function
   b. Some indicators should be selected (and deployed) in combination
   c. Indicators should be selected using the guidance of criteria acceptable to all involved parties.

Finally, the assessment of the endpoints for each indicator must be measured against values established as objectives, criteria, guidelines or standards. This approach has been recently successfully applied by EPA (EPA, 2001) in their assessment of coastal condition; each of the indicators (water clarity, dissolved oxygen, coastal wetlands, eutrophic conditions, sediment, benthos, fish tissue) were ranked using carefully chosen ranking criteria, for good, fair or poor assessments. The overall national coastal condition was then established based on an average of the seven rankings, hence completing the loop from monitoring and indicators to assessments of coastal condition or marine ecosystem health.

17.3.3   Status and Trends Analysis

Indicators of marine ecosystem health should be deployed so as to answer the question: is the health (short-term) and quality (long-term) of the Fundy marine ecosystem (or parts of it) getting better or worse? This requires substantial databases for a status and trends analysis. A number of monitoring programs for coastal waters have selected measures of health and quality that allow for such analysis (e.g. Harding, 1990; Chase et al., 2001; EPA, 2001). That is, measurements of selected variables in water, sediments, and biota are conducted over space and time to provide a picture of the magnitude of the response and the direction of change.

Examples include: Gulfwatch for trace chemical contaminants (Chase et al., 2001, Jones et al., 2001); algal toxins (as per the harmful algal blooms (HAB) program); input of aerial contaminants (Clair et al., 2002, acid rain; Cox et al. pers comm., toxic chemicals in fog; Sunderland, pers. comm., mercury in air and marine sediments); bacterial levels along the coasts, especially at shellfish beds; condition — that is, catch statistics, of specific fisheries; state of the water and benthic environments around aquaculture operations; numbers and locations of tidal barriers or obstructions; location of dykes, maintained and unmaintained; sediment quality at selected sites (Tay, pers. comm.); nutrients, especially nitrogen; sewage treatment and effects, etc.

Measurements of the same suite of indicators and endpoints over time and space, as in the US Coastal condition reports (EPA, 2001), will provide the best opportunity to assess the trends of key indicators and identify problems in specific locations, again completing the loop between monitoring of key indicators and management action.

17.4 SUMMARY AND CONCLUSIONS

This chapter discusses the concepts of health and marine ecosystem health, and describes indicators useful in monitoring and assessment, in the context of the Gulf of Maine and Bay of Fundy. At this time, there are two urgent needs in this field of marine ecosystem health — one, to reach consensus on the terminology and simplify it so that it is useful for practitioners and understandable by policy makers and environmental managers; two, to reach consensus on the choice of indicators and their endpoints for monitoring key issues affecting the oceans, and maintain rigorous monitoring and assessment programs. Progress is being made in this regard in the Gulf of Maine — Bay of Fundy region, with monitoring and assessment now focussed on six issues — land use, contaminants and pathogens, fisheries and aquaculture, eutrophication (nutrients), aquatic habitat change/loss, and climate change. A consensus is finally being reached on the ecological indicators for assessing marine ecosystem health for this region, and periodically reporting on progress.

ACKNOWLEDGMENTS

This work was supported by Environment Canada (Atlantic Region) over many years. Many scholars contributed to the ideas and facts summarized in this paper, and are duly thanked.

REFERENCES

Adams, S.M. *Biological Indicators of Aquatic Ecosystem Stress.* American Fisheries Society, Bethesda, MD, 2002, 644 p.

Adey, W.H. Coral reef ecosystems and human health: biodiversity counts! *Ecosys. Health* 6, 227–236, 2000.

AMAP Arctic Pollution Issues. A State of the Arctic Environment Report. AMAP, Oslo, Norway. 188 p.

Atlantic Geoscience Society. *The Last Billion Years. A Geological History of the Maritime Provinces of Canada.* Nimbus Publishing Ltd., Halifax, NS, 1997, 212 p.

Backus, R.H. and Bourne, D.W. *Georges Bank.* MIT Press, Cambridge, MA, 1987, 593 p.

Baird, D.J. and Burton, G.A. *Ecological Variability: Separating Natural from Anthropogenic Causes of Ecosystem Impairment.* SETAC Press, Pensacola, FL, 2001, 307 p.

Bigelow, H.B. and Schroedet, W.C. Fishes of the Gulf of Maine. Fish. Bull. 74. Fish. Bull. Fish. Wildlife serv. Vol. 53, US Govt Printing office, Wash. DC. 1953. 577 p.

Bilyard, G.R. The value of benthic infauna in marine pollution monitoring studies. *Mar. Poll. Bull.* 18, 581–585, 1987.

Buckley, D.E. "Sediments and environmental quality of the Miramichi Estuary: new perspectives," in *Water, Science and the Public: the Miramichi Ecosystem,* Chadwick, E.M.P., Ed. Can. Spec. Publ. Fish. Aquat. Sci. Rept. 123, NRCC, Ottawa, ON, 1995, pp. 179–190.

Burger, J. and Gochfeld, M. On developing bioindicators for human and ecological health. *Environ. Mon. Assess.* 66, 23–46, 2001.

Burt, M.D.B. and Wells, P.G. Coastal Monitoring and the Bay of Fundy. *Proceedings of the Maritime Atlantic Ecozone Science Workshop,* 1997. Huntsman Marine Science Center, St. Andrews, and Environment Canada, Dartmouth, NS, 1998, 196 p.

Busch, D.E. and Trexler, J.C. *Monitoring Ecosystems. Interdisciplinary Approaches for Evaluating Ecoregional Initiatives.* Island Press, Washington, 2003, 447 p.

Butler, R.W. *The Great Blue Heron. A Natural History and Ecology of a Seashore Sentinel.* UBC Press, Vancouver, 1997, 167 p.

Cairns, J., Jr. "Biotic community response to stress." in *Biological Response Signatures. Indicator Patterns Using Aquatic Communities,* Simon, T.P., Ed. CRC Press, Boca Raton, FL, 2003, pp. 13–21.

Calow, P. Can ecosystems be healthy? Critical considerations of concepts. *J. Aqua. Ecosys. Health* 1, 1–5, 1992.

Calow, P. Ecosystems not optimized. *J. Aqua. Ecosys. Health* 2, 55, 1993.

Calow, P. "Ecosystem health — a critical analysis of concepts," in *Evaluating and Monitoring Health of Large Scale Ecosystems.* NATO ASI Series 1. Global Environmental Change, Vol. 28. Springer Verlag, Berlin, 1995, pp. 33–41.

Calow, P. Critics of ecosystem health misrepresented. *Ecosys. Health* 6, 3–4, 2000.

Campbell, D.E. Using energy systems theory to define, measure, and interpret ecological integrity and ecological health. *Ecosys. Health* 6, 181–204, 2000.

Chandler, H. Marine Monitoring Programs in the Gulf of Maine: An Inventory. Internal report prepared for the Gulf of Maine Council on the Marine Environment. Maine State Planning Office and the Gulf of Maine Council, Augusta, ME, 2001, 115 p.

Chang, C.P. A Marine Environmental Quality Framework: Managing the Marine Ecosystem by Choosing Appropriate Guidelines, Objectives and Standards. MSc thesis, Dalhousie University, Halifax, NS, 1999, 88 p.

Chang, C.P. and Wells, P.G. A marine environmental quality (MEQ) framework and the Bay of Fundy. *Opportunities and Challenges for Protecting, Restoring and Enhancing Coastal Habitats in the Bay of Fundy: Proceedings of the 4th Bay of Fundy Science Workshop*, Saint John, NB, 2000, Chopin, T., and Wells, P.G., Eds. Environment Canada — Atlantic Region, Occasional report No. 17, Sackville, NB, 2001, 127 p.

Chase, M.E., Jones, S.H., Hennigar, P., Sowles, J., Harding, G.C.H., Freeman, K., Wells, P.G., Krahforst, C., Coombs, K., Crawford, R., Pederson, J., and Taylor, D. Gulfwatch: Monitoring spatial and temporal trends of trace metal and organic contaminants in the Gulf of Maine (1991–1997) with the blue mussel. *Mytilus edulis. Mar. Pollut. Bull.* 42, 491–505, 2001.

Chopin, T. and Wells, P.G., Eds. Opportunities and Challenges for Protecting, Restoring and Enhancing Coastal Habitats in the Bay of Fundy. *Proceedings of the 4th Bay of Fundy Science Workshop*, Saint John, NB, 2000. Environment Canada — Atlantic Region, Occasional report No. 17. Environment Canada, Sackville, NB, 2001, 237 p.

Clair, T.A. Ehrman, J.M., Ovellet, A.J., Brun, G.L., Lockerbie, D., Ro, C.V. Changes in freshwater acidification trends in Canada's Atlantic provinces: 1983–1997. *Wat. Air, Soil Pollut.* 135, 335–354, 2002.

Clark, R.A. and Frid, C.L.J. Long-term changes in the North Sea ecosystem. *Environ. Rev.* 9, 131–187, 2001.

Coastal Research and Monitoring Strategy Workgroup (CRMSW). Clean Water Action Plan: Coastal Research and Monitoring Strategy. Ocean and Coastal Protection Division, USEPA, Washington, D.C., 2001, 34 p. plus appendices.

Costanza, R. "Toward an operational definition of ecosystem health," in *Ecosystem Health*, Costanza, R., Notton, B.G., and Haskell, B.D., Eds. Island Press, Washington, D.C., 1992, pp. 239–256.

Costanza, R. and Mageau, M. "What is a healthy ecosystem?" in *The Gulf of Mexico Large Marine Ecosystem: Assessment, Sustainability, and Management*, Kumpf, H., Sherman, K., and Steidinger, K., Eds. Blackwell Science, Malden, MA, 1999.

Costanza, R., Norton, B.G., and Haskell, B.D. *Ecosystem Health. New Goals for Environmental Management*. Island Press, Washington, D.C., 1992, 269 p.

Côté, R.P. The Many Dimensions of Marine Environmental Quality. Science Council of Canada, Discussion Paper. Science Council of Canada, Ottawa, ON, 1989, 37 p.

Di Giulio, R.T. and Benson, W.H. *Interconnections Between Human Health and Ecological Integrity.* SETAC Press, Pensacola, FL, 2002, 110 p.

Di Giulio, R.T. and Monosson, E. *Interconnections Between Human and Ecosystem Health.* Ecotoxicology Series 3, Chapman and Hall, London, 1996, 275 p.

Downs, T.J. and Ambrose, R.F. Syntropic ecotoxicology: a heuristic model for understanding the vulnerability of ecological systems to stress. *Ecosyst. Health* 7, 266–283, 2001.

Duarte, C.M., Cebrian, J., and Marba, N. Uncertainty of detecting sea change. *Nature* 356, 190, 1992.

Earle, S.A. *Sea Change. A Message of the Oceans.* G.P. Putnam, New York, 1992, 361 p.

Ehrenfeld, D. Ecosystem health and ecological theories. *Ecosystem Health,* Costanza, R., Norton, B.G., and Haskell, B.D., Eds. Island Press, Washington, D.C., 1992, pp. 135–143.

Environment Canada. The Ecosystem Approach: Getting Beyond the Rhetoric. Task Group on Ecosystem Approach and Ecosystem Science, Environment Canada, Ottawa, ON, 1996, 21 p.

Environment Canada. Environmental Signals. Canada's National Environmental Indicator Series 2003. National Indicators and Reporting Office, Environment Canada, Ottawa/Hull, ON, 2003. K1A 0H3, 78 p.

EPA. Condition of the Mid-Atlantic Estuaries. USEPA Office of Research and Development, Washington, D.C., 1998. EPA 600-R-98-147, 50 p.

EPA. The Ecological Condition of Estuaries in the Gulf of Mexico. USEPA Office of Research and Development, Washington, D.C., 1999, EPA 620-R-98-004, 71 p.

EPA. Stressor Identification Guidance Document. USEPA, Office of Research and Development, Washington, D.C., 2000a. EPA/822/B-00/025, 242 p.

EPA. Evaluation Guidelines for Ecological Indicators. EMAP, USEPA, Office of Research and Development, Washington, D.C., 2000b. EPA/620/R-99/005, 97 p.

EPA. National Coastal Condition Report. Office of Research and Development, Office of Water, EPA, Washington, D.C., 2001. EPA-620/R-01/005, 204 p.

EPA. Mid-Atlantic Integrated Assessment (MAIA) Estuaries 1997–98. Summary report. Environmental Conditions in the Mid-Atlantic Estuaries. US EPA, Narragansett, RI, 2000b. EPA/620/R-02/003, 115 p.

Epstein, P.R. Climate, ecology, and human health. *Consequences* 3, 1–14, 1997. See http://www.gerio.org/consequences/vol3no2/climhealth.html.

Epstein, P.R. "Large marine ecosystem health and human health," in *The Gulf of Mexico Large Marine Ecosystem: Assessment, Sustainability, and Management,* Kumpf, H., Sherman, K., and Steidinger, K., Eds. Blackwell Science, Boston, MA, 1999, pp. 417–438.

Epstein, P.R. Marine ecosystem health. in Sherman, B.H., 2000, p. 249.

Epstein, P.R. and Rapport, D.J. Changing coastal marine environments and human health. *Ecosys. Health* 2, 166–176, 1996.

Fairweather, P.G. Determining the 'health' of estuaries: priorities for ecological research. *Austral. J. Ecol.* 24, 441–451, 1999.

Fisheries and Oceans Canada. Ensuring the Health of the Oceans and Other Seas. Sustainable Development in Canada Monograph Series: no. 3. Department of Fisheries and Oceans, Ottawa, ON, 1997, 12 p.

Fisheries and Oceans Canada. Working Together to Protect and Promote Canada's Oceans. Booklet, Department of Fisheries and Oceans, DFO, Ottawa, ON, 2000, 9 p.

Fisheries and Oceans Canada. Canada's Oceans Strategy. Our Oceans, Our Future. Oceans Directorate, DFO, Ottawa, ON, 2002a, 30 p.

Fisheries and Oceans Canada. Canada's Oceans Strategy. Our Oceans, Our Future. Policy and Operational Framework for Integrated Management of Estuarine, Coastal and Marine Environments in Canada. Oceans Directorate, DFO, Ottawa, ON, 2002b, 36 p.

Frith, K. Ocean and Human Health: How the Health of the Oceans Affects our Well-Being. Currents, Bermuda Biological Station for Research, 1999, pp. 7–8.

Galloway, T., Brown, R.J., Browne, M.A., Dissanayake, A., Lowe, D., Jones, M.B., and Depledge, M.H. A multibiomarker approach to environmental assessment. *Environ. Sci. Technol.* 38, 1723–1731, 2004.

Galvin, J., Moser, F., Dewailly, E., and Knap, A. A review of indicators of ocean and human health. International Centre for Ocean and Human Health, Bermuda Biological Station for Research, Bermuda. Manuscript report. Nov, 1999. 74 p.

GESAMP. Biological Indicators and Their Use in the Measurement of the Condition of the Marine Environment. GESAMP Reports and Studies No. 55. UNEP, Nairobi, 1994a, 56 p.

GESAMP. A Sea of Troubles. GESAMP Reports and Studies No. 70. UNEP, Nairobi. 2001a, 35 p.

GESAMP. Protecting the Oceans from Land-based Activities. GESAMP Reports and Studies No. 71, UNEP, Nairobi, 2001b. 162 p.

Goldberg, E.D. *The Health of the Oceans.* Unesco Press, Paris, 172 p.

GOMCME. Gulf of Maine Council on the Marine Environment Action Plan 2001–2006, 2002, See http://www.gulfofmaine.org.

Gordon, D.C., Jr., Griffiths, L.D., Hurley, G.V., Muecke, A.L., Muschenheim, D.K., and Wells, P.G. Understanding the environmental effects of offshore hydrocarbon development. *Can. Tech. Rep. Fish. Aquat. Sci.* No. 2311, 82 pages plus appendices, 2000.

GPAC. (Global Programme of Action Coalition). 2004. Summary Reports and Matrices from United States and Canadian Regional Forums and Meetings. The Gulf of Maine Summit Conference, St. Andrews, NB, October 2004. Manuscript report, on CD, in Pesch and Wells (2004).

Hall, J. and Wadleigh, M. The Scientific Challenge of Our Changing Environment. Canadian Global Change Program, Incidental report series no. IR93-2. The Royal Society of Canada, Ottawa, ON, 1993, 89 p.

Harding, G.C.H. A review of major marine environmental concerns off the east coast in the 1980s. *Can. Tech. Rept. Fish. Aquat. Sci.* 1885, 38 p., 1992a.

Harding, L.E. Monitoring status and trends in marine environmental quality. *Proceedings of a Symposium in conjunction with the 17th Annual Aquatic Toxicity Workshop*, Vancouver, BC, 1990. Environment Canada, Vancouver, BC, 1990, 200 p.

Harding, L.E. Measures of marine environmental quality. *Mar. Pollut. Bull.* 25, 23–27, 1992b.

Harris, H.J., Sager, P.E., Regier, H.A., and Francis, G.R. Ecotoxicology and ecosystem integrity: the Great Lakes examined. *Environ. Sci. Technol.* 24, 598–603, 1990.

Harvell, C.D., Kim, K., Burkholder, J.M., Colwell, R.R., Epstein, P.R., Grimes, D.J., Hofmann, E.E., Lipp, E.K., Osterhaus, A.D.M.E., Overstreet, R.M., Porter, J.W., Smith, G.W., and Vasta, G.R. Emerging marine diseases — climate links and anthropogenic factors. *Science* 285, 1505–1510, 1999.

Harvey, J., Coon, D., and Abouchar, J. *Habitat Lost: Taking the Pulse of Estuaries in the Canadian Gulf of Maine*. Conservation Council of New Brunswick, Fredericton, NB, 1998, 79 p.

Haskell, B.D., Norton, B.G., and Costanza, R. "Introduction. What is ecosystem health and why should we worry about it?" in *Ecosystem Health. New Goals for Environmental Management*, Costanza, R., Norton, B.G., Haskell, B.D., Eds. Island Press, Washington, D.C., 1992, pp. 3–20.

HEED. *Marine Ecosystems: Emerging Diseases as Indicators of Change*. Health of the Oceans from Labrador to Venezuela. Health, Ecological and Economic Dimensions (HEED) of Global Change Program, Center for Health and the Global Environment, Harvard Medical School, Boston, MA, 1998, 85 p.

Hinch, P.R., Bryon, S., Hughes, K., and Wells., P.G. *Sewage Management in the Gulf of Maine: Workshop Proceedings*. Gulf of Maine Council on the Marine Environment, Concord, NH, 2002, 52 p. See http://www.gulfofmaine.org.

Holling, C.S. Cross-scale morphology, geometry, and dynamics of ecosystems. *Ecol. Monogr.* 62, 447–502, 1992.

Huntsman, A.G. 1922. The fishes of the Bay of Fundy. *Contrib. Can. Bibl.* 3, 49–72, 1921.

International Joint Commission. A Proposed Framework for Developing Indicators of Ecosystem Health for the Great Lakes Region. Council of Great Lakes Research Managers, IJC, U.S. and Canada, 1991, 47 p.

Intergovernmental Oceanographic Commission. A strategic plan for the assessment and prediction of the health of the ocean: a module of the Global Ocean Observing System. IOC/UNESCO, IOC/INF-1044. IOC, Paris, 1996, 39 p.

IUCN. An ecosystem strategy for the assessment and management of international coastal ocean waters. Brochure. IUCN, NOAA and IOC/UNESCO, Paris, 1998, 8 p.

Jackson, J.B.C. What was natural in the coastal oceans? *Proc. Nat. Acad. Sci.* 98, 5411–5418, 2001.

Jackson, J.B.C., Kirby, M.X., Berger, W.H., Bjorndal, K.A., Botsford, L.W., Bourgue, B.J., Bradbury, R.H., Cooke, R., Erlandson, J., Estes, J.A., Hughes, T.P., Kidwell, S., Lange, C.B., Lenihan, H.S., Pandolfi, J.P., Peterson, C.H., Steneds, R.S., Tegner, M.J., and Warner, R.R. Historical over-fishing and the recent collapse of coastal ecosystems. *Science* 293, 629–638, 2001.

Jakeman, A.J., Beck, M.B., and McAleer, M.J., Eds. *Modelling Change in Ecological Systems*. Wiley, Chichester, 1993, 584 p.

Jamieson, G., O'Boyle, R., Arbour, J., Cobb, D., Courtenay, S., Gregory, R., Levings, C., Munro, J., Perry, I., and Vandermeulen, H. (Eds.) *Proceedings of the National Workshop on Objectives and Indicators for Ecosystem-based Management*, Sidney, BC, 2001. Can. Sci. Advisory Secretariat Proceedings Series 2001/09. Fisheries and Oceans, Ottawa, ON, 2001, 140 p.

Jickells, T.D. Nutrient bio-geochemistry of the coastal zone. *Science* 281, 217–222, 1998.

Jones, S.H. 2004. Chapter 5. Contaminants and Pathogens. In: Tides of Change Across the Gulf. G.G. Pesch and P.G. Wells, eds. Gulf of Maine Council on the Marine Environment, Augusta, ME., and Concord, NH. p. 33–41.

Jones, S.H. and Wells, P.G. Gulf of Maine Environmental Quality Monitoring Workshop, April 3–May 1, 2001. Summary report. Gulf of Maine Council on the Marine Environment, Durham NH, 2002, 37 p.

Jones, S.H. Chapter 5. Contaminants and Pathogens. In: Tides of Change Across the Gulf. Posch, G.G. and Wells, P.S., Eds. Gulf of Maine Council on the Marine Environment, Angusta, ME, and Concord, NH. 2004, pp. 33–41.

Jones, S.H., Chase, M., Sowles, J., Hennigar, P., Landry, N., Wells, P.G., Harding, G.C.H., Krahforst, C., and Brun, G.L. Monitoring for toxic contaminants in *Mytilus edulis* from New Hampshire and the Gulf of Maine. *J. Shellfish Res.* 20, 1203–1214, 2001.

Jørgensen, S.E. *Ecosystem health. Integration of Ecosystem Theories: A Pattern*, 2nd ed., Jorgensen, S.E. Kluwer, Dordrecht, 1997, pp. 265–280.

Karr, J.R. Assessment of biotic integrity using fish communities. *Fisheries* 6, 21–27, 1981.

Karr, J.R. Biological integrity: a long-neglected aspect of water resource management. *Ecol. Applic.* 1, 66–84, 1991.

Karr, J.R. "Ecological integrity. Protecting earth's life support systems." in *Ecosystem Health*, Costanza, R., Norton, B.G., and Haskell, B.D., Eds. Island Press, Washington, D.C., 1992, pp. 223–238.

Karr, J.R. Defining and assessing ecological integrity: beyond water quality. *Environ. Toxicol. Chem.* 12, 1521–1531, 1993.

Karr, J.R., Fausch, K.D., Angermeier, P.L., Yant, P.R., and Schlosser, I.J. Assessment of Biological Integrity in Running Water: A Method and its Rationale. Spec. Public. 5, Illinois Natural History Survey, Champaign, IL, 1986, 28 p.

Kay, J.J. "The concept of ecological integrity, alternative theories of ecology, and implications for decision-support indicators," in *Economic, Ecological, and Decision Theories. Indicators of Ecologically Sustainable Development*, P.A. Vicot, Kay, J.J., and Ruitenbeek, H.J., Eds. Canadian Environmental Advisory Council, Environment Canada, Ottawa, ON, 1991, pp. 23–58.

Kiddon, J.A., Paul, J.F., Buffum, H.W., Strobel, C.S., Hale, S.H., Cobb D., and Brown, B.S. Ecological condition of US mid-Atlantic estuaries, 1997–1998. *Mar. Pollut. Bull.* 46, 1224–1244, 2003.

Knap, A.H., Dewailly, E., Furgal, C., Galvin, J., Baden, D., Bowen, R.E., Depledge, M., Duguay, L., Fleming, L.E., Ford, T., Moser, F., Owen, R., Suk, W.A., and Unlvata, V. Indicators of ocean health and human health: developing a research and monitoring framework. *Environ. Health Perspect.* 110, 839–845, 2002.

Kullenberg, G. The Review of the Health of the Oceans. GESAMP Reports and Studies No. 15, UNESCO, Paris, 1982, 108 p.

Kutchenberg, T.C. Measuring the health of the ecosystem. *Environment* 27, 32–37, 1985.

Lane, P.A. "Assessing cumulative health effects in ecosystems," in *Ecosystem Health*, Rapport, D., Costanza, R., Epstein, P.R., Gaudet, C., Levins, R., Eds. Blackwell Science, Oxford, 1998, pp. 129–153.

Leopold, A. *A Sand County Almanac, and Sketches Here and There.* Oxford University Press, London, 1949, 226 p.

Livingston, R.J. Trophic Organization in Coastal systems. CRC Press, Boca Raton, FL. 2003, 388 p.

Lomborg, B. *The Skeptical Environmentalist. Measuring the Real State of the World.* Cambridge University Press, Cambridge, 2001, 515 p.

Long, E.R. and M.F. Buchman. An evaluation of candidate measures of biological effects for the National Status and Trends Program. NOAA Tech. Mem., NOS OMA 45. Seattle, 1989 (exec. summ.).

Lotze, H. and Milewski, I. *Two Hundred Years of Ecosystem and Food Web Changes in the Quoddy Region, Outer Bay of Fundy*. Conservation Council of New Brunswick, Fredericton, NB, 2002, 188 p.

MacDonald, D.D., Smith, S.L., Wong, M.P., and Murdoch, P. The Development of Canadian Marine Environmental Quality Guidelines. Marine Environmental Quality Series No. 1. Conservation and Protection, Environment Canada, Ottawa, 1992, 121 pp.

MacDonald, R.W., Morton, B., and Johannessen, S.C. A review of marine environmental issues in the North Pacific: the dangers and how to identify them. *Environ. Rev.* 11, 103–139, 2003.

Mann, K.H. *Ecology of Coastal Waters, with Implications for Management*. Blackwell, Cambridge, UK, 2000, 406 p.

Martens, P. and McMichael, A.J. *Environmental Change, Climate and Health*. Cambridge University Press, Cambridge, 2002, 352 p.

McGinn, A.P. *Safeguarding the Health of Oceans*. Worldwatch Paper 145, March 1999. Worldwatch Institute, Washington, D.C., 1999, 87 p.

McGowan, J.A., Cayan, D.R., and Dorman, L.M. Climate-ocean variability and ecosystem response in the Northeast Pacific. *Science* 281, 210–217, 1998.

McIntyre, A.D. and Pearce, J. Biological Effects of Marine Pollution and the Problems of Monitoring. *Rapp. Proces-Verbaux Réun. Cons. Int. Explor. Mer* 179, 1–346. ICES, Copenhagen, Denmark, 1980.

McMichael, A.J. *Planetary Overload. Global Environmental Change and the Health of the Human Species*. Cambridge University Press, Cambridge, 1993, 352 p.

McMichael, A.J. Human culture, ecological change, and infectious disease: are we experiencing history's fourth great transition? *Ecosystem Health* 7, 107–115, 2001.

McMichael, A.J. The biosphere, health, and "sustainability." *Science* 297, 1093, 2002.

Mills, K. 2004. Chapter 6. Fisheries and Aquaculture. In: Tides of Change Across the Gulf. An Environmental Report on the Gulf of Maine and Bay of Fundy. G.G. Pesch and P.G. Wells, eds. Gulf of Maine Council on the Marine Environment, Augusta, ME, and Concord, NH. p. 42–50.

Myers, N. Environmental unknowns. *Science* 269, 358–360, 1995.

Montevecchi, W.A. "Birds as bio-indicators in marine and terrestrial ecosystems." in The Scientific Challenge of Our Changing Environment. Canadian Global Change Program. Incidental report series no. IR93-2. The Royal Society of Canada, Ottawa, ON, 1993, pp. 60–62.

National Ocean Service. NS&T Program. National Status and Trends Program for Marine Environmental Quality. Brochure, National Ocean Service, National Centers for Coastal Ocean Science, Center for Coastal Monitoring and Assessment. NOAA, Washington, D.C., 1998, 32 p.

Nettleship, D.N. "Long-term monitoring of Canada's seabird populations," in Monitoring Bird Populations: the Canadian Experience, Dunn, E.H., Cadman, M.D. and Falls, J.B., Eds. Occasional Paper Number 95, Canadian Wildlife Service, Environment Canada, Ottawa, 1997, pp. 16–23.

Nielsen, N.O. The meaning of health. *Ecosystem Health* 5, 65–66, 1999.

NOAA. National Status and Trends Program for Marine Environmental Quality. South Florida. NOAA, CCMA, Regional report series 2, 1999. 40 p.

Norse, E.A. (Ed.). 1993. Global Marine Biological Diversity. Island Press, Washington, D.C., 383 p.

Norton, B.G., Ulanawicz, R.E., and Haskell, B.D. Scale and Environmental Policy Goals. Report to the EPA, Office of Policy, Planning and Evaluation. EPA, Washington, D.C., 1991.

Noss, R. *Maintaining Ecological Integrity in Representative Reserve Networks*. WWF Canada, WWF USA, Toronto and Washington, D.C., 1995, 77 p.

O'Connor, J.S. and Dewling, R.T. Indices of marine degradation: their utility. *Environ. Man.* 10, 335–343, 1986.

Ollerhead, J., Hicklin, P.W., Wells, P.G., and Ramsey, K. Understanding Change in the Bay of Fundy Ecosystem. Environment Canada — Atlantic Region. Occasional report no. 12, 1999, 143 p.

Pauly, D. Anecdotes and the shifting baseline syndrome of fisheries. *Trends in Ecology and Evolution* 10, 430, 1995.

Pauly, D., Christensen, V., Dalsgaard, J., Froese, R., and Torres, F., Jr. Fishing down food webs. *Science* 279, 860–863, 1998.

Pauly, D. and Maclean, J., 2003. In a Perfect Ocean. The State of Fisheries and Ecosystems in the North Atlantic Ocean. Island Press, Wash., D.C. 175 p.

Pearce, J.B. The New York Bight. *Mar. Pollut. Bull.* 41, 44–55, 2000.

Pearce, J.B. 2004. Taking the pulse of the seas. Can it be done? Pages 309–319 in Wells, P.G., et al., eds. Health of the Bay of Fundy: Assessing Key Issues. Environment Canada — Atlantic Region, Occasional Report No. 21. March 2004. Environment Canada, Dartmouth, NS.

Pearce, J.B. and Wells, P.G. Key(s) to marine ecology and understanding pollution impacts. *Mar. Pollut. Bull.* 44, 179–180, 2002.

Percy, J.A. and Wells, P.G. Taking Fundy's Pulse: Monitoring the Health of the Bay of Fundy. BoFEP Fundy Fact Sheet #22. Environment Canada, Dartmouth, NS, 2002, 8 pp

Percy, J.A., Wells, P.G., and Evans, A.J. Bay of Fundy Issues: A Scientific Overview. Environment Canada — Atlantic Region, Occasional report no. 8, Sackville, NB, 1997. Reprinted April 2002.

Pesch, G. Establishing a framework for effective monitoring of the Gulf of Maine. RARGOM report 00-1, 2000, Univ. of New Hampshire, Durham, NH, 9 p.

Pesch, G.G. and P.G. Wells. (Eds.). 2004. Tides of Change Across the Gulf. An Environmental Report on the Gulf of Maine and Bay of Fundy. Gulf of Maine Council on the Marine Environment, Augusta, ME, and Concord, NH, 81 p. (www.gulfofmaine.org).

Plant, S. Bay of Fundy environmental and tidal power bibliography. *Can. Tech. Rept. Fish. Aquat. Sci.* 1339, 159 p. plus index, 1985.

Rapport, D.J. *State of Canada's Environment*. Environment Canada, Ottawa, ON, 1986.

Rapport, D.J. What constitutes ecosystem health? *Perspect. Biol. Med.* 33, 120–132, 1989.

Rapport, D.J. Challenges in the detection and diagnosis of pathological change in aquatic ecosystems. *J. Great Lakes Res.* 16, 609–618, 1990.

Rapport, D.J. Evolution of indicators of ecosystem health. 1991, 34 p. (Unpublished manuscript).

Rapport, D.J. "Evaluating ecosystem health," in *Assessing Ecosystem Health: Rationale, Challenges and Strategies*, Munawar, M., Ed. Kluwer, Amsterdam, 1992, pp. 15–24.

Rapport, D.J. "Dimensions of ecosystem health," in *Ecosystem Health*, Rapport, D., Costanza, R., Epstein, P.R., Gaudet, C., Levins, R., Eds. Blackwell, Oxford, 1998, pp. 34–40.

Rapport, D.J. Epidemiology and ecosystem health: natural bridges. *Ecosyst. Health* 5, 174–180, 1999.

Rapport, D.J., Thorpe, C. and Regier, H.A. "Commentary. Ecosystem medicine," in *Perspectives on Adaptation, Environment and Population*, Calhoun, J.C. Ed. Praeger, New York, 1980, pp. 180–189.

Rapport, D.J., Regier, H.A., and Hutchinson, T.C. Ecosystem behaviour under stress. *Amer. Naturalist* 125, 617–640, 1985.

Rapport, D.J., Costanza, R., and McMichael, A.J. Assessing ecosystem health. *Trends Ecol. Evol.* 13, 397–401, 1998a.

Rapport, D.J., Costanza, R., Epstein, P.R., Gaudet, C., Levins, R. *Ecosystem Health*. Blackwell Science, Oxford, 1998b, 372 p.

Rapport, D.J., Böhm, G., Buckingham, D., Cairns, J., Jr., Costanza, R., Karr, J.R., de Kruiff, H.A.M., Levins, R., McMichael, A.J., Nielson, N.O., Whitford, W.G. Ecosystem health: the concept, the ISEH, and the important tasks ahead. *Ecosys. Health* 5, 82–90, 1999.

Rapport, D.J. and Whitford, W.G. How ecosystems respond to stress. Common properties to arid and aquatic systems. *Bioscience* 49, 193–203, 1999.

Rose, G.A., deYoung, B., Kuika, D.W., Goddard, S.V., and Fletcher, G.L. Perspective. Distribution shifts and overfishing the northern cod (*Gadus morhua*): a view from the ocean. *Can. J. Fish. Aquat. Sci.* 57, 644–663, 2000.

Salanki, J. Biological Monitoring of the State of the Environment: Bioindicators. An Overview of the IUBS Programme on Bioindicators 1985. IUBS, Paris, 1986.

Schaeffer, D.J. Diagnosing ecosystem health. *Ecotoxicol. Environ. Safety* 34, 18–34, 1996.

Schaeffer, D.J., Herricks, E.E., and Kerster, H.W. Ecosystem health: 1. Measuring ecosystem health. *Environ. Manag.* 12, 445–455, 1988.

Schindler, D.W. Detecting ecosystem responses to anthropogenic stress. *Can. J. Fish. Aquat. Sci.* 44 (Suppl.1), 6–25, 1987.

Schwaiger, J., Wanke, R., Adams, S., Pawert, M., Honnen, W., Triebskorn, R. The use of histopathological indicators to evaluate contaminant-related stress in fish. *J. Aquatic Ecosystem Stress and Recovery* 6, 75–86, 1997.

Selye, H. *Stress Without Distress*. Lippincott, Philadelphia, PA, 1974, 171 p.

Shaw, S. Seals as sentinels for the Gulf of Maine ecosystem. 2003, 4 p. (Unpubl. man.).

Sherman, B.H. Marine ecosystem health as an expression of morbidity, mortality and disease events. *Mar. Pollut. Bull.* 41, 232–254, 2000.

Sherman, K. Sustainability, biomass yields, and health of coastal ecosystems: an ecological perspective. *Mar. Ecol. Progr. Ser.* 112, 277–301, 1994.

Sherman, K. Why regional coastal monitoring for assessment of ecosystem health? *Ecosystem Health* 6, 205–216, 2000.

Sherman, K. The use of indicators in international large marine ecosystem programs and a baseline for the US Northeast Shelf. Paper prepared for the Northeast Coastal Indicators Summit, 6–8th January 2004, UNH, Durham, NH. 24 p. (Unpubl. man.). (see www.gulfofmaine.org)

Sherman, K., Jaworski, N.A., and Smayda, T.J. *The North-East Shelf Ecosystem. Assessment, Sustainability and Management*. Blackwell Science, Cambridge, MA, 1996, 564 p.

Sherman, K. and Skjoldal, H.R. *Large Marine Ecosystems of the North Atlantic. Changing States and Sustainability* (in press).

Simon, T.P. *Biological Response Signatures. Indicator Patterns Using Aquatic Communities.* CRC Press, Boca Raton, FL, 2003a, 576 p.

Simon, T.P. "Biological response signatures: toward the detection of cause-and-effect and diagnosis in environmental disturbances," in *Biological Response Signatures. Indicator Patterns Using Aquatic Communities*, Simon, T.P., Ed. CRC Press, Boca Raton, FL, 2003b, pp. 3–12.

Simon, T.P., Rankin, E.T., Dufour, R.L., and Newhouse, S.A. "Using biological criteria for establishing restoration and ecological recovery endpoints," in *Biological Response Signatures. Indicator Patterns Using Aquatic Communities*, Simon, T.P., Ed. CRC Press, Boca Raton, FL, 2003, pp. 83–96.

Smiley, B., Thomas, D., Duvall, W., and Eade, A. State of the Environment Reporting. Selecting Indicators of Marine Ecosystem Health: A Conceptual Framework and an Operational Procedure. Environment Canada, Occasional report series no. 9, Environment Canada, Ottawa, ON, 1998, 33 p.

Soule, D.F. and Kleppel, G.S. *Marine Organisms as Indicators.* Springer-Verlag, New York, 1998.

Spellerberg, I.F. *Monitoring Ecological Change.* Cambridge University Press, Cambridge, 1991, 334 p.

Stewart, P. and White, L. A review of contaminants on the Scotian Shelf and in adjacent waters: 1970–1995. *Can. Tech. Rept. Fish. Aquat. Sci.* 2351, 158, 2001.

Stokes, P. "Discussion," in *Ecotoxicology and the Aquatic Environment*, Stokes, P.M., Ed. Pergamon Press, Oxford, 1981, pp. 85–89.

Strain, P.M. and Macdonald, R.W. Design and implementation of a program to monitor ocean health. *Ocean Coast. Manage.* 45, 325–355, 2002.

Suter, G.W. A critique of ecosystem health concepts and indexes. *Environ. Toxicol. Chem.* 12, 1533–1539.

Sutter, G.C. Can ecosystem health be ecocentric? *Ecosys. Health* 7, 77–78, 2001.

Sykes, J.B. *The Concise Oxford Dictionary of Current English*, 6th ed. Oxford University Press, Oxford, 1976.

Tabor, G.M., Ostfeld, R.S., Poss, M., Dobson, A.P. and Aguirre, A.A. "Conservation biology and the health sciences. Defining the research priorities of conservation medicine," in *Conservation Biology. Research Priorities for the Next Decade*, Soule, M.E. and Orians, G.H., Eds. Island Press, Washington, D.C., 2001, pp. 155–173.

Tait, J.T.P., Cresswell, I.D., Lawson, R., and Creighton, C. Auditing the health of Australia's ecosystems. *Ecosys. Health* 6, 149–163, 2000.

Thompson, D.R. and Hamer, K.C. Stress in seabirds: causes, consequences and diagnostic value. *J. Aquat. Ecosys.m Stress Recov.* 7, 91–110, 2000.

Ulanowicz, R.E. "Ecosystem health and trophic flow networks," in *Ecosystem Health*, Costanza et al., Eds. Island Press, Washington, D.C., 1992.

Vandermeulen, H. The development of marine indicators for coastal zone management. *Ocean Coast. Manage.* 39, 63–71, 1998.

Vandermeulen, H. and Cobb, D. Objectives, indicators and reference points related to marine environmental quality. In *Proceedings of the National Workshop on Objectives and Indicators for Ecosystem-based Management*, Sidney, BC, 2001. Can. Sci. Advisory Secretariat Proceedings Series 2001/09. Fisheries and Oceans, Ottawa, ON, 2001, pp. 54–55.

Van Dolah, F.M. Marine algal toxins: origins, health effects, and their increased occurrence. *Environ. Health Perspect.* 108(Suppl. 1), 133–141, 2000.

Wade, T.L. et al. NOAA's mussel watch project: current use of organic compounds in bivalves. *Mar. Pollut. Bull.* 37, 20–26, 1998.

Wallace, G.T. and Braasch, E.F. *Proceedings of the Gulf of Maine Ecosystem Dynamics, Scientific Symposium and Workshop*, 1996, St. Andrews, NB. RARGOM Report 97-1, 352 p.

Webster's Dictionary. *Webster's Third New International Dictionary*, Unabridged. Merriam-Webster, Springfield, MA, 1993.

Wells, P.G. "Assessment," in *Health of the Oceans. A Status Report on Canadian Marine Environmental Quality*, Wells, P.G. and Rolston, S., Eds. Environment Canada, Ottawa and Dartmouth, 1991, pp. 115–122.

Wells, P.G. "Measuring ocean health: resetting a cornerstone of Canadian marine policy, science and management," in *Annexes of the Canadian Ocean Assessment. A Review of Canadian Ocean Policy and Practice*. International Ocean Institute, Halifax, NS, 1996.

Wells, P.G. Biomonitoring the health of coastal marine ecosystems — the roles and challenges of microscale toxicity tests. *Mar. Pollut. Bull.* 39, 39–47, 1999a.

Wells, P.G. Understanding change in the Bay of Fundy ecosystem. Ollerhead, J., Hicklin, P.W., Wells, P.G., and Ramsey, K. Understanding Change in the Bay of Fundy Ecosystem. Environment Canada — Atlantic Region. Occasional report no. 12, 1999a, pp. 4–11.

Wells, P.G. Assessing health of the Bay of Fundy — concepts and framework. *Mar. Pollut. Bull.* 46, 1059–1077, 2003.

Wells, P.G. and Côté, R.P. Protecting marine environmental quality from land-based pollutants. *Mar. Pol.* 12, 9–21, 1988.

Wells, P.G., Daborn, G.R., Percy, J.A., Harvey, J., and Rolston, S.J. (Eds.). Assessing health of the Bay of Fundy. *Proceedings of the 5th BoFEP Bay of Fundy Science Workshop, and Coastal Forum, Environment Canada-Atlantic Region.* Occasional Report No. 21, Environment Canada, Dartmouth, NS. 2004, March 2004. 402 p.

Wells, P.G., Depledge, M.H., Butler, J.N., Manock, J.J., and Knap, A.H. Rapid toxicity assessment and biomonitoring of marine contaminants — exploiting the potential of rapid biomarker assays and microscale toxicity tests. *Mar. Pollut. Bull.* 42, 799–804, 2001.

Wells, P.G. and Gratwick, J. *Proceedings of the Canadian Conference on Marine Environmental Quality.* ITTOPS, Dalhousie University, Halifax, NS, 1988.

Wells, P.G. and Rolston, S.J. Health of the Oceans. A Status Report on Canadian Marine Environmental Quality. Environment Canada, Ottawa and Dartmouth. 1991, 187 p.

Wells, P.G. et al. 2004. Nine-year review of Gulfwatch: trends in tissue contaminant levels in the blue mussel, *Mytilus edulis*, with special emphasis on the Bay of Fundy. Summary of poster, in press, Percy, J.A. et al. The Changing Bay of Fundy — Beyond 400 Years. The 6th BOFEP Bay of Fundy Workshop, Proceedings, Environment Canada, Dartmouth, NS.

Westhead, M.C. and T.B. Reynoldson. 2004. The reference condition approach: on trial in the Minas Basin. Pages 262–266 in Wells, P.G. et al. (eds.). Health of the Bay of Fundy: Assessing Key Issues. Environment Canada — Atlantic Region, Occasional Report No. 21, Environment Canada, Dartmouth, NS.

Wichert, G.A. and Rapport, D.J. Fish community structure as a measure of degradation and rehabilitation of riparian systems in an agricultural drainage basin. *Environ. Manage.* 22, 425–443, 1998.

Wilcox, B.A. Ecosystem health in practice: emerging areas of application in environment and human health. *Ecosys. Health* 7, 317–325, 2001.

Wilson, R.C.H. and Addison, R.F. Health of the Northwest Atlantic. A report to the Interdepartmental Committee on Environmental Issues. Environment Canada, Department of Fisheries and Oceans, Department of Energy, Mines and Resources, Ottawa, ON, 1984, 174 p.

Wood, N. and Lavery, P. Monitoring seagrass ecosystem health — the role of perception in defining health and indicators. *Ecosys. Health* 6, 134–148, 2000.

Woodley, S., Kay, J. and Francis, G. *Ecological Integrity and the Management of Ecosystems.* St. Lucie Press, 1993, 220 p.

Woodwell, G.M. On purpose in science, conservation and government. *Ambio* 31, 432–436, 2002.

Yoder, C.O. and DeShon, J.E. "Using biological response signatures within a framework of multiple indicators to assess and diagnose causes and sources of impairments to aquatic assemblages in selected Ohio rivers and streams," in *Biological Response Signatures. Indicator Patterns Using Aquatic Communities*, Simon, T.P., Ed. CRC Press, Boca Raton, FL, 2003.

Zabel, R.W., Harvey, C.J., Katz, S.L., Good, T.P., and Levin, P.S. Ecologically sustainable yield. *Amer. Sci.* 91, 150–157, 2003.

Zaitsev, Y. and Mamaev, V. *Marine Biological Diversity in the Black Sea. A Study of Change and Decline.* GEF Black Sea Environmental Programme. United Nations Publications, New York, NY, 1997, 208 p.

Zelazny, D.E. Introduction. *Great Lakes Research Review* 5, i–ii, 2001.

Index